Flora of Tirunelveli Hills
(Southern Western Ghats)

Volume III
MONOCHLAMYDEAE

[Text]

V. S. Manickam
G. Jeya Jothi
S. Ignacimuthu

Centre for Biodiversity and Biotechnology,
Xavier Research Foundation, St Xavier's college,
Palayamkottai – 627 002, Tamil Nadu, India

Bishen Singh Mahendra Pal Singh
23-A, New Connaught Place,
Dehra Dun - 248 001 (INDIA)
2022

Flora of Tirunelveli Hills (Southern Western Ghats) Volume III Monochlamydeae (Text)

ISBN : 978-81-943323-6-7

Published by Abhimanyu Gahlot for M/s. Bishen Singh Mahendra Pal Singh, 23-A, New Connaught Place, Dehra Dun, India and Printed by him at Shiva Offset Press, 14, Old Connaught Place, Dehra Dun, India.
phone +91 135 2715748 | fax +91 135 2715107 | bsmpsbooks@gmail.com

FOREWORD

Owing to their significant geographic position in the southern end of the Western Ghats, varied climate and altitude, the 'Tirunelveli hills' possess a very rich and diverse flora. A reliable, up-dated floristic inventory is a pre-requisite for a just assessment of the plant wealth of a given area, more so in a tropical country like India, wherein the population pressure continues to lead to over-exploitation and consequent disappearance of the precious forests. Since Tropical rain forest has entered a period of rapid decline as a world natural resource, the "Tirunelveli Hills" must be regarded as a prime example of this ecosystem in Southern India.

This treatise (Volume III - Monochlamydeae) comprising 278 taxa with illustrations, forms a welcome contribution to the present day knowledge of the flora, that too attempted by a local educational institution of repute, viz., Centre for Biodiversity and Biotechnology, Xavier Research Foundation, St Xavier's College, Palayamkottai, Tamil Nadu.

It is hoped that this work will be of real service to students and teachers of Biology and other local practical plant users in the better understanding and utilization of the rich biodiversity of the area, leading to rural development.

A.N. Henry

Joint Director (Retd.)
Botanical Survey of India

PREFACE

The current emergence of awareness and interest on Biodiversity inventory and conservation can be traced to the publication of a book entitled "Tragedy of the Commons" by Barry Commoner, a British Scientist. Since then, several international organizations such as IUCN, UNESCO, UNEP, WWF, UNCED, BGCI have been playing an active role in disseminating the awareness of the biodiversity conservation to individuals and groups of people in all the areas of human activities. The Biological Diversity Convention held at Rio de Janerio in 1992, roused both an interest and a sense of responsibility in scientists and political leaders all over the world in order to make the biodiversity lasting for use by future humankind and to prevent further global warming which today threatens the very existence of humankind on earth.This awareness, indirectly, has led both the animal and plant taxonomists to take interest in taxonomic inventories of various hilly regions and plains.

When compared with other countries both in the temperate and tropical belts, with regard to taxonomic inventories of biodiversity rich areas, India still remains backward. With regard to the Western Ghats, so far only Flora of Khandala published long ago by Santapau, Flora of Coorg by K.R. Keshava Murthy and S.N. Yoganarasimhan, Flora of Hassan District by Cecil J. Saldanha & Nicolson, Flora of Thiruvananthapuram by M. Mohanan and A.N. Henry, Flora of Palghat by Vajravelu, Flora of Cannanore by V.S. Ramachandran and V.J. Nair, Flora of the Palni Hills by K.M. Matthew, Flora of Thrissur Forest by Sasidharan and Flora of Agasthyamala (Kerala) by N. Mohanan and Sivadasan are available. Agasthiyamalai and Nilgiris are two important hot spots of the Western Ghats, but only a check-list of the Flora of Nilgiris has been published. Manickam and Irudayaraj (1992) published the Pteridophytic Flora of the Western Ghats between Kanyakumari and Palghat. Manickam and Irudayaraj (2003) brought out Pteridophytic Flora of Nilgiris. There has been no taxonomic account of Bryophytes, algae and fungi of the Western Ghats. Under the AICOPTAX project supported by Ministry of Environment, Forest and Climate Change, Government of India, Delhi, Manickam and his five collaborators worked out the pteridophytic and gymnosperm flora of the entire India. In view of the fast depleting floral diversity, more active interest on the part of the plant taxonomists is needed to know the endemic and vulnerable species in order to prevent their extinction by prevalent anthropogenic pressure on plant diversity.

Out of 25 hot spots of the world, two are present in India viz. the Western Ghats and the Himalayas. The Western Ghats are one of the mega endemic centres in India, which run north-south for about 1,600 km and a rich treasure house of flowering plants and ferns. Western Ghats are compared to oceanic island (Subramaniam & Nayar, 1974) due to the natural barrier, such as the Arabian sea on the western side, Vindhya and Satpura ranges on the northern side and semi-arid Deccan plateau on the eastern side. In the Western Ghats endemic taxa are spread over 58 families with major representations from Euphorbiaceae (34 species), Rubiaceae (34 species), Lamiaceae (33 species) and Myrtaceae

(30 species). In all, 60 genera and 1500 species are considered to be endemic to the Western Ghats (Nayar 1996), which means 70% of the endemic taxa of the peninsular India are confined to the Western Ghats. Raghavan and Singh (1983) indicated that 103 species are confined to the Western Ghats of Maharashtra, 46 to Karnataka region, 93 exclusively confined to the Nilgiris (Blasco 1970), 18 to Palnis, 13 to Anamalais and 150 to Agasthiyamalai and its environs (Tirunelveli District, Tamil Nadu: Thiruvananthapuram District, Kerala) (Henry *et al*., 1984). The above data clearly indicate that the Agasthiyamalai region has greatest floristic richness compared to other regions of the Western Ghats.

The total area of the Agasthiyamalai is approximately 2,500 sq. km. and falls within the hilly tracts of Tirunelveli and Kanyakumari districts of Tamil Nadu and Thiruvananthapuram district of Kerala. The Agasthiyamalai peak (1868 m) is locally known as Pothigaimudi, Agasthiyarkudam, Eka Pothigai or Periya Pothigai. These hills form a very compact block comprising Papanasam R.F., Kottur R.F., Mahendragiri R.F., Kalamalai R.F., Veerapuli R.F., Nattukaltheri R.F. and Ashamboo R.F. This region is drained by several small perennial rivers, such as Tambaraparani, Neyyar, Karamanayar and Kothayar. Since the ecosystem diversity is high, almost all vegetational types known from the Western Ghats are found in this region, such as southern tropical thorn forest (about 350 m), Grassland at low altitudes (below 500 m), southern tropical moist deciduous forest (above 500 m), tropical semi-evergreen forest (above 700 m), southern tropical wet evergreen forest (800-1,500 m), subtropical montane forest (above 1,500 m) and grassland at high altitudes (above 1,200 m).

Agasthiyamalai is a remarkable microendemic centre and is one of the hot spots of Indian flora (Nayar 1996), attested by IUCN as CPD site IS7 (Davis et al., 1994). The hills harbour about 2,000 species of flowering plants which include about 150 strict endemics. According to previous surveys, the 10 families of flowering plants which dominate the flora are: Poaceae (420 species in 121 genera), Fabaceae (340 species in 72 genera), Acanthaceae (165 species in about 14 genera), Euphorbiaceae (140 species in about 55 genera), Asteraceae (135 species in about 57 genera), Lamiaceae (115 species in 25 genera), Rubiaceae (109 species in 40 genera) and Asclepiadaceae (92 species in 30 genera). But recently numerous new species have been described and reported. According to Gopalan and A.N. Henry (2000), there are 139 endemic species of different habits viz., medium to large trees - 38, large shrubs - 28: shrubs/undershrubs/large herbs - 20, herbs - 30, lithophytic/epiphytic tree - 1, epiphytic/lithophytic herbs - 7 and climbers - 9.

The Agasthiyamalai was explored by the earlier botanists viz, Hooker (1872), Ramaswamy (1914), Rangacharya (1919), Mudaliar and Sundararaj (1954), Govindarajalu and Swamy (1958) and Shankarnarayan (1960). After the establishment of the Botanical Survey of India, Southern circle at Coimbatore, there has been an excellent progress of floristic exploration and many new species have been described. Among them, notable contributions are from Sebastine and

A.N. Henry (1960), Sharma *et al.* (1973), Henry *et al.* (1984) and Vajravelu *et al.* (1987). A.N. Henry and Gopalan (1995) described more than 20 new species from this area.

The present taxonomic account is based entirely on field work in all the hilly tracts of the Western Ghats in the district of Tirunelveli, from Sivagiri in the north to Mahendragiri in the south. Also included are certain small areas of hills like Maramalai, Korimani and Balamore, which fall within the district of Kanyakumari. The project was funded by UGC during 1996-99. The authors carried out field work for over three years, collecting about 13,000 field numbers in 60 field trips extended over 300 days. In 2003, a check-list of the flora of Tirunelveli hills was published. In this check-list, 1872 species belonging to 669 genera and 153 families were enumerated. After this publication, a careful and critical look on the unidentified species revealed more species, thus increasing the total number of species to 2000.

The striking peculiarity of Agasthiyamalai lies in the large preponderance of several typical Sri Lankan plants, thus showing marked affinity with the island flora of Ceylon (Sri Lanka). Some such species are *Ampelocissus wightiana, Antidesma alexiteria, Gastrochilus acaulis, Eugenia roxburghii, Paracroton pendulus* subsp. *zeylanicus*, etc.

Agasthiyamalai houses the richest plant diversity in the Western Ghats and so is a priority hot spot area for conservation. The threatened species are conserved by Tropical Botanic Garden and Research Institute (TBGRI), Palode, Thiruvananthapuram. The Foundation for Revitalization of Local Health Tradition (FRLHT) concentrates on the conservation of medicinal plants. In the green house attached to our research centre, there are germplasm collections of rare and vulnerable species. Some of these are *Acranthera grandiflora, Aeschynanthus perrottetii, Begonia floccifera, Exacum courtallense, Geophila repens, Decalepis arayalpathra, Justicia beddomei, Medinilla beddomei, Rauvolfia serpentina, R. tetraphylla, Trichopus zeylanicus* etc. The herbal garden in the centre includes about 150 species, collected both from the hills and plains.

In a project on the conservation of biodiversity of rare, economic and endemic species of Agasthiyamalai, funded by Department of Biotechnology, Government of India, Delhi, we multiplied through tissue culture, three tree species, namely *Gluta travancorica, Calophyllum austroindicum* and *Poeciloneuron pauciflorum*. So far, 15 populations of *Gluta travancorica*, 13 populations of *Calophyllum austroindicum* and 11 populations of *Poeciloneuron pauciflorum* have been located in the Tirunelveli Hills. We have attained some success in the multiplication of these three tree species through embryo, nodal and shoot tip culture.

It is worth mentioning that the Tirunelveli hills are also rich in pteridophytes. In Gujarat, there are 26 Species, in Maharashtra 62 species, in Goa 51 species, in the entire Karnataka 186 species, in Nilgiris 144 species, in Anamalais 130 species, in Palnis 136 species and in Tirunelveli hills 176 species.

Tirunelveli hills are known for a large number of plants which are being used by tribals and also by professional and medical practitioners. The list of medicinal plants collected during three years of field work includes 567 species in polypetalae 131 species in gamopetalae and monochlamydeae and 75 species in monocotyledons. A few very important medicinal plants are cultivated in the green house and herbal garden belonging to the research centre.

The present volume deals with the taxonomic account of monochlamydeae including 278 species, subspecies and varieties distributed in 107 genera belonging to 22 families.

We have prepared illustrations for all species of the flora of Tirunelveli hills and plains except a few for which we could not collect the flowers. The entire taxonomic account of the flora of the hills and plains of Tirunelveli district will be completed in near future.

The present taxonomic account is in continuation with the earlier published volume (2008). It is the first comprehensive and illustrated one for Tirunelveli and stands as a significant milestone in the history of the Western Ghats of India. In this volume, 22 families of Monochlamydeae covering 278 taxa have been treated with illustrations for almost all the taxa. Since, nearly all the species are illustrated, students and scientists will be able to identify taxa not only of Tirunelveli hills, but also of other areas, particularly of Tamil Nadu. Earlier taxonomists have called the flora of Tirunelveli district as the epitome of the flora of Tamil Nadu as specialities belonging to other hill tracts and plains of other districts are also found in this district. We hope that this flora will stimulate interest in both research scholars and scientists to take up similar work in other unexplored parts of Tamil Nadu and the rest of India.

Centre for Biodiversity and Biotechnology,
Xavier Research Foundation,
St Xavier's College,
Palayamkottai - 627 002.

V.S. Manickam
G. Jeya Jothi
S. Ignacimuthu

ACKNOWLEDGEMENTS

We thank Dr A.N. Henry for the Foreward. We express our heartfelt gratitude to the University Grants Commission, New Delhi for financial support. We are deeply grateful to the French Jesuit provincials, Paris, to Rev. P.J. Ubelmesser, S.J., Nurnberg, Germany and to Rev. Hubert Hanggi, S.J., Zurich, Switzerland for generous financial support needed to complete the Flora of Tirunelveli Hills. We express our deep indebtedness to Dr K.P.S. Katwal, LF.S. and to Dr Sukh Dev, I.F.S., Chief Wildlife Warden, Chennai, Dr V.K. Melkani and Dr R. Annamalai, Field Directors of Kalakad - Mundanthurai Tiger Reserve and to the successive district Forest Officers, Tirunelveli, for generous permissions to carry out field-work and for encouragement. We express our sincere thanks to Dr P. Daniel, ex Joint Director, BSI, Southern Regional Centre, Coimbatore for giving permission to consult the herbarium.

We also express our gratitude to Dr R. Gopalan and Dr V.J. Nair, BSI, Coimbatore and to Dr V. Chelladurai, Siddha Medicinal Plants Research Unit (CCRAS), Palayamkottai for help in identification of species. We thank Dr V. Sundaresan and Dr V. Irudayaraj for their help in various ways. Mrs Shanthi Jesuraj and Miss T. Latha have meticulously and patiently carried out computational works. Our deep gratitude is due to them. Mr H. Sankar and Mrs G. Venkateswari took great efforts to draw line drawings for all the species of Tirunelveli District. We thank them whole heartedly. Mr P. John Peter, the driver has taken great risks in driving the jeep in very rugged and narrow roads in the Tirunelveli hills. Our most sincere thanks are due to him. We also thank I. Jegan, A. Rex and M. Sundar, Field men for untiring assistance in field studies. We also thank Dr P. Pandikumar and Dr S. Mutheeswaran, Scientists and Mr M. Muthu Kanagavel, technician at Xavier Research Foundation for their assistance in compilation of this book.

We are grateful to the Editor for his suggestions to improve the text. We thank M/S Bishen Singh Mahendra Pal Singh, Publishers and Distributors for bringing out the book elegantly.

Above all, the authors express their sense of deep indebtedness to the Jesuit Management, St Xavier's college for all their support and help. The authors thank all the members of St Xavier's College Jesuit Community for encouragement and appreciation of the scientific contributions.

CONTENTS

INTRODUCTION

The study area (about 3,200 sq. km.) falls within the hilly tracts of Tirunelveli and Kanyakumari districts of Tamil Nadu, India. Three seasons viz., winter (December to March), summer (April to June) and rainy seasons (July to September; October to November) are recognizable in the study area. The red ferruginous sandy loam soil has been observed in the lower elevations; on the hilly slopes it appears yellow or red due to high level of soil erosion. In the wet evergreen and semi-evergreen forests, the soil is rich in humus. Starting from Wight (1835-36), many explorations were taken to study the flora of Tirunelveli hills.

For this work, Tirunelveli hills were explored periodically, between 1996-2001 covering various seasons. A total of 60 field trips were taken and 13,000 field numbers were collected. Specimens were identified using Gamble's Flora of the Presidency of Madras (1915-1936), Hooker's Flora of British India (1872-1897), K.M Matthew's Flora of the Tamil Nadu Carnatic (1983) and other related works. These were later confirmed through matching with authentic specimens at Madras Herberium (MH), Botanical Survey of India, Southern Regional Centre, Coimbartore, India. All the specimens were deposited in XCH (Xavier's College Herbarium), St Xavier's College, Palayamkottai, Tamil Nadu.

All the forests types of Western Ghats are found in the Tirunelveli Hills (N.C. Nair & P. Daniel, 1986; Gopalan & A.N. Henry, 2000). The area represents diverse ecosystems with almost all types of vegetation known to occur in the Western Ghats. This includes Southern tropical dry deciduous forests. Southern tropical wet ever green forests, Southern subtropical montane forests and grasslands at lower and higher altitudes.

It is reported that Agasthiamalai area in Tirunelveli and Travancore hills has about 150 localised endemic species (Henry et al., 1983). Among Monochlamydeae, 2 species of Aristolochiaceae, 3 species of Piperaceae, one species of Myristicaceae, 19 species of Lauraceae, 7 species of Loranthaceae, 2 species of Urticaceae and 3 species of Moraceae are found in Tirunelveli hills. Most of the endemic tree species of Monochlamydeae distributed in Tirunelveli hills are from the family Euphorbiaceae and Lauraceae. Some Monochlamydae members such as *Claoxylon anomalum, Euphorbia santapaui*, *Euphorbia vajravelui* and *Phyllanthus singampattianus* are the stenoendemics.

In Tirunelveli hills Monochlamydeae is represented by 22 families, among which 8 families are represented by single genus. Trees are dominant followed by shrubs and herbs. The climbers are poorly represented. Detailed analysis of the study area, geology, soil, previous works, general pattern of vegetation, endemism and floristic analysis are given in volume 1 (Manickam et al., 2008).

MONOCHLAMYDEAE

Trees, shrubs or herbs, rarely climbers. Mostly chlorophyllous and autotrophic, sometimes achlorophyllous and parasitic (partial or total). Leaves mostly simple, rarely compound, alternate, opposite or whorled, sometimes absent. Perianth usually 1 seriate, rarely more seriate, sometimes absent, herbaceous, rarely petaloid and corolline; imbricate, valvate or connate into a tube. Stamens few, rarely numerous, inserted on the thalamus or rarely at the base of the perianth. Carpel 1 - ∞, free or connate.

Key to the families

1a. Flowers very often 2-sexual or if 1-sexual then perianth lobes 2 seriate 2
1b. Flowers 1-sexual or if 2-sexual then perianth 0 or 1 seriate 14

2a. Ovary inferior 3
2b. Ovary superior 6

3a. Ovary 4-6 celled; ovules very many, 1-2 seriate; seeds with a distinct testa ....6. **Aristolochiaceae**
3b. Ovary 1 celled; ovules 1-4; seeds without distinct testa 4

4a. Parasitic herbs, undershrubs or shrubs; ovule solitary, basal, erect 5
4b. Terrestrial herbs, shrubs or trees; ovules 2-4, pendulous, placentation free central ....17. **Santalaceae**

5a. Anthers not adnate to perianth lobes, opening longitudinally ....15. **Loranthaceae**
5b. Anthers adnate to perianth lobes, opening by many pores ....16. **Viscaceae**

6a. Seeds usually with copious albumen, or if albumen scanty or 0, then embryo curved, excentric 7
6b. Seeds without albumen, or if albuminous then embryo straight 12

7a. Stipules 0 8
7b. Stipules present, connate in a tube around the nodes, persistent or rarely deciduous, leaving a circular scar ....5. **Polygonaceae**

8a. Perianth lobes completely or partially united 9
8b. Perianth lobes free ....4. **Phytolaccaceae**

9a. Perianth tubular .... 1. **Nyctaginaceae**
9b. Perianth not tubular 10

10a. Perianth tube constricted above the ovary, subcorolline .... 14. **Elaeagnaceae**
10b. Perianth tube not constricted above the ovary, lobes not corolline 11

11a. Perianth scarious, dry ....2. **Amaranthaceae**
11b. Perianth membranous, herbaceous, coriaceous or fleshy ....3. **Basellaceae**

12a. Perianth lobes valvate; stamens 4 in 1 series; ovules basal or lateral, ascending ....12. **Proteaceae**

12b. Perianth lobes imbricate; stamens in 2, 3 or 4 series; ovules pendulous 13

13a. Perianth tube very short, lobes 2-seriate; stamens usually accompanied by glands, anthers opening by as many upcurved valvular lids .... 10. **Lauraceae**

13b. Perianth tube rather long, lobes 1-seriate; stamens not accompanied by glands, anthers opening by slits .... 13. **Thymelaeaceae**

14a. Leafy herbs, shrubs or trees, leaves sometimes much reduced, sap milky............. 15

14b. Leafless herbs..........................18. **Balanophoraceae**

15a. Ovary 1-locular, 1-celled, or if syncarpous, then 2 or more celled, ovules 1-2 collateral in each cell 16

15b. Ovary 2 - more celled, rarely 1 celled with 2 ovules in each cell.... 19. **Euphorbiaceae**

16a. Stipules 0 17

16b. Stipules present.......................... 18

17a. Flowers bracteolate; perianth 3, sometimes 4 lobed; stamens 6-30, connate in a column; staminodes 0; fruit fleshy, at length 2-4 valved; seeds enclosed in a thin or fleshy aril .... 9. **Myristicaceae**

17b. Flowers ebracteolate; perianth 4-7, partite in male, 2 lobed in female; stamens 4-7 free, alternate with clavate staminodes; fruit a drupe like; crowned with elongate, spathulate, wing like perianth lobes; seeds without aril .... 11. **Hernandiaceae**

18a. Seeds with copious albumen and a minute embryo; perianth 0.......................... 19

18b. Seeds with little or no albumen; perianth present.......................... 20

19a. Stamens 2-4, filaments short, free; ovule erect.......................... 7. **Piperaceae**

19b. Stamens 1 or 3, confluent; ovule pendulous .... 8. **Chloranthaceae**

20a. Filaments not inflexed in bud, anthers not reversed .... 21. **Ulmaceae**

20b. Filaments inflexed in bud with reversed anthers 21

21a. Styles or style branches 2; stamen 1; ovules pendulous; herbs, shrubs and trees usually with milky sap.......................... 22. **Moraceae**

21b. Style single, undivided; stamens 3-5; ovule basal; erect; herbs, shrubs or trees with watery sap .... 20. **Urticaceae**

1. NYCTAGINACEAE Juss.

Herbs, shrubs or trees. Leaves usually opposite; stipules 0. Flowers bisexual, rarely unisexual, actinomorphic, sometimes zygomorphic, often involucrate. Perianth monophyllous, petaloid; tube persistent, campanulate, funnel or salver shaped, enveloping the fruit; limb 3-5 lobed, persistent or deciduous; lobes plicate in bud. Stamens 1-39, hypogynous, sometimes unilateral; filaments usually unequal, inflexed in bud; anthers included or exserted, dorsifixed, didymous. Ovary free, 1 locular; ovule solitary, basal, erect; style filiform, involute in bud; stigma small. Fruit indehiscent, enclosed in the hardened perianth tube; seeds erect; testa thin.

There are about 27 genera and 390 species in the tropics and subtropics; in India 4 genera and 15 species are reported.

Key to the genera

1a. Armed shrubs 2
1b. Unarmed herbs 3
2a. Bracts green, smaller than the flower; stamens exserted .... 4. **Pisonia**
2b. Bracts other than green, larger than the flower; stamens included .. 2. **Bougainvillea**
3a. Branchlets succulent; leaves in pairs, equal .... 3. **Mirabilis**
3b. Branchlets not succulent; leaves in pairs, unequal .... 1. **Boerhavia**

1. BOERHAVIA L.

This genus is represented by ca 50 species, mostly distributed in tropics and sub tropics. In India, 4 species have been reported.

Key to the species

1a. Anthocarp club shaped, glandular hairy, top rounded; perianth lobes glandular within .... 1. **B. diffusa**
1b. Anthocarp obconic, glabrous, top-truncate; perianth lobes not glandular within .... 2. **B. erecta**

1. **Boerhavia diffusa** L., Sp. Pl. 3. 1753; Gamble, Fl. Pres. Madras 2: 1162. 1925; Matthew & Rani in Matthew, Fl. Tamil. Carnatic 3: 1287. 1983; Kumari in A.N. Henry *et al*., Fl. Tamil Nadu 2: 188. 1987; Matthew, Fl. Palni Hills 2: 1021. 1999. *B. repens* L., Sp. Pl. 3: 1753; Hook.f., Fl. Brit. India 4: 709. 1885. **(Fig. 1)**.

Diffuse herbs to 30 cm. Leaves suborbicular or ovate or lanceolate, 1-2 x 0.8-1.2 cm, chartaceous, glabrous to minutely puberulous above, glaucous beneath, base subtruncate, margin entire or undulate, apex obtuse; petioles 0.3-1 cm.

Panicles axillary, branched; branches ca 8 flowered. Pedicels to 2 mm. Perianth tube ovoid, to 2.5 mm, prominently ribbed, upper part pinkish, campanulate, 5 lobed; lobes shortly acuminate at apex. Stamens 2; filaments to 1.75 mm; anthers 0.25 mm, exserted. Ovary ca 1 x 0.5 mm, ribbed, glandular-pubescent; style to 2 mm; stigma capitate. Anthocarp club shaped, ca 6 mm, ribbed, glandular; stalk to 1 cm; seeds oblong, to 4 mm.

Common in foot hills.

Fl. & Fr.: Throughout the year.

Specimens examined: XCH 16333, 16470 (Thirukurungudi); 11535 (Cheranmahadevi).

Distribution: Pantropical, throughout Malaysia, Australia, Pacific.

2. **Boerhavia erecta** L., Sp. Pl. 3. 1753; Nair in Bull. Bot. Surv. India 9: 283. 1967; Matthew & Rani in Matthew, Fl. Tamil. Carnatic 3: 1288. 1983; Kumari in A.N. Henry *et al.*, Fl. Tamil Nadu 2: 188. 1987; Matthew, Fl. Palni Hills 2: 1021. 1999. **(Fig. 2).**

Herbs to 50 cm, puberulous. Leaves elliptic-ovate, 1.5-4 x 0.5-1.5 cm, chartaceous, glabrescent to puberulous, base rounded to subacute, margin undulate, apex acute to apiculate; petioles 0.5-1.5 cm, puberulous. Panicles axillary, terminal or at the forks of dichotomy, 2-4 flowered; peduncles ca 3.5 cm. Pedicels ca 5 mm. Perianth tube ca 2 mm, lower part with 5 faint ribs, upper part campanulate, white with pink stripes, ca 2.75 mm. Stamens 3, ca 1.8 mm. Ovary ca 1 mm; style to 1.5 mm. Anthocarps obconic, ca 4 x 1.3 mm, ribbed; ribs puberulous.

Common in foot hills.

Fl. & Fr.: Throughout the year.

Specimen examined: XCH 11498 (Cheranmahadevi).

Distribution: Pantropical weed.

2. BOUGAINVILLEA Commers. ex A.L. Juss.

This genus is represented by ca 18 species worldwide; 2 in India.

Bougainvillea glabra Choisy in DC., Prodr. 13(2): 437. 1849; Gamble, Fl. Pres. Madras 2: 1164. 1925; Matthew & Rani in Matthew, Fl. Tamil. Carnatic 3: 1288. 1983; Kumari in A.N. Henry *et al.*, Fl. Tamil Nadu 2: 188. 1987; Matthew, Fl. Palni Hills 2: 1021. 1999.

Tamil: Kahithappoo.

Armed, scandent shrubs reaching 4-6 m. Leaves simple, alternate, upto 10 cm. Flowers in panicles; bracts larger than the flowers, varying in colour. Achenes elongate, ribbed.

Commonly found in the plains and occasionally in the hills.

Fr.: Throughout the year.

Specimen examined: XCH 18105 (Courtallam).

Distribution: Native of Brazil, widely cultivated in tropics and subtropics.

3. MIRABILIS L.

This genus is represented by ca 63 species mostly found in the America; 2 from India.

Mirabilis jalapa L., Sp. Pl. 177. 1753; Gamble, Fl. Pres. Madras 2: 1164. 1925; Matthew & Rani in Matthew, Fl. Tamil. Carnatic 3: 1291. 1983; Kumari in A.N. Henry *et al.*, Fl. Tamil Nadu 2: 189. 1987; Matthew, Fl. Palni Hills 2: 1022. 1999. **(Fig. 3).**

Tamil: Anthimantharai.

Herbs to 50 cm; branchlets succulent, swollen at nodes, minutely puberulous. Leaves broadly triangular-ovate to elliptic, 4.5-9.5 x 2.5-5 cm; chartaceous, sparsely appressed-pubescent on nerves, base truncate, apex acute to acuminate; lateral nerves ca 6 pairs; petioles 1-3 cm, puberulous. Flowers ca 5, in terminal, sessile corymbs, subtended by an involucre of 5, triangular-ovate, shortly connate bracts, accrescent in fruits. Perianth funnel shaped; tube to 3 cm, lower part green, globose, upper part dark purple or yellow. Stamens 3-6, exserted, ca 3 cm. Ovary globose, ca 1 mm across; style to 2.5 mm; stigma fimbriate. Anthocarp globose, 4 mm across, black.

Occasional in deciduous forests, escape.

Fl. & Fr.: Throughout the year.

Specimen examined: XCH 14020 (Courtallam).

Distribution: Native in Peru, now cultivated as an ornamental or medicinal plant, occasionally an escape, in all tropical regions.

4. PISONIA L.

This genus is represented by ca 40 species distributed in tropical and subtropical regions, mostly in America; 3 in India.

Pisonia aculeata L., Sp. Pl. 1026. 1753; Wight, Ic. t. 1763 & 1764. 1852; Hook.f., Fl. Brit. India 4: 711. 1885; Gamble, Fl. Pres. Madras 2: 1163. 1925; Matthew & Rani in Matthew, Fl. Tamil. Carnatic 3: 1291. 1983; Kumari in A.N. Henry *et al.*, Fl. Tamil Nadu 2: 189. 1987; Matthew, Fl. Palni Hills 2: 1023. 1999. **(Fig. 4).**

Armed, scandent shrubs to 1 m; prickles axillary, to 0.8 cm. Leaves elliptic-ovate, 2-6.5 x 1-3 cm, chartaceous, glabrous or puberulous on nerves, base acute, margin entire, apex obtusely acute to shortly acuminate; lateral nerves 6-10 pairs, obscure; petioles ca 1 cm. Flowers bisexual or polygamous, sessile, aggregated in axillary, stalked umbel. Perianth funnel-shaped; tube ca 2.5 mm, puberulous without, 4 lobed; lobes triangular-ovate, ca 1 mm, cream in colour. Stamens ca 10, exserted; filaments connate below in a tube to 4 mm; anthers oblong, to 1 mm, didymous. Ovary oblong-globose, ca 0.75 mm; style to 1.5 mm; stigma fimbriate. Anthocarp clavate, ca 1.7 x 0.4 cm, with 5 rows of 2 seriate viscid glands; seeds oblong, to 1.3 cm.

Occasional in open forests and along the stream banks in deciduous forests.

Fl.: Throughout the year; *Fr.:* March

Specimens examined: XCH 11099, 15460, 18061 (Courtallam); 13322, 16450 (Thirukurungudi); 12398, 12404 (Papanasam); 18302 (Sivagiri).

Distribution: (Sub) tropical America, Africa, Madagascar, Mauritius, Seychelles, Sri Lanka, India, Myanmar, Indo-china, Hainan, Taipei, Malaysia.

2. AMARANTHACEAE Juss.

Herbs, rarely shrubs, erect or climbing. Leaves opposite or alternate; stipules 0. Flowers usually hermaphrodite, rarely polygamous or dioecious; small, usually in terminal and axillary, simple or panicled spikes, cymes or clusters. Bracts hyaline or scarious; bracteoles 2, scarious. Tepals dry, calycine, persistent, usually of 5, free or slightly connate, hyaline or scarious, imbricate segments. Stamens 1-5, opposite to the tepals; filaments free or connate below into a membranous, hypogynous cup, often alternating with membranous staminodes; anthers dorsifixed, 1-2 celled. Ovary 1 locular, ovoid-ellipsoid or globose; ovule 1 or more, amphitropous, erect or suspended from basal funicle; styles simple sometimes with capitellate stigma, sometimes 2-3 fid with acute stigmas or styles 2-3 papillosely stigmatic on the lower face. Fruit a membranous utricle, rarely a berry or a capsule, breaking irregularly or circumscissly; seeds erect or inverted, orbicular-ovate or reniform, compressed; testa crustaceous; tegmen, if present membranous; embryo horseshoe shaped or annular.

A large and almost exclusively tropical family of some 65 genera and over 1000 species, including many cosmopolitan weeds and a large number of xerophytic plants. 46 genera and 151 species are reported from India.

Key to the genera

1a. Leaves alternate 2
1b. Leaves opposite or whorled 5

2a. Fertile flowers subtended by 2 lateral, sterile, modified flowers .... 7. **Digera**
2b. Fertile flowers not subtended by lateral, sterile modified flowers 3

3a. Ovary multiovulate .... 5. **Celosia**
3b. Ovary uniovulate 4

4a. Staminodes present; ovule pendulous from long basal funicle ....2. **Aerva**
4b. Staminodes absent; ovule erect..4. **Amaranthus**

5a. Flowers surrounded by other imperfect flowers 6
5b. Flowers all perfect 7

6a. Staminodes present .... 6. **Cyathula**
6b. Staminodes absent .... 10. **Pupalia**

7a. Filaments connate almost to the top ....8. **Gomphrena**
7b. Filaments connate at the base only or free.. 8

8a. Staminodes absent ....9. **Psilotrichum**
8b. Staminodes present 9

9a. Leaves very narrow, linear-filiform ....11. **Trichuriella**
9b. Leaves other than linear-filiform 10

10a. Flowers in dense, sessile, axillary heads; anthers 1 celled ....3. **Alternanthera**
10b. Flowers in elongate, pedunculate spikes; anthers 2 celled .... 1. **Achyranthes**

1. ACHYRANTHES L.

Ca 15 species in the warm temperate and tropical regions of the world; 3 have been reported from India.

Key to the species

1a. Wings of the bracteoles broad, nearly half as long as the spine; staminodes truncate, fimbriate..1. **A. aspera**
1b. Wings of the bracteoles merely auricles at the base; staminodes truncate, toothed ....2. **A. bidentata**

1. **Achyranthes aspera** L., Sp. Pl. 204. 1753; Hook.f., Fl. Brit. India 4: 730. 1885; Gamble, Fl. Pres. Madras 2: 1176. 1925; C.Towns. in Dassan. & Fosberg, Rev. Handb. Fl. Ceylon, 1: 38. 1980; Rani & Matthew in Matthew Fl. Tamil. Carnatic 3: 1296. 1983; Kumari in A.N. Henry *et al.*, Fl. Tamil Nadu 2: 189. 1987; Matthew, Fl. Palni Hills 2: 1025. 1999. **(Fig. 5).**

Tamil: Nayuruvi.

Herbs, 0.5-1 m, stiffly erect to subscandent, simple to much-branched; stem stout or weak, 4 angled, striate, subglabrous to densely tomentose. Leaves elliptic, oblong or ovate, 2-10(15) x 1-5 cm; puberulous above, densely appressed-pubescent beneath or more or less densely tomentose on both surfaces, attenuate at apex; nerves obscure; petioles 0.2-2.5 cm. Inflorescence at first dense, finally elongating to 25 cm, pedunculate. Bracts lanceolate, membranous, ca 3 mm; bracteoles linear-lanceolate, ca 5 mm, basal one adnate to spine. Tepals white or pale green to red, 5 lobed; lobes lanceolate, ca 6 mm, with distinct lateral nerves. Stamens 5, alternate with staminodes. Style ca 3 mm.

Very common in the plains and foot hills to deciduous forests.

Fl. & Fr.: Throughout the year.

Specimens examined: XCH 11532, 14636 (Papanasam); 14024, 15804 (Courtallam); 18417, 18375 (Sivagiri); 12378 (Mahendragiri).

Distribution: Throughout the world in tropical and warmer regions.

2. **Achyranthes bidentata** Blume, Bijdr. 545. 1826; Wight, Ic. t. 1779. 1852; Hook.f., Fl. Brit. India 4: 730. 1885; Gamble, Fl. Pres. Madras 2: 1176. 1925; C.Towns. in Dassan. & Fosberg, Rev. Handb. Fl. Ceylon 1: 40. 1980; Rani & Matthew in Matthew, Fl. Tamil. Carnatic 3: 1298. 1983; Kumari in A.N. Henry *et al.*, Fl. Tamil Nadu 2: 189. 1987; Matthew, Fl. Palni Hills 2: 1025. 1999. **(Fig. 6)**.

Tamil: Sivappu Nayuruvi.

Herbs to 0.5-1 m, much branched; stem and branches striate or sulcate. Leaves variable, elliptic-ovate to oblong or broadly to narrowly lanceolate, 4-15 x 1-3 cm, chartaceous, glabrescent to pubescent, acuminate, often mucronate; nerves 5-10 pairs, obscure to faint, looping; petioles 0.5-2 cm, glabrescent or appressed pubescent. Spikes to 20 cm; peduncles ca 4 cm. Flowers at first dense, finally lax. Bracts membranous, linear-lanceolate, ca 4.5 mm, apex minutely awned; bracteoles broadly ovate, ca 2 x 3 mm, sharply pointed at apex, membranous, basally adnate to midrib. Perianth 5 lobed; lobes linear-lanceolate, to 6 mm, appressed-pubescent without. Stamens 5, ca 2.8 mm, alternate with minute appendage like pseudo-staminodes, often obscure; staminal sheath fimbriate. Ovary oblong-globose, ca 2.5 mm long; style ca 1 mm; stigma capitate.

Rare in shady places mostly along paths in evergreen forests between 800-1200 m.

Fl. & Fr.: Throughout the year.

Specimens examined: XCH 10486, 11642, 11915, 15085, 17180 (Kakachi); 10999, 14323 (Kalakad); 18289 (Courtallam); 18340 (Sivagiri).

Distribution: Asia from W. Pakistan, India and Sri Lanka, east to China, Japan, Indonesia and Solomon islands; east and west Tropical Africa.

2. AERVA Forssk.

About 10 species in the warm and tropics of old world; ca 3 have been reported from India.

Key to the species

1a. Leaves well-developed throughout; spikes paniculate, axillary or terminal .... 1. **A. javanica**

1b. Leaves gradually smaller towards apices; spikes clustered, all axillary..2. **A. lanata**

1. **Aerva javanica** (Burm.f.) Juss. ex Schult. in Roem. & Schult., Syst. 5: 565. 1819; Wight, Ic. t. 876. 1844-1845; Hook.f., Fl. Brit. India 4: 727. 1885; C.Towns. in Dassan. & Fosberg, Rev. Handb. Fl. Ceylon 1: 31. 1980; Rani & Matthew in Matthew, Fl. Tamil. Carnatic 3: 1299. 1983. *Iresine javanica* Burm.f., Fl. Ind. 212. 1768. *Aerva tomentosa* Forssk., Fl. Aeg.-Arab. 122: 727. 1775; Gamble, Fl. Pres. Madras 2: 1178. 1925. *Aerva persica* (Burm.f.) Merr. in Philipp. J. Sci. 19: 348. 1928; Kumari in A.N. Henry *et al.*, Fl. Tamil Nadu 2: 190. 1987; Matthew, Fl. Palni Hills 2: 1027. 1999. **(Fig. 7).**

Subshrubs to 1 m, erect to scandent; stem and branchlets terete, striate, more or less densely whitish or yellowish tomentose. Leaves alternate, very variable, linear-oblong, or oblanceolate or spathulate, 1-5 x 0.3-1.2 cm, subcoriaceous, more or less densely white or yellowish tomentose, more so beneath, base cuneate to attenuate, obtusely apiculate at apex; petioles 2-6 mm. Spikes terminal, paniculate, cylindrical, densely white woolly, sessile. Flowers unisexual. Bracts deltoid-ovate, ca 2 x 3 mm, hyaline, densely lanate; bracteoles like bracts, persistent. Perianth 5 lobed; lobes spathulate, ca 2.5 x 1.5 mm, tomentose without; outer 2 larger, with yellowish midrib. Stamens 5, ca 2 mm. Ovary obovoid, ca 1 x 1 mm; style ca 0.3 mm; stigma deeply forked.

Rare in open dry places of foot hills.

Fl.: Throughout the year, peak in December – February.

Specimen examined: XCH 14541 (Surankudi).

Distribution: Tropics and subtropics of the old world, from Myanmar, India and Sri Lanka westwards through S.W. Asia, across N. Africa to Morocco and South to Cape Verde Islands and Cameroon through Uganda and Tanzania to Madagascar. Introduced in Australia and elsewhere.

2. **Aerva lanata** (L.) Juss. ex Schult. in Ann. Mus. Hist. Nat. 11: 131. 1808; Wight, Ic. t. 723. 1840; Hook.f., Fl. Brit. India 4: 728. 1885; Gamble, Fl. Pres. Madras 2: 1178. 1925; C.Towns. in Dassan. & Fosberg, Rev. Handb. Fl. Ceylon 1: 32. 1980; Rani & Matthew in Matthew, Fl. Tamil. Carnatic 3: 1299. 1983; Kumari in A.N. Henry *et al.*, Fl. Tamil Nadu 2: 189. 1987; Matthew, Fl. Palni Hills 2: 1026. 1999. *Achyranthes lanata* L., Sp. Pl. 204. 1753. *Illecebrum lanatum* (L.) L., Mant. 344. 1771. *Aerva elegans* Moq. in DC., Prodr. 13(2): 303. 1849. **(Fig. 8).**

Tamil: Pongappoo, Poolaipoo.

Herbs to 60 cm, more or less densely lanate. Leaves alternate, suborbicular or ovate-elliptic or elliptic-lanceolate, 1-3 x 0.5-2 cm, gradually smaller towards apices, subcoriaceous, lanate or canescent, more so beneath, cuneate to attenuate at base, obtusely acute at apex, often mucronate; petioles to 0.5 cm. Spikes clustered, axillary, subsessile, cylindrical, white to cream in colour, often silky. Flowers bisexual. Perianth 5 lobed; outer 2 hyaline, ovate-oblong, 1 nerved, mucronate, ca 1 mm; inner 3 slightly shorter, linear, with green midrib, acute. Stamens 5, ca 6 mm. Ovary ca 0.2 mm across; style ca 0.1 mm; stigma bifid.

Very common in the plains, occasional in foot hills.

Fl.: Throughout the year, peak during October – January.

Specimens examined: XCH 12365 (Papanasam); 11506 (Cheranmahadevi).

Distribution: Widespread in the drier parts of the tropics and subtropics of the old world in Africa from Sierra Leone across to Egypt, South to Africa (rare) and Madagascar, also in the Seychelles; in Asia from Arabia east to Malaysia, Indonesia, the Philippines and New Guinea.

3. ALTERNANTHERA Forssk.

A large genus of over 80 species, the greatest number occurring in the American tropics and subtropics; 7 in India.

Alternanthera pungens Kunth in H.B.K., Nov. Gen. Sp. 2: 206. 1818; C.Towns. in Dassan. & Fosberg, Rev. Handb. Fl. Ceylon 1: 47. 1980; Rani & Matthew in Matthew, Fl. Tamil. Carnatic 3: 1303. 1983; Kumari in A.N. Henry *et al.*, Fl. Tamil Nadu 2: 191. 1987; Matthew, Fl. Palni Hills 2: 1028. 1999. *A. echinata* J.E. Smith in Rees, Cyclop. 39(Suppl.): n. 10. 1819. **(Fig. 9)**.

Prostrate herbs to 25 cm, much branched; branches terete, more or less villous with long hairs, often glabrescent. Leaves opposite, opposite pairs unequal, rhomboid-ovate or obovate or broadly ovate, 0.5-2.3 x 0.3-2 cm, chartaceous, glabrous or thinly appressed pilose on both surfaces, base cuneate to attenuate, narrowed into petiole, margin entire, obtuse or acute at apex, mucronate; petioles 0.2-1 cm, with appressed pilose hairs. Spikes sessile, axillary, solitary or 2-3 together. Bracts membranous, lanceolate, ca 4 mm, denticulate at apex; bracteoles like bracts, ca 3 mm. Perianth 5 lobed; outer 2 lobes deltoid-lanceolate, to 5 mm, rigid, 5 nerved beneath; inner pair faintly visible; lobes oblong-lanceolate, ca 3 mm, serrulate, obtusely apiculate; lateral lobe ca 2.5 mm, sinuate, sharply mucronate, all 3 lobes with a tuft of glochidiate, barbellate bristles at base. Stamens 4, ca 1 mm, alternate with staminodes; staminodes membranous. Ovary depressed globose, ca 0.75 mm across; ovule pendulous; style very short, wider; stigma capitate.

Rare in foot hills.

Fl.: Throughout the year.

Specimen examined: XCH 16469 (Thirukurungudi).

Distribution: A native of the American continent from the southern parts of USA to Argentina and Peru, now widespread, used as food in the tropics and subtropics of the Old World.

4. AMARANTHUS L.

About 70 species, chiefly in the tropical and subtropical regions of the world, about a dozen species being more or less cosmopolitan weeds in the tropics also; ca 13 species have been reported from India.

Amaranthus spinosus L., Sp. Pl. 991. 1753; Hook.f., Fl. Brit. India 4: 718. 1885; Gamble, Fl. Pres. Madras 2: 1170. 1925; C.Towns. in Dassan. & Fosberg, Rev. Handb. Fl. Ceylon 1: 9. 1980; Rani & Matthew in Matthew, Fl. Tamil. Carnatic 3: 1307. 1983; Kumari in A. N. Henry *et al.*, Fl. Tamil Nadu 2: 191. 1987; Matthew, Fl. Palni Hills 2: 1030. 1999; Mohanan & Sivadasan, Fl. Agasthyamala 541. 2002. **(Fig. 10).**

Tamil: Mullukkeerai.

Armed, erect or decumbent herbs to 1 m; stem angular stout; spines axillary, paired or clustered, subtending the inflorescence. Leaves ovate or rhomboid-ovate or elliptic or lanceolate, 1.5-11 x 0.5-5 cm, base acute to cuneate, margin entire or undulate, obtusely acute or retuse at apex; petioles filiform, flexuous, 1.5-8 cm, sometimes longer than the leaves. Flowers in fascicles in lower axils, towards the end of the stem and branches, the clusters forming simple or branched leafless spikes. Bracts deltoid-ovate, ca 1.5 mm, hyaline, with an erect awn formed by excurrent green midrib; bracteoles like bracts, shorter than or subequalling or little exceeding the tepals, mostly smaller than the bracts. Male: Perianth 5 lobed; lobes subequal, broadly lanceolate or lanceolate-oblong, ca 1.5 x 5 mm; midrib green, mucronate. Stamens 5; filaments slightly connate at base, ca 0.25 mm; anthers 4 celled. Female: Perianth 5 lobed, unequal; 2 outer narrowly oblong or oblanceolate, to 1.5 x 0.75 mm, mucronate; midvein prominent; inner 3 ovate, to 0.75 mm, acute. Ovary depressed-ovoid, 1 locular; ovule solitary, erect; style very short; stigmas 3, ca 1 mm, reflexed.

Frequent in foot hills.

Fl.: August – November.

Specimens examined: XCH 10665 (Mancholai); 11311 (Courtallam); 10067 (Kothayar); 17386 (Mahendragiri).

Distribution: In the warmer regions of the world, of presumed American origin.

5. CELOSIA L.

About 65 species in the warmer regions of both old and new world, especially America and Africa; ca 3 species from India.

Key to the species

1a. Flowers in dense, stout, imbricate spikes .... 1. **C. argentea**

1b. Flowers in small clusters, on slender, interrupted spikes 2

2a. Leaves obtuse or subacute, scarcely longer than broad .... 2. **C. polygonoides**
2b. Leaves acuminate, much longer than broad .... 3. **C. pulchella**

1. **Celosia argentea** L., Sp. Pl. 205. 1753; Wight, Ic. t. 1767. 1852; Hook.f., Fl. Brit. India 4: 714. 1885; Gamble, Fl. Pres. Madras 2: 1166. 1925; C.Towns. in Dassan. & Fosberg, Rev. Handb. Fl. Ceylon 1: 3. 1980; Rani & Matthew in Matthew, Fl. Tamil. Carnatic 3: 1309. 1983; Kumari in A.N. Henry *et al.*, Fl. Tamil Nadu 2: 192. 1987; Matthew, Fl. Palni Hills 2: 1031. 1999. **(Fig. 11)**.

Herbs to 1 m. Leaves elliptic-ovate or linear-lanceolate, 5-10 x 0.2-1 cm, base cuneate to attenuate, acute at apex; petioles to 1 cm long. Spikes silvery to pink, terminal, long pedunculate; peduncles up to 35 cm. Bracts and bracteoles lanceolate, ca 5 mm, hyaline, more or less aristate with excurrent midrib, persistent. Perianth 5 lobed, narrowly elliptic-oblong, ca 6 x 2 mm, mucronate with excurrent midrib, margins widely hyaline. Stamens 5; filaments delicate, ca 0.2 mm, subequalling the staminal sheath, cream or pink in colour; anthers linear-oblong, to 1.3 mm. Ovary obovoid, ca 1x1 mm; style to 5 mm, filiform; stigma bifid.

Occasional in foot hills.

Fl.: March.

Specimen examined: XCH 18483 (Sivagiri).

Distribution: Pantropical weed.

2. **Celosia polygonoides** Retz., Obs. Bot. 2: 12. 1781; Hook.f., Fl. Brit. India 4: 715. 1885; Gamble, Fl. Pres. Madras 2: 1166. 1925; C.Towns. in Dassan. & Fosberg, Rev. Handb. Fl. Ceylon 1: 5. 1980; Rani & Matthew in Matthew, Fl. Tamil. Carnatic 3: 1310. 1983; Kumari in A.N. Henry *et al.*, Fl. Tamil Nadu 2: 192. 1987; Matthew, Fl. Palni Hills 2: 1031. 1999. **(Fig. 12)**.

Herbs to 30 cm, woody at base; stem and branches more or less furnished with pubescence, often glabrous. Leaves deltoid-ovate or ovate, 1.4-3 x 1-2 cm, membranous, pilose, more so on nerves beneath or subglabrous, base truncate, apex obtuse or subacute; petioles filiform, to 1.2 cm, pubescent. Spikes slender, terminal, unbranched; flowers 2-3 in distant, sessile clusters, axis hairy; peduncles 2-4 cm. Bracts and bracteoles lanceolate or deltoid-ovate, ca 1.5 mm. Pedicels very short. Perianth pinkish, 5 lobed; lobes oblong-ovate, ca 2 x 1 mm, mucronate. Stamens 5, ca 1.5 mm, basal sheath fimbriate. Ovary obovoid, ca 1 x 0.75 mm; ovules 9; style to 0.25 mm; stigma trifid.

Occasional in foot hills.

Fl.: October.

Specimen examined: XCH 11534 (Cheranmahadevi).

Distribution: India, Sri Lanka.

3. **Celosia pulchella** Moq. in DC. Prodr., 13(2): 238. 1849; Wight, Ic. t. 1768. 1852; Hook.f., Fl. Brit. India 4: 715. 1885; Gamble, Fl. Pres. Madras 2: 1166. 1925; C.Towns. in Dassan. & Fosberg, Rev. Handb. Fl. Ceylon 1: 4. 1980; Kumari in A.N. Henry *et al.*, Fl. Tamil Nadu 2: 192. 1987; Matthew, Fl. Palni Hills 2: 1032. 1999. **(Fig. 13).**

Herbs to 30 cm; branchlets striate, glabrous or minutely pilose about nodes. Leaves deltoid-ovate to lanceolate, 2.5-7 x 0.7-2.7 cm, membranous, glabrous, base cuneate, apex acute to shortly acuminate; lateral nerves and reticulations punctiform; petioles 1-3 cm, filiform. Spikes terminal, slender, unbranched. Flowers 2-3, in distant, sessile clusters. Bracts and bracteoles lanceolate, ca 1.5 mm. Perianth with excurrent midrib, 5 lobed, straw coloured; lobes oblong-lanceolate, ca 5 x 1 mm. Stamens 5, connate at base, free part of filaments much longer, ca 2 mm. Ovary depressed globose, ca 5 x 5 mm; style to 5 mm; stigma trifid.

Occasional from foot hills to dry deciduous forests.

Fl.: Throughout the year.

Specimens examined: XCH 14135 (Papanasam); 15403 (Kalakad); 13860, 19250 (Courtallam).

Distribution: India, Sri Lanka.

6. CYATHULA Lour.

About 25 species in the tropics of both hemispheres; 3 from India.

Cyathula prostrata (L.) Blume, Bijdr. 549. 1826; Hook.f., Fl. Brit. India 4: 723. 1885; Gamble, Fl. Pres. Madras 2: 1172. 1925; C.Towns. in Dassan. & Fosberg, Rev. Handb. Fl. Ceylon 1: 26. 1980; Kumari in A.N. Henry *et al.*, Fl. Tamil Nadu 2: 192. 1987. Mohanan & Sivadasan, Fl. Agasthyamala 541. 2002. *Achyranthes prostrata* L., Sp. Pl. ed. 2. 296. 1762. *Desmochaeta prostrata* (L.) DC., Cat. Hort. Monsp. 102. 1813. *Pupalia prostrata* (L.) Mart. in Nov. Act. Acad. Caes. Leop.-Carol. 13: 321. 1826. **(Fig. 14).**

Slender herbs, prostrate to ascending or erect, rooting at nodes. Leaves opposite, mostly rhomboid-ovate or rhomboid-elliptic or elliptic, 1-4.5 x 0.8-2 cm, subglabrous or prominently pilose on both surfaces, cuneate at base, acute or obtuse at apex; petioles ca 0.5 mm, pilose. Spikes terminal, at first dense, later elongating to ca 25 cm. Flowers 2-3 in distant, sessile fascicles, with imperfect flowers, the fascicles often in heads or racemes. Bracts persistent; lanceolate, later deflexed; bracteoles 2, lanceolate, 1.5 mm, ciliate, with strongly hooked arista. Perianth 5 lobed, calycine, oblong-lanceolate, ca 3 x 0.7 mm, mucronate, surrounding imperfect flowers reduced to hooked, rigid awns. Stamens 5, reddish; filaments connate below into hypogynous, membranous cup, with intervening lacerate or 2-3 fid staminodes; anthers 2 celled, ca 1.25 mm. Ovary obovoid, ca 1 x 1 mm, 1 locular; ovule solitary, pendulous; style simple, filiform, ca 2 mm; stigma bifid.

Rare in semi-evergreen forests between 500-800 m.

Fl.: Throughout the year.

Specimens examined: XCH 15497, 18246 (Courtallam), 18885 (Papanasam).

Distribution: Widespread in the tropics of both old and New World.

7. DIGERA Forssk.

Monotypic genus found in the old world tropics.

Digera muricata (L.) Mart. in Nov. Act. Acad. Caes. Leop.-Carol. 13(1): 285. 1826; C.Towns. in Dassan. & Fosberg, Rev. Handb. Fl. Ceylon 1: 23. 1980; Rani & Matthew in Matthew, Fl. Tamil. Carnatic 3: 1311. 1983; Kumari in A.N. Henry *et al.*, Fl. Tamil Nadu 2: 192. 1987; Matthew, Fl. Palni Hills 2: 1032. 1999. *Achyranthes muricata* L., Sp. Pl. ed. 2: 295. 1762. *Achyranthes alternifolia* L., Mant. 50. 1767. *Digera arvensis* Forssk., Fl. Aeg.-Arab. 65. 1775; Hook.f., Fl. Brit. India 4: 717. 1885; Gamble, Fl. Pres. Madras 2: 1169. 1925. **(Fig. 15).**

Herbs to 30 cm, prostrate or ascending to erect; branchlets striate, glabrous or sparsely pilose, with faint ridges. Leaves alternate, variable, deltoid-ovate, 2.5-6 x 0.6-2.8 cm, membranous, sparsely appressed pilose on nerves, straw colour when dry, base truncate to cuneate, obtusely acute or acuminate at apex, apiculate; lateral nerves ca 7 pairs, faint, looping; petioles filiform, 2-2.5 cm, appressed pubescent. Flowers 3, in interrupted, ca 15 cm long spike; central flower fertile; two lateral flowers subtended by bracts. Bracts obovate; bracteoles 2, lanceolate, ca 1.5 cm. Perianth pinkish, 5 lobed; lobes subequal; outer 2 oblong-lanceolate, ca 3 mm, nerved; midrib prominent; inner 3 slightly shorter, delicate, hyaline. Stamens 5, basally connate, subequalling or shorter than style, ca 2.5 mm. Ovary obovoid, ca 1 x 1 mm; ovule 1; style filiform; stigmas 2. Fruits subglobose, slightly compressed; style persistent.

Rare in disturbed deciduous forests.

Fl.& Fr.: Throughout the year, peak in December – February.

Specimens examined: XCH 18418 (Sivagiri); 11988 (Mancholai).

Distribution: Asia from tropical Arabia and the Yemen to Afghanistan, W. Pakistan, India, Malaysia and Indonesia; South, Central and east tropical Africa, Madagascar.

8. GOMPHRENA L.

A rather large genus of about 90 species, centred in tropical America; 3 in India.

Key to the species

1a. Bracteoles with a small, dorsal crest only at apex; stamens included.....1. **G. celosioides**

1b. Bracteoles with a large, conspicuous crest throughout; stamens subexserted ....2. **G. globosa**

1. **Gomphrena celosioides** Mart. in Nova Acta Phys.-Med. Acad. Caes. Leop. Carol. 13: 301. 1826; C.Towns. in Dassan. & Fosberg, Rev. Handb. Fl. Ceylon 1: 53. 1980; Rani & Matthew, in Matthew. Fl. Tamil. Carnatic 3: 1312. 1983; Matthew, Fl. Palni Hills 2: 1033. 1999. *G. decumbens* sensu Gamble, Fl. Pres. Madras 2: 1179. 1925, non Jacq. **(Fig. 16).**

Ascending or erect herbs to 30 cm, much branched; branchlets spreading, lanate. Leaves oblanceolate or oblong-elliptic, 1.5-3 x 0.5-1 cm, chartaceous, lanate when young, base attenuate, narrowed into petiole, apex obtuse, mucronate; petioles 0-5 mm. Spikes terminal and axillary, sessile in uppermost pair of leaves, at first globose, to ca 1 x 1 mm, later cylindrical, to 3.5 x 1 cm, more or less lanate. Bracts ovate, membranous, ca 3 mm, mucronate with excurrent midrib; bracteoles ca 4 mm, mucronate with excurrent midrib. Perianth 5 lobed; lobes broadly lanceolate, ca 6 mm, lanate at base, apex acuminate; inner 3 acute, densely lanate at apex. Stamens 5; staminal tube subequalling tepals, included, ca 5 mm. Ovary oblong-globose, ca 2 x 1 mm; ovule 1, compressed; style ca 0.2 mm; stigma divergent, ca 0.3 mm.

Occasional in disturbed deciduous forests.

Fl.: Throughout the year.

Specimens examined: XCH 10571, 15159 (Mancholai); 13785 (Courtallam).

Distribution: Native of S. America, spreading rapidly in the warmer regions of the world.

2. **Gomphrena globosa** L., Sp. Pl. 224. 1753; Wight, Ic. t. 1784. 1852; Hook.f., Fl. Brit. India 4: 732. 1885; Gamble, Fl. Pres. Madras 2: 1179. 1925; C.Towns. in Dassan. & Fosberg, Rev. Handb. Fl. Ceylon 1: 55. 1980; Rani & Matthew in Matthew, Fl. Tamil. Carnatic 3: 1313. 1983; Kumari in A.N. Henry *et al.*, Fl. Tamil Nadu 2: 192. 1987; Matthew, Fl. Palni Hills 2: 1033. 1999. **(Fig. 17).**

Tamil: Vaadamalligai.

Erect or ascending herbs to 50 cm; branches appressed white hairy when young. Leaves elliptic-ovate or oblong, 2.5-7.5 x 1-2 cm, thinly pilose on both surfaces, base cuneate, apex acute, apiculate; petioles 2-5 mm. Spikes terminal, globose or depressed-globose, pinkish. Bracts ovate, membranous, mucronate; bracteoles strongly laterally compressed, mucronate with excurrent midrib. Perianth 5 lobed, lanceolate, ca 3 mm, pilose without. Stamens 5, ca 3.5 mm with staiminal sheath. Ovary oblong-globose, ca 2 mm. Ovule compressed; style short.

Planted, rare in foot hills.

Fl.: Throughout the year.

Specimen examined: XCH 14787 (Pechiparai).

Distribution: A native of tropical America, cultivated in the warmer regions of the world.

9. PSILOTRICHUM Blume

About 14 species in tropical Asia and Africa, chiefly in the latter; ca 5 from India.

Key to the species

1a. Branches tomentose; leaves scaberulous on both surfaces; spikes many-flowered ....2. **P. scleranthum**

1b. Branches pubescent or glabrous; leaves sparsely hairy; spikes few-flowered .... 1. **P. elliotii**

1. Psilotrichum elliotii Baker & C.B.Clarke in Dyer, Fl. Trop. Africa 6(1): 58. 1909; C.Towns. in Dassan. & Fosberg, Rev. Handb. Fl. Ceylon 1: 45. 1980; Rani & Matthew, Fl. Tamil. Carnatic 3: 1315. 1983; Kumari in A.N. Henry *et al.*, Fl. Tamil Nadu 2: 193. 1987; Matthew, Fl. Palni Hills 2: 1034. 1999. *P. calceolatum* Moq. ex Hook.f., Fl. Brit. India 4: 725. 1885, nom. illeg.; Gamble, Fl. Pres. Madras 2: 1174. 1925. *P. nudum* sensu Wight, Ic. t. 1775. 1852, non (B. Heyne ex Wallich) Moq. 1849. **(Fig. 18)**.

Spreading, matty herbs; stem and branches quadrangular and sulcate when young later terete, glabrous. Leaves deltoid to elliptic-ovate, sparsely white-hairy, sometimes glabrous, truncate to cuneate at base, apex acute, mucronate, 1-3 x 0.5-1.2 cm; petioles 0-4 mm. Spikes axillary, few flowered. Bracts deltoid-ovate, ca 2 mm, white-pilose; bracteoles broadly ovate, ca 1.7 mm, pilose. Perianth 5 lobed; outer 2 larger, ca 3 x 0.8 mm, 3-5 nerved, pilose without; inner 3 smaller, ca 2.5 x 0.6 mm, pilose without; middle one only ciliate along margin. Stamens 5, connate only at base, delicate, ca 3 mm. Ovary obong-globose, ca 1 x 0.75 mm; ovule 1, on curved funicle; style ca 0.5 mm; stigma capitate.

Occasional in foot hills to deciduous forests.

Fl. & Fr.: Throughout the year.

Specimens examined: XCH 11758 (Kalakad); 16606, 16605 (Papanasam); 16250 (Thirukurungudi).

Distribution: Sri Lanka, Peninsular India, Tropical Africa.

2. Psilotrichum scleranthum Thwaites, Enum. Pl. Zeyl. 248. 1861; Hook.f., Fl. Brit. India 4: 725. 1885; Gamble, Fl. Pres. Madras 2: 1174. 1925; C.Towns. in Dassan. & Fosberg, Rev. Handb. Fl. Ceylon 1: 44. 1980. **(Fig. 19)**.

Woody herbs, erect or scandent, much branched; branches spreading, tomentose, quadrangular, striate, often terete when mature. Leaves elliptic-ovate or elliptic-oblong, 2-4.6 x 1-1.8 cm, scabrous, more so on nerves beneath, base truncate-acute, apex acute, mucronate; petioles 0-3 mm. Spikes many flowered, branched, terminal and generally 2 at each node in axils of opposite stem or branches, up to 2.5 cm, sessile or pedunculate; peduncles upto 3 cm. Bracts ovate, ca 1.5 cm, appressed pubescent, mucronate with excurrent midrib; bracteoles whitish, cordate-ovate, ca 2.5 mm, mucronate with excurrent midrib, pilose along midrib and margins. Perianth 5 lobed; lobes lanceolate, ca 4 x 1.5 mm; outer 2 oblong-lanceolate, ca 4 mm, shortly pilose without; inner 3 lanceolate, ca 3.5 mm, pilose mostly in centre and on margins. Stamens 5, ca 2 mm. Ovary obovoid, ca 1 x 1 mm; ovule 1, on curved funicle; style ca 0.8 mm; stigma capitate. Utricle oblong, ca 2.5 mm.

Occasional in deciduous forests.

Fl. & Fr.: March.

Specimen examined: XCH 18465 (Sivagiri).

Distribution: Sri Lanka, Peninsular India, widespread in tropical Africa.

10. PUPALIA Juss.

4 species in the tropics of the old World from W. Africa to Malaysia and the Philippines; 1 in India.

Pupalia lappacea (L.) Juss. in Ann. Mus. Natl. Hist. Nat. 2: 132. 1803; Hook.f., Fl. Brit. India 4: 724. 1858; Gamble, Fl. Pres. Madras 2: 1173. 1925; C.Towns. in Dassan. & Fosberg, Rev. Handb. Fl. Ceylon 1: 29. 1980; Kumari in A.N. Henry *et al.*, Fl. Tamil Nadu 2: 192. 1987; Matthew, Fl. Palni Hills 2: 1035. 1999. *Achyranthes lappacea* L., Sp. Pl. 204. 1753. *A. atropurpurea* Lam., Encycl. 1: 546. 1785. *Pupalia atropurpurea* (Lam.) Moq. in DC., Prodr. 13(2): 331. 1849; Hook.f., Fl. Brit. India 4: 723. 1885; Gamble, Fl. Pres. Madras 2: 1173. 1925. **(Fig. 20).**

Tamil: Adai-otti.

Suberect herbs to 60 cm; branchlets pubescent, angular when young, later terete. Leaves opposite, decussate, elliptic-ovate or elliptic-lanceolate, 4.5-8.3 x 1.8-2.5 cm, chartaceous, sericeous, more so on nerves beneath, base truncate to cuneate, apex acute to shortly acuminate; lateral nerves ca 8 pairs, faint, punctiform; petioles filiform, 0.2-0.6 cm. Flowers lax, in slender, terminal, ca 18 cm long spikes. Bracts lanceolate, persistent, mucronate with midrib, subtending on each side by a triad of one solitary, central bisexual flower with two modified flowers on either side; spines of modified flowers accrescent into hooked bristles, in fruits bibracteolate, outer bracteoles ovate, inner lanceolate. Perianth 5 lobed; lobes lanceolate, ca 5 x 1 mm, sericeous, aristate, with excurrent midrib; outer 2 tomentose without, 3 nerved. Stamens 5, ca 2 mm, basally connate. Ovary obovoid, ca 1 x 1 mm; ovule 1, pendulous; style ca 0.8 mm; stigma capitate. Capsules ovoid, ca 7 mm, subtended by more than 15 hooked bristles.

Occasional in disturbed sites of deciduous forests.

Fl. & Fr.: September.

Specimen examined: XCH 14010 (Courtallam).

Distribution: Peninsular India, Sri Lanka, South and East Africa.

11. TRICHURIELLA Bennet

Monotypic genus native to India, Hainan and Indo-China.

Trichuriella monsoniae (L.f.) Bennet in Indian J. Forest. 8: 86. 1985. *Illecebrum monsoniae* L.f., Suppl. Pl. 161. 1781. *Celosia monsoniae* Retz., Obs. Bot. 2: 13. 1781, "monsonia". *Aerva monsoniae* (Retz.) Mart. in Nova Acta Phy. Med. Acad. Caes. Leop.-Carol. 13: 291. 1826; Wight, Ic. t. 725. 1843; Hook.f., Fl. Brit. India 4: 728. 1885; Gamble, Fl. Pres. Madras 2: 1178. 1925. *Trichurus monsoniae* (L.f.) C.Towns. in Kew Bull. 29: 466. 1974 et in Dassan. & Fosberg, Rev. Handb. Fl. Ceylon 1: 38. 1980; Rani & Matthew in Matthew, Fl. Tamil. Carnatic 3: 1317. 1983; Kumari in A.N. Henry *et al*., Fl. Tamil Nadu 2: 194. 1987. **(Fig. 21)**

Herbs to 50 cm, ascending or prostrate, much branched; branches lanate. Leaves linear-filiform, 1-2 cm, crowded on lower internodes, in upper parts mostly in pairs or in whorls. Flowers bisexual, numerous, in globose or ovoid, axillary and terminal spikes; spikes 0.5-1 cm, pedunculate; peduncles to 1 cm. Bracts and bracteoles linear-lanceolate, ca 1.25 mm, hyaline; midrib prominent, excurrent into awn. Perianth 4 lobed; lobes lanceolate, ca 1.75 x 0.25 mm, lanate without, hyaline. Stamens 4, connate at base, alternating with membranous, bifid staminodes; filaments ca 0.25 mm; anthers oblong, to 0.15 mm. Ovary subglobose, ca 0.5 mm; ovule 1, pendulous, on a long, curved funicle; style and stigma minute. Fruit a thin-walled utricle, circumcissile, ca 3 x 2 mm, enclosed by persistent bracteoles and tepals.

Occasional in foot hills.

Fl. & Fr.: February.

Specimen examined: XCH 14544 (Surankudi).

Distribution: Sri Lanka, India, Thailand.

3. BASELLACEAE Raf.

Twiners, Leaves alternate, entire, somewhat fleshy, estipulate. Flowers regular, bisexual, spicate or racemose. Bracts small; bracteoles 2, connate at the base of calyx. Tepals 5, free or basally united to form a tube. Stamens 5. ovary superior, 1 locular; ovule solitary. Fruit indehiscent, often surrounded by persistent, fleshy bracteols and perianth.

4 genera and 19 species in tropical and warm parts especially in the America; 2 from India.

BASELLA L.

5 species have been reported, mostly from Madagaskar; 1 from India.

Basella alba L., Sp. Pl. 272. 1753; Matthew & Rani in Matthew, Fl. Tamil. Carnatic 3: 1327. 1983; Harriman in Dassan. & Fosberg, Rev. Handb. Fl. Ceylon 7: 1. 1991; Kumari in A.N. Henry *et al*., Fl. Tamil Nadu 2: 194. 1987; Matthew, Fl. Palni Hills 2: 1038. 1999. **(Fig. 22)**.

Tamil: Paslakkirai.

Twining herbs; branchlets succulent, terete, glabrous. Leaves distally alternate, elliptic-ovate, glabrous, subsucculent, base rounded to acute, decurrent, margin entire, apex acute, ca 4.5 x 3 cm; petioles filiform, to 1 cm. Spikes axillary, ca 4 cm, unbranched, bearing flowers distally. Flowers bisexual, actinomorphic, with 3 small basal bracts to 1.25 mm, distally with 2 larger bracteoles exceeding perianth. Perianth 5 lobed; lobes oblanceolate, ca 3 x 1 mm, fleshy, connate at base. Stamens 5, opposite to perianth lobes and attached to them; filaments very minute, ca 1 mm. Ovary superior, ovoid-globose, ca 0.5 mm across, 1 locular; ovule 1, basal; style arms 3, ca 2 mm.

Occasional in foot hills.

Fl.: October – December.

Specimens examined: XCH 11203 (Courtallam); 14236 (Papanasam); 19849 (Sivagiri).

Distribution: Probably native to tropical Asia and Africa, often cultivated.

4. PHYTOLACCACEAE R.Br.

Erect or scandent, herbs or shrubs. Leaves alternate, entire, pinnately nerved, estipulate. Flowers bisexual, in axillary or leaf opposed spikes or racemes. Bracts and bracteoles small. Perianth green or somewhat coloured, 11-10 partite. Stamens 5-30, inserted on fleshy disc; filaments free or connate at base. Carpels 11-12, free or fused; ovule 1 in each carpel. Fruits fleshy or drupe like.

12 genera and about 100 species, mostly in tropical America and Africa; 4 genera and 7 species in India.

RIVINA L.

About 2 species in the world; 1 in India.

Rivina humilis L., Sp. Pl. 121. 1753; Matthew & Rani in Matthew, Fl. Tamil. Carnatic 3: 1329. 1983; Kumari in A.N. Henry *et al*., Fl. Tamil Nadu 2: 196. 1987; Philcox in Dassan. & Clayton, Rev. Handb. Fl. Ceylon 10: 325. 1996. **(Fig. 23)**.

Herbs to 30 cm, pubescent. Leaves broadly elliptic, membranous, puberulous on nerves beneath, otherwise glabrous, base acute, apex acute to acuminate, 3-6.3 x 1-3 cm; lateral nerves ca 6 pairs, obscure; petioles 0.5-4 cm, filiform. Racemes ca 10 cm, terminal and axillary, laxly many flowered. Flowers actinomorphic; pedicels ca 3 mm. Perianth 4 lobed; lobes oblong-obovate, ca 3 x 1 mm, free. Stamens 4, alternate with perianth lobes, connate at base; filaments ca 2 mm; anthers oblong, ca 5 mm. Ovary ovoid-globose, superior, 1 mm across, 1 locular; ovule 1, basal; style lateral, bent, upper half erect, to 1 mm; stigma peltate. Fruit pseudoberry, ovoid, ca 4 x 3 mm, with persistent style and connivent perianth lobes.

Occasional weed in deciduous forests.

Fl. & Fr.: May.

Specimen examined: XCH 12776 (Mahendragiri).

Distribution: Native in tropical America, naturalized in Sri Lanka, India and Malaysia.

5. POLYGONACEAE Juss.

Herbs, rarely shrubs, very rarely trees, sometimes climbing. Leaves alternate, rarely opposite, entire or serrulate; stipules scarious or membranous, usually sheathing the stem. Flowers bisexual, rarely polygamous, regular, solitary or in small, bracteate clusters; pedicels usually articulate. Perianth simple, calycine or corolline, inferior 3-6 lobed, free or connate, imbricate in bud. Stamens 5-8, rarely more or fewer, opposite to perianth-lobes. Disc annular, glandular or 0. Ovary free, sessile, 1 locular; ovule solitary, orthotropous, sessile or on a distinct funicle; styles 3 or 2, rarely 4; stigma capitate, peltate or fimbriate. Fruit a small, hard, trigonous or biconvex nutlet, enclosed in the perianth; seed erect; testa membranous; albumen floury, sometimes ruminate, embryo various; radicle superior.

Family of about 50 genera, occurring worldwide and mostly in northern temperate regions; 12 genera and 121 species have been reported from India.

Key to the genera

1a. Flowers in terminal racemes or spikes or heads; ochreae membranous 1. **Persicaria**
1b. Flowers in axillary clusters; ochreae hyaline..2. **Polygonum**

PERSICARIA (L.) P. Miller

Ca 130 species, cosmopolitan in distribution; nearly 30 species have been reported from India.

Key to the species

1a. Inflorescence of sessile or pedicellate flowers in branched spikes or racemes......... 2
1b. Inflorescence capitate, the heads single or in corymbs.. 4

2a. Bracts ciliate, hispid; flowers white................. 3
2b. Bracts not ciliate, glabrous; flowers pink ....3. **P. glabra**

3a. Leaves silky-pubescent on both surfaces..5. **P. pulchra**
3b. Leaves appressed hairy, more so on nerves beneath............................ 1. **P. barbata**

4a. Heads in the axils of an involucral leaf; perianth 4 fid .... 4. **P. nepalensis**
4b. Heads not in the axils of an involucral leaf; perianth 5 fid .... 2. **P. chinensis**

1. Persicaria barbata (L.) H. Hara, Fl. E. Himal. 1: 70. 1966; Matthew & Rani in Matthew, Fl. Tamil. Carnatic 3: 1334. 1983; Matthew, Fl. Palni Hills 2: 1044. 1999. *Polygonum barbatum* L., Sp. Pl. 362. 1753; Wight, Ic. t. 1798. 1852; Hook.f., Fl. Brit. India 5: 37. 1886; Gamble, Fl. Pres. Madras 2: 1189. 1925; Chandrasek. in A.N. Henry *et al.*, Fl. Tamil Nadu 2: 197. 1987; Philcox in Dassan. & Clayton, Rev. Handb. Fl. Ceylon 11: 397. 1997; Mohanan & Sivadasan, Fl. Agasthyamala 542. 2002. **(Fig. 24)**.

Herbs to 60 cm, glabrous or subglabrous, with appressed pubescence. Leaves lanceolate or linear-lanceolate, chartaceous, appressed hairy on both surfaces, more so on nerves, acute to cuneate at base, acute at apex, 6-11 x 0.8-2 cm. Ochreae tubular, coarsely ciliate with cilia to 1.7 cm, often longer than tube. Spikes terminal and axillary, to 7 cm, densely, closely flowered. Flowers deciduous. Bracts closely imbricate, oblong-obtuse, ciliate, ca 3.5 mm. Perianth white, ca 2 mm, 5 lobed; lobes ovate, subequal, ca 2 x 1 mm, obtuse. Stamens 5, ca 1.5 mm. Ovary obovoid, ca 0.5 mm across; style to 0.5 mm, 3 armed; stigma capitate. Nutlets triangular, ca 2 x 1.3 mm.

Rare near stream banks in deciduous forests.

Fl. & Fr.: Throughout the year.

Specimen examined: XCH 14618 (Kallar).

Distribution: Throughout the eastern hemisphere.

2. Persicaria chinensis (L.) H.Gross in Bot. Jahrb. Syst. 49: 26. 1913; Matthew & Rani in Matthew, Fl. Tamil. Carnatic 3: 1334. 1983; Matthew, Fl. Palni Hills 2: 1045. 1999. *Polygonum chinense* L., Sp. Pl. 363. 1753; Wight, Ic. t. 1806. 1852; Hook.f., Fl. Brit. India 5: 44. 1886 incl. var. *ovalifolia* Meisner; Gamble, Fl. Pres. Madras 2: 1190. 1925; Chandrasek. in A.N. Henry *et al.*, Fl. Tamil Nadu 2: 197. 1987; Philcox in Dassan. & Clayton, Rev. Handb. Fl. Ceylon 11: 400. 1997; Mohanan & Sivadasan, Fl. Agasthyamala 543. 2002. **(Fig. 25)**.

Straggling shrubs to 2 m; branches long, cylindrical, swollen at nodes, glabrous or puberulous or glandular pubescent. Leaves ovate-oblong or elliptic-oblong, glabrous or pubescent on nerves, base rounded or acute, tapering into petiole, margin entire to crenulate, apex obtuse to acute, 4-10 x 1-4 cm; lateral nerves ca 5 pairs; petioles to 1.5 cm. Ochreae 1-3.5 cm, membranous, narrowly lanceolate, abruptly acuminate, extending to the base of upper node. Inflorescence corymbose, ultimate branches with peduncle terminated by 1-3 pedunculate capitula; peduncles glandular-hispid. Bracts oblong, ca 2 mm. Perianth white, or pinkish-white, 5 lobed; lobes ovate, ca 3 x 1.5 mm, concave. Stamens 7-8, subexserted. Ovary globose, ca 0.5 mm across; style 3 armed, stout; stigmas capitellate. Nutlets trigonous, smooth.

Very common on the borders or on thickets in evergreen forests.

Fl. & Fr.: Throughout the year.

Specimens examined: XCH 18115, 19189, 18112, 15910 (Courtallam), 16043 (Naraikadu); 1537, 10964, 19649 (Kalakad); 19056, 14460 (Papanasam); 13483, 10406, 10458 (Kakachi); 10328, 11620, 10182, 10039, 10094, 10497, 10507, 12198, 12530, 10166, 14916, 15031, 10073, 13629 (Kothayar); 17518, 17414, 12765 (Mahendragiri).

Distribution: India, Sri Lanka, Myanmar, Malaysia, Philippines, China, Japan.

3. **Persicaria glabra** (Willd.) M.Gomez in Ann. Inst. Segunda Ensef., Habana. 2: 278. 1896; Matthew & Rani in Matthew, Fl. Tamil. Carnatic 3: 1336. 1983; Matthew, Fl. Palni Hills 2: 1045. 1999. *Polygonum glabrum* Willd., Sp. Pl. 2: 447. 1799; Wight, Ic. t. 1797. 1852; Hook.f., Fl. Brit. India 5: 34. 1886; Gamble, Fl. Pres. Madras 2: 1189. 1925; Chandrasek. in A.N. Henry *et al.*, Fl. Tamil Nadu 2: 198. 1987; Philcox in Dassan. & Clayton, Rev. Handb. Fl. Ceylon 11: 400. 1997. **(Fig. 26).**

Shrubs to 2 m, stout, glabrous, swollen above nodes, often reddish. Leaves linear-lanceolate, thick-chartaceous, glabrous, coppery-brown above when dry, paler beneath, base cuneate, decurrent into petioles, apex gradually acuminate, 13.5-21 x 1.4-2.8 cm; lateral nerves 20-35 pairs, close, obscure; midrib prominently raised above; petioles to 1 cm. Ochreae tubular, 2-4 cm, membranous, glabrous. Racemes terminal or in upper axils, ca 10 cm, pedunculate. Flowers many, closely arranged. Perianth rose in colour, 5 lobed; lobes oblong, ca 5 x 1.5 mm, connate at base, acute. Stamens 5; filaments to 3 mm; anthers oblong, ca 3 mm. Ovary ovoid, ca 1.5 mm across; styles 3, ca 1 mm, often divergent; stigmas capitate.

Rare by the stream sides in deciduous forests.

Fl.: July.

Specimen examined: XCH 16440 (Thirukurungudi).

Distribution: Old World tropics and America.

4. **Persicaria nepalensis** (Meisner) H.Gross in Bot. Jahrb. Syst. 49: 277. 1913; Matthew & Rani in Matthew, Fl. Tamil. Carnatic 3: 1337. 1983; Matthew, Fl.

Palni Hills 2: 1046. 1999. *Polygonum nepalense* Meisner, Monog. Polygon. 84. t. 7. f. 2. 1826; Wight, Ic. t. 1804. 1852; Chandrasek. in A.N. Henry *et al.*, Fl. Tamil Nadu 2: 198. 1987; Philcox in Dassan. & Clayton, Rev. Handb. Fl. Ceylon 11: 402. 1997. *P. punctatum* Buch.-Ham. ex D. Don, Prodr., Fl. Nep. 72. 1825, non Rafin. 1820; Gamble, Fl. Pres. Madras 2: 1190. 1925. *P. alatum* Buch.-Ham. ex Spreng., Syst. Cur. Post. 4(2): 154. 1827; Hook.f., Fl. Brit. India 5: 41. 1886 incl. var. *metzianum*. **(Fig. 27).**

Spreading herbs to 50 cm, glabrous or pubescent; stem pinkish. Leaves deltoid-ovate or rhomboid, 1-2.3 x 0.5-1.3 cm, appressed pubescent on nerves, base truncate to acute, decurrent into petiole, apex acute, often mucronate; lateral nerves ca 5 pairs, faint; petioles ca 2 mm, pubescent. Ochreae tubular, membranous, ca 0.5 cm, hairy. Flowers few, crowded in slender, pedunculate capitulum; capitula subtended by cordate, sessile, leafy bracts. Perianth pinkish or white, 5 lobed; lobes oblong-ovate, ca 3.5 x 0.8 mm, connate at base. Stamens 10, in two rows, one row alternate with perianth another one opposite to perianth, ca 1.5 mm. Ovary ovoid, ca 0.5 mm across; style to 1.4 mm, 3 armed; stigmas 3, capitate. Nutlets triangular-ovoid, ca 2 x 1.3 mm, compressed.

Rare in rocky crevices in evergreen forests, above 1000 m.

Fl.& Fr.: Throughout the year.

Specimens examined: XCH 12580, 17239 (Kothayar).

Distribution: Afghanistan, India and Sri Lanka, Java, Japan and Ethiopia.

5. **Persicaria pulchra** (Blume) Sojak in Preslia 46(2): 154. 1974; Matthew & Rani in Matthew, Fl. Tamil. Carnatic 3: 1339. 1983. *Polygonum pulchrum* Blume, Bijdr. 530. 1826; Chandrasek. in A.N. Henry *et al.*, Fl. Tamil Nadu 2: 199. 1987. *P. tomentosum* Willd., Sp. Pl. ed. 4, 2: 447. 1799, non Schrank 1789; Hook.f., Fl. Brit. India 5: 30. 1886; Gamble, Fl. Pres. Madras 2: 1189. 1925; Chandrasek. in A.N. Henry *et al.*, Fl. Tamil Nadu 2: 199. 1987; Philcox in Dassan. & Clayton, Rev. Handb Fl Ceylon 11; 400.1997. **(Fig. 28).**

Herbs to 1 m. Leaves linear-lanceolate, 6-11 x 1.5-2.5 cm, chartaceous, silky-tomentose, more so beneath, attenuate to decurrent at base, apex acute to acuminate; petioles very short. Ochreae ca 2.5 cm, tubular, membranous, softly ciliate. Panicles terminal, to 8 cm, pedunculate; peduncles to 3 cm, tomentose. Perianth whitish, 5 lobed; lobes oblanceolate, ca 2.5 x 1 mm, connate at base. Stamens 5, ca 4 mm. Ovary globose, ca 1 mm across; style arms ca 1 mm. Nutlets globose, compressed, ca 4 mm across, with persistent style.

Rare in foot hills.

Fl. & Fr.: May.

Specimen examined: XCH 15825 (Courtallam).

Distribution: Tropical and South Africa, Sri Lanka, India east to Malaysia.

POLYGONUM L.

About 157 species of cosmopolitan distribution; ca 15 species have been reported from India.

Polygonum plebeium R.Br., Prodr. 420. 1810, "*plebejum*"; Hook.f., Fl. Brit. India 5: 27. 1886 incl. var. *indica*; Gamble, Fl. Pres. Madras 2: 1188. 1925; Matthew & Rani in Matthew, Fl. Tamil. Carnatic 3: 1339. 1983; Chandrasek. in A.N. Henry *et al.*, Fl. Tamil Nadu 2: 199. 1987; Matthew, Fl. Palni Hills 2: 1048. 1999. *P. indicum* B. Heyne ex Roth, Nov. Pl. Sp. 208. 1821; Wight, Ic. t. 1808. 1852. **(Fig. 29)**.

Prostrate herbs, puberulous. Leaves small, oblong, ca 1.6 cm, coriaceous, obtusely acute. Ochreae obliquely tubular, lacerate, ca 1.5 mm. Flowers in axillary fascicles. Perianth rose in colour, ca 5 lobed; lobes broadly ovate, ca 1.4 x 1 mm. Stamens 8, included, alternate with basal disc lobes, ca 0.4 mm. Ovary 3-gonous, ovoid, ca 0.4 mm; style ca 0.2 mm. Nutlets ovoid, ca 3 mm, with persistent style.

Rare in foot hills.

Fl. & Fr.: March.

Specimen examined: XCH 20311 (Kalakad)

Distribution: Africa, W. Asia, Himalaya, India, east to China and Ryukyu, Malaysia, Australia.

6. ARISTOLOCHIACEAE Juss.

Herbs or shrubs, often climbing, usually with more or less unpleasant odour. Leaves alternate, petioled, entire or lobed; stipules 0. Flowers hermaphrodite, often large, usually lurid, terminal, axillary or lateral, solitary, racemose or cymose. Bracts various, often 0. Perianth simple, corolline, superior, regular or irregular, 3 lobed or tubular and variously shaped; lobes valvate. Stamens 6 or more, adnate in 1 or 2 series to the style column; anther cells parallel, extrorse, opening dorsally. Ovary 4-6 celled; placenta parietal, free or meeting at the axis; ovules numerous, anatropous; style columnar, ending in stigmatic lobes. Fruit a capsule or berry. Seeds numerous, horizontal or pendulous; albumen fleshy, copious; embryo minute.

Eight genera and 660 species worldwide; 4 genera and 30 species in India.

Key to the genera

1a. Climbing shrubs; ovary 6 celled .... 1. **Aristolochia**

1b. Herbs, shrubs or subshrubs; ovary 4 celled .... 2. **Thottea**

1. ARISTOLOCHIA L.

About 500 species with pantropical distribution; 18 species have been reported from India.

Key to the species

1a. Leaves cordate to sagittate at base, 5-7 nerved.. 2
1b. Leaves truncate to subcordate at base, 3 nerved.. 3

2a. Leaves cordiform-reniform, apex obtuse or acute.......................... 2. **A. grandiflora**
2b. Leaves broadly ovate or oblanceolate, apex acuminate.................. 1. **A. acuminata**

3a. Ovary glabrous; fruits not winged ....3. **A. indica**
3b. Ovary hairy; fruits winged................................ ....4. **A. krisagathra**

1. **Aristolochia acuminata** Lam., Encycl. 1: 254. 1783; *A. tagala* Cham. in Linnaea 7: 207. t. 5. f. 3: 1832; Gamble, Fl. Pres. Madras 2: 1202. 1925; Matthew & Rani in Matthew, Fl. Tamil. Carnatic 3: 1345. 1983; Kumari in A.N. Henry *et al*., Fl. Tamil Nadu 2: 202. 1987; Matthew, Fl. Palni Hills 2: 1052. 1999. *Aristolochia roxburghiana* Klotzsch in Monatsber Koenigl. Preuss. Akad. Wiss. Berlin. 596. 1859; Hock.f., Fl. Brit. India 5: 75. 1886. **(Fig. 30).**

Twining robust shrubs; stems grooved. Leaves broadly ovate or oblanceolate, 13-18 x 6-9 cm, membranaceous, glabrous, 5 nerved at base; nerves converging towards apex, base deeply cordate, margin entire, apex acuminate; lateral nerves 2-3 pairs, prominent beneath; petioles 4-6 cm, glabrous or pubescent, slightly dilated at base. Flowers 4-5, in axillary raceme; pedicels to 3 mm. Bracts ovate-lanceolate, ciliate. Perianth dark green, ca 7 cm, base villose within, 5 villous pits on each side within; tube short, to 1 cm, sparsely hairy within; mouth ca 1 cm across; limb entire, 3-5 cm, oblong, apex acute-emarginate. Stamens 6, glabrous; anthers oblong, ca 2 mm. Ovary ca 1 cm across, puberulous; style ca 5 mm; stigmatic lobes 6, obtuse. Capsules glabrous, ca 4 x 3 cm, grooved, winged; ribs prominent.

Rare, in semi-evergreen to evergreen forests between 700-900 m.

Fl.: June – September; *Fr.:* August.

Specimens examined: XCH 19036 (Papanasam); 19910 (Sivagiri).

Distribution: Myanmar, Sri Lanka, India and Java.

2. **Aristolochia grandiflora** Sw., Nov. Pl. 126. 1788; Bor & Raiz. Beautif. Ind. Climbers & Shrubs 15. 1954; Matthew & Rani in Matthew, Fl. Tamil. Carnatic 320. 1981; Kumari in A.N. Henry *et al*., Fl. Tamil Nadu 2: 201. 1987. **(Fig. 31).**

Vines to 6 m; stem faintly grooved. Leaves cordiform or reniform, 3-11 x 4.5-13.5 cm, glaucous beneath, membranous, glabrous, 5 nerved from base, base deeply cordate, sinus broad, margin entire, apex obtuse to acute; petioles to 4 cm, glabrous; stipules like prophyll, auriculate. Flowers solitary, 15-25 cm. Bracts auriculate; pedicels to 10 cm; tube ca 8 cm, bent above, with 2 patches of white glandular hairs at the base within; mouth ca 25 cm across; limb 2 lipped; upper lip straight, folded, ca 14 x 3 cm, densely scabrous within; lower lip broadly cordate, ca 4 x 7 cm, emarginate, separated from mouth by an elongate stalk. Stamens 6; anthers linear, ca 0.5 mm. Ovary oblong-lanceolate, 3-5 cm; ovules numerous; style ca 0.4 cm; stigma 6 lobed, erect, lanceolate, spreading, margin rolled back. Capsules oblong-lanceolate, pubescent, ca 6 x 2 cm, winged.

Occasional in moist deciduous forests.

Fl.: February – April; *Fr.:* October – November.

Specimens examined: XCH 17825, 20496 (Papanasam).

Distribution: India, Central America, Columbia.

3. **Aristolochia indica** L., Sp. Pl. 960. 1753; Hook.f., Fl. Brit. India 5: 75. 1886; Gamble, Fl. Pres. Madras 2: 1201. 1925; Matthew & Rani in Matthew, Fl. Tamil. Carnatic 3: 1345. 1983; Kumari in A.N. Henry *et al.*, Fl. Tamil Nadu 2: 201. 1987; Matthew, Fl. Palni Hills 2: 1052. 1999. *Aristolochia lanceolata* Wight, Ic. t. 1858. 1852. **(Fig. 32)**

Twining shrubs; stem faintly grooved. Leaves very variable, linear or ovate-lanceolate or oblong or subpanduriform, membraneous, glabrous or pubescent, 3 nerved from base, rounded or truncate or subcordate, margin entire, apex acute, 2-6.5 x 0.7-4 cm; lateral nerves 2-4 pairs, prominent beneath; petioles to 5 mm long, pubescent. Flowers 3-5, in axillary or terminal racemes. Bracts lanceolate, ca 1 mm. Pedicels ca 1.2 cm; perianth oblong-lanceolate, pale green, ca 4.8 cm, base globose, ca 5 mm, 2 villous hairy pits on each side within; tube funnel shaped, ca 1 cm, hairy or glabrous within; mouth oblique; limb 1 lipped, elliptic, ca 2 x 0.8 cm, glabrous or pubescent, apex acute, slightly emarginate. Stamens 6; anthers glabrous, ca 1 mm. Ovary oblong or lanceolate, ca 7.5 x 2.5 mm; ovules numerous; style linear, to 5 mm, thick; stigma 6 lobed, erect, incurved, ca 1.2 mm. Capsules 6 valved, oblong-ovoid, ca 3.5 x 2 cm, glabrous; seeds triangular, ca 6 x 5.2 mm, winged, flattened.

Common in moist deciduous to evergreen forest between 300-900 m.

Fl.: August to December; *Fr.:* Throughout the year.

Specimens examined: XCH 10808, 11065, 11655, 14302, 14360, 14475, 19542, 19639, 16734, 15347 (Kalakad); 13981, 15535, 18018, 18427 (Courtallam); 10349, 13564, 13604, 17258, 17346, 17552, 19409 (Kothayar).

Distribution: Sri Lanka, India, Nepal, Bangaladesh.

4. **Aristolochia krisagathra** Sivarajan & Pradeep in Pl. Syst. Evol. 163: 31. 1989; Mohanan & Sivadasan, Fl. Agasthyamala 546. 2002. **(Fig. 33).**

Twining shrubs; stems faintly grooved. Leaves simple, alternate, oblong-ovate; 3-8 x 1.5-3 cm, chartaceous, sparsely hairy, densely beneath, less so above, 3 nerved at base, base obtuse-shallowly cordate, margin entire, apex acute or retuse; lateral nerves 3-4 pairs; petioles slender, 1-2 cm, slightly puberulous. Flowers 5-7 in axillary or terminal racemes, irregular, bisexual. Bracts ovate, hairy, to 1.2 mm. Pedicels to 5 mm. Perianth ca 2.5 cm. base globose, to 6 mm, villous, 2 pits on each side within; tube short, funnel shaped, slightly puberulous without, villous within; mouth oblique; limb 1 lipped, greenish-purplish black, apex acute, puberulous without. Stamens 6; anthers 2 celled, extrorse, ca 1 mm. Ovary oblong, ca 6 mm, pubescent, 6 locular; ovules numerous; style linear, thick, ca 7 mm; stigma 6 lobed, connate, ca 7 mm. Capsules 6 valved, 6 ribbed, winged; pedicels 2 or 3-lated at ends, ca 2 cm, pubescent; seeds triangular, ca 7 x 6 mm, flat, winged; albumen fleshy.

Occasional, in dry deciduous-moist deciduous forests.

Fl. & Fr.: February – April.

Specimens examined: XCH 18301, 18424, 18439 (Sivagiri).

Distribution: India.

2. THOTTEA Rottb.

Ca 43 species in the Bangladesh, Borneo, Hainan, India, Jawa, Laos, Malaya, Myanmar, Philippines, Sri Lanka, Sulawesi, Sumatera, Thailand and Vietnam; 11 species have been reported from India.

Key to the species

1a. Stamens uniseriate 2
1b. Stamens biseriate .... 2. **T. dinghoui**
2a. Leaves hairy; stamens, ovary and capsules glabrous; fruits ca 2.5 cm long .... 1. **T. barberi**
2b. Leaves glabrous; stamens, ovary and capsules hairy; fruits > 5 cm long. .... 3. **T. siliquosa**

1. **Thottea barberi** (Gamble) Ding Hou in Blumea 27: 331-332. 1981; Kumari in A.N. Henry *et al.*, Fl. Tamil Nadu 2: 102. 1987. *Apama barberi* Gamble in Kew Bull. 1924: 386. 1924 et Fl. Pres. Madras 2: 1200.1925. **(Fig. 34).**

Shrubs with branches covered with lenticels, 30-60 cm, hairy when young. Leaves elliptic or oblong or oblanceolate, ca 5-11 x 3-6 cm, coriaceous, base rounded, margin entire, apex acuminate, 3 ribbed at base; lateral nerves 3-5 pairs, prominent beneath; petioles 2-3 mm, pubescent. Flowers in axillary cymes, dark purple. Bracts ca 2 mm, glabrous; pedicels short, glabrous. Perianth ovate-elliptic, densely pubescent without, slightly pubescent within, gland dotted within, ca 4.2 x 3.5 mm, apex acute-subobtuse. Stamens 6; anthers elliptic, introrse, glabrous, ca 1.2 x 0.5 cm; filaments short. Ovary oblong, pubescent, ca 5 mm long; ovules many; stigma 3-5 lobed. Capsules lanceolate, glabrous, ca 2.5 cm long, greenish.

Rare, in dense evergreen forests between 1000-1200 m.

Fl. & Fr.: June – August.

Specimens examined: XCH 18881, 17961 (Papanasam).

Distribution: India, Sri Lanka.

2. **Thottea dinghoui** Swarup in Blumea 28: 407-411. 1983; Manickam *et al.* in J. Bombay Nat. Hist. Society 98(2): 319-320. 2001. **(Fig. 35).**

Erect herbs; stems rusty pubescent, 30-50 cm. Leaves elliptic-oblanceolate, 4.5-7 x 1.8-3 cm, base acute, margin entire, apex acute-mucronate, chartaceous, dorsal surface puberulous, ventral surface densely pubescent; lateral nerves 3-4 pairs, prominent beneath; petioles ca 1.0 cm, pubescent. Flowers 1-3, radical, pale yellow. Bracts and bracteoles small. Perianth free, ovate-elliptic, apex acute, pubescent without, glabrous within, 5 x 5 cm. Stamens 15-20; anthers elliptic, introrse, glabrous, ca 1 x 0.6 cm; filaments long. Ovary oblong, pubescent, ca 1 cm; ovules many; stigma 3-6 lobed. Capsules lanceolate, pubescent, ca 4 x 0.5 cm, quadrangular, pale green, winged.

Rare, in dense evergreen forests between 400-600 m.

Fl. & Fr.: June – September.

Specimens examined: XCH 16673, 17174 (Papanasam).

Distribution: Southern Western Ghats (India).

3. **Thottea siliquosa** (Lam.) Ding Hou in Blumea 27: 327. 1981. Mohanan & Sivadasan, Fl. Agasthyamala 548. 2002. *Apama siliquosa* Lam., Encycl. 1: 91. 1783. *Bragantia wallichii* R.Br. ex Wight & Arn. in Edinburgh New Philos. J. 15: 181. 1833; Hook.f., Fl. Brit. India 5: 73. 1886; Gamble, Fl. Pres. Madras 2: 1200. 1925. **(Fig. 36).**

Erect shrubs; stem slightly pubescent, to 5 m. Leaves variable, normally oblong or elliptic-oblanceolate, ca 5 x 14 cm, base acute to slightly obtuse, chartaceous, rarely coriaceous, pubescent beneath, 3-5 ribbed at base, prominently beneath, 3-5 pairs from the midrib; petioles to 5 mm, glabrous. Flowers in axillary cymes, or lateral cymes, deep-purple or pale yellow. Bracts lanceolate, small, pubescent within, ca 2.7 x 1.5 mm, apex acute; pedicels short, hairy. Perianth ovate or elliptic, apex acute, 0.5-1 x 0.3-0.5 cm, hairy without. Stamens 6-12, pubescent, ca 2.5 mm; anthers short pubescent. Ovary hairy; lanceolate, 2-8.5 cm; ovules many; styles 3-4 fid, column short; stigma 6 lobed. Capsules lanceolate, densely pubescent, apex acute, 6-11 x 0.3-0.4 cm, greenish or purple; seeds oblong, trigonous, ca 2.4 x 1.1 mm.

Frequent, in the undergrowth of evergreen forests in shaded moist localities between 700-1500 m.

Fl. & Fr.: April.

Specimens examined: XCH 12132, 12450, 13551, 14076, 14900, 15078, 17209, 17264, 19307 (Kothayar); 10995, 12796, 12940, 14355, 15791, 16751, 17168, 18801, 18988, 19271, 19584 (Papanasam); 15768, 15840, 15895, 15916 (Courtallam); 17478, 18648 (Mahendragiri).

Distribution: India and Sri Lanka.

7. PIPERACEAE Giseke

Herbs or shrubs, usually aromatic; branches often swollen at the nodes. Leaves alternate, opposite or whorled, often gland dotted; stipules none or 2, connate or adnate to the petiole. Flowers minute, hermaphrodite or unisexual, bracteate, in axillary or terminal, catkin like spikes. Perianth 0. Stamens 1-10, hypogynous or connate with the ovary at its base; anthers often jointed on the filaments, the cells dehiscing longitudinally. Ovary usually 1 celled, with a single, terminal orthotropous ovule, sometimes of 3-4 carpels; stigma sessile, simple or penicillate. Fruit an indehiscent berry in species with 1 cell, otherwise of separate cocci or follicles; seed adherent to the endocarp, globose, ovoid or oblong; testa thin; albumen floury; embryo in a cavity remote from the hilum; cotyledons minute or obsolete; radicle superior.

5 genera and 3500 species worldwide; 2 genera and 95 species in India.

Key to the genera

1a. Herbs; flowers hermaphrodite...1. **Peperomia**

1b. Erect or climbing shrubs; flowers dioecious..2. **Piper**

1. PEPEROMIA Ruiz & Pav.

Ca. 1300 species in the tropical and subtropical regions, especially in the America; around 12 species have been reported from India.

Key to the species

1a. Leaves thin, membranous, cordate at base .... 3. **P. pellucida**

1b. Leaves thick, succulent, cuneate to broad at base 2

2a. Stem glabrous 3

2b. Stem crisped hairy .... 1. **P. blanda**

3a. Plants robust; lamina >2 cm long 4

3b. Plants slender; lamina < 1.5 cm long.. 5

4a. Leaves 3 ribbed at base; spike stout ....4. **P. portulacoides**
4b. Leaves 5 ribbed at base; spike slender...6. **P. thomsonii**
5a. Spikes hairy; leaves subsessile, generally whorled ....5. **P. tetraphylla**
5b. Spikes glabrous; leaves petioled, opposite, ternate towards tip ....2. **P. heyneana**

1. Peperomia blanda (Jacq.) Kunth, Nov. Gen. Sp. 1: 67. t. 13. 1815. *P. dindigulensis* Miq., Syst. Piperac. 122. 1843; Hook.f., Fl. Brit. India 5: 98. 1886; Gamble, Fl. Pres. Madras. 2: 1210. 1925; Matthew & Rani in Matthew, Fl. Tamil. Carnatic 3: 1348. 1983; Kumari in A.N. Henry *et al.*, Fl. Tamil Nadu 2: 202. 1987; Matthew, Fl. Palni Hills 2: 1055. 1999. *P. blanda* (Jacq.) Kunth var. *floribunda* (Miq.) H.Huber in Dassan. & Fosberg, Rev. Handb. F1. Ceylon 6: 294. 1987. **(Fig. 37).**

A small, erect herb, rooting at lower nodes, to 6-14 cm; stem crisped hairy. Leaves opposite, pellucid-punctate, ovate-elliptic or obovate, 1.5-4 x 1-2.5 cm, chartaceous, glabrous, base cuneate, margin entire, apex acute-subobtuse, 3-5 nerved at base; petioles to 1.2 cm, hairy. Spikes terminal, linear, 1-2, pubescent, 2-5 cm. Bracts ovate, fleshy, apex subacute, ca 0.4 mm. Stamens 2, ca 0.1 mm; filaments ca 0.4 mm; anthers oblong, ca 0.6 mm, papillate without. Berries globose, ca 0.9 mm across, punctate.

Occasional, on wet rocks or rotten trees above 700 m.

Fl. & Fr.: July – November.

Specimens examined: XCH 11216, 13985 (Courtallam); 10933, 17358 (Kalakad).

Distribution: Peninsular and Sri Lanka.

2. Peperomia heyneana Miq., Syst. Piperac. 123. 1843 et in London J. Bot. 5: 550 1846; Hook.f., Fl. Brit. India 5: 99. 1886; Gamble, Fl. Pres. Madras 2: 1210. 1925; Kumari in A.N. Henry *et al.*, Fl. Tamil Nadu 2: 202. 1987;. Matthew, Fl. Palni Hills 2: 1055. 1999. *Peperomia haycockii* C.DC. in DC., Prodr. 16(1): 404. 1869. **(Fig. 38).**

A small, slender, epiphytic, ascending herb to 4-8 cm, rooting at nodes; stem glabrous, tender parts soft puberulous. Leaves opposite at base, ternate towards tip, ciliolate, oblong-obovate, 1-1.5 x 0.4-0.9 cm, coriaceous, pubescent on both surfaces, base cuneate, margin entire, apex emarginate-obtuse; petioles to 1.5 mm, puberulous. Bracts peltate, spathulate, ca 0.5 x 0.5 mm; stalk ca 0.5 mm. Stamens minute. Ovary ellipsoid, globose, puberulous within, ca 0.4 x 0.2 mm; stigma minute. Berries globose-ellipsoid, ca 0.6 x 0.4 mm, minutely stipitate, finely verrucose, with pseudocupule at base.

Rare, in shade of trees in evergreen forests at 900 m.

Fl.: & Fr.: March – April.

Specimens examined: XCH 19267 (Courtallam); 18637, 17391, 18637 (Mahendragiri); 19020, 17924, 13386, 18462 (Kalakad).

Distribution: Sri Lanka, India.

3. **Peperomia pellucida** (L.) Kunth, Nov. Gen. Sp. 1: 64. 1816: Gamble, Fl. Pres. Madras 2: 1210. 1925; Huber in Dassan. & Fosberg, Rev. Handb. Fl. Ceylon 6: 299. 1987; Mohanan & Sivadasan, Fl. Agasthyamala 551. 2002. *Piper pellucidum* L., Sp. Pl. 30. 1753. *Piper exiguum* Blume in Verh. Batav. Genootsch Kunsten 11: 232. 1826. *Peperomia exigua* (Blume) Miq., Syst. Piperac. 77. 1843. **(Fig. 39).**

A delicate, entirely glabrous, annual, succulent herb to 15-20 cm; branchlets transluscent, striate when dry; stem weak, procumbent. Leaves alternate, membranaceous, broadly ovate-deltoid, 1-2.5 x 0.8-2.3 cm, base cordate, with sinus, margin entire, apex acute, palmately 5-7 nerved from base, pellucid, shining above, very pale beneath; petioles 0.8 to 1.2 cm, thin, striate, decurrent towards base. Spike solitary, terminal but frequently leaf-opposed, 2-2.5 cm; peduncle 1-2 cm, as thick as spike axis, pellucid. Flowers scarcely immersed in slender spikes. Bracts peltate, sessile, brownish. Stamens 2-3, ca 0.25 cm; anthers ca 0.2 mm. Ovary oblong, glabrous, ca 0.28 x 0.16 cm; stigma 3 lobed, sessile, fleshy, oblique, penicillate, subapical. Berries globose-ellipsoid, minute, 3-gonous, ca 1 x 0.75 mm; seeds warty.

Occasional, in wet rocky places in semievergreen forests around 600 m.

Fl. & Fr.: Throughout the year.

Specimen examined: XCH 17538 (Mahendragiri).

Distribution: Pantropical of American origin, India, Sri Lanka.

4. **Peperomia portulacoides** (Lam.) Dietr., Sp. Pl. ed. 6. 1: 172. 1831; Hook.f., Fl. Brit. India 5: 98. 1886; Gamble, Fl. Pres. Madras 2: 1210. 1925; Kumari in A.N. Henry *et. al.*, Fl. Tamil Nadu 2: 202. 1987; Mohanan & Sivadasan Fl. Agasthyamala 552. 2002. *Piper portulacoides* Lam., Tabl. Encycl. 1: 82. 1791. **(Fig. 40).**

Epiphytic, succulent herbs, erect, ascending, to 25-45 cm; stem glabrous, fleshy, rooting at lower nodes, striate when dry. Leaves alternate below, opposite or ternately whorled above, obovate or elliptic or rhomboid, membranaceous, glabrous on both surfaces, dark above, pale beneath, base acute or cuneate, margin undulate, apex emarginate or obtuse. 2-4 x 2-3 cm; petioles 0.5-1 cm. Spikes terminal and axillary, one or two, glabrous, to 5 cm. Bracts ovate, peltate, sessile, apex acute, ca 8 x 0.2 cm; peduncle to 2 cm, striate, glabrous. Stamens 2, ca 1 mm; filaments ca 0.5 mm; anthers oblong, ca 0.4 mm. Ovary obovoid, beaked, 0.5 x 0.25 mm; stigma penicillate. Berries small, glabrous, ca 0.75 x 0.5 mm, partly immersed in rachis.

Common, on rocks and tree trunks between 600-800 m.

Fl. & Fr.: March – August.

Specimen examined: XCH 19101 (Kalakad).

Distribution: Madagascar to South West India.

5. **Peperomia tetraphylla** (Forst.f.) Hook. & Arn., Bot. Bech. Voy. 96. 1832; Kumari in A.N. Henry *et al.*, Fl. Tamil Nadu 2: 203. 1987; Matthew, Fl. Palni Hills 2: 1056. 1999. *Piper tetraphyllum* Forst.f., Insul. Austr. Prodr. 5. 1786. *Peperomia reflexa* (L.f.) Dietr., Sp. Pl. ed. 6. 1: 180. 1831, non H.B.K. 1816; Wight, Ic.t. 1923. f. 1. 1853; Hook.f., Fl. Brit. India 5: 99. 1886; Gamble, Fl. Pres. Madras 2: 1209. 1925. **(Fig. 41).**

Slender, perennial herbs to 20 cm, main axis creeping, branched; branches ascending or erect; stems glabrous, shallowly channelled when fresh, deeply furrowed and acutely angled when dry. Leaves whorled, rhomboid or orbicular, 0.5-1.25 x 0.3-0.6 cm, coriaceous, glabrous above, glabrate beneath, base cuneate, margin entire, apex obtuse; veins obscure; petioles to 2 cm, densely pubescent. Spikes solitary, terminal, 1-2 x 0.2 cm, axis hairy, much exceeding the leaves; peduncle to 1 cm, thinly pubescent. Bracts orbicular, ca 0.5 x 0.5 mm. Filaments ca 0.15 mm; anthers 0.1-0.25 mm, exserted. Ovary ovoid, ca 0.5 x 0.3 mm, sessile, smooth, basal one third forming mucilaginous pseudocupule; style 0; stigma penicillate. Berries ovoid, ca 1 x 0.5 mm.

Occasional, in shady, rocky places in evergreen forests between 1000-1300 m.

Fl. & Fr.: December – April.

Specimens examined: XCH 12498, 12895 (Papanasam); 19095 (Kalakad); 19155, 19163 (Courtallam); 15324, 19891, 19902 (Sivagiri); 20102 (Kothayar).

Distribution: Africa, China, India, Sri Lanka, Indo-China and Malaysia

6. **Peperomia thomsonii** Hook.f., Fl. Brit. India 5: 97. 1886; Gamble, Fl. Pres. Madras 2: 1210. 1925; Kumari in A.N. Henry *et al.*, Fl. Tamil Nadu 2: 203. 1987. **(Fig. 42).**

Epiphytic, 4.5-8 x 2.3-4 cm, succulent herbs. Leaves alternate, acute to cuneate at base, prominently 5 ribbed at base; apex acute. Spikes very slender, ca 10 cm. Flowers minute, sunk in the rachis. peduncles ca 5 cm. Bracts sessile, peltately attached, suborbicular, ca 0.5 mm across. Stamens 2, ca 0.25 mm; filaments very short; anther cells confluent. Ovary ovoid, ca 0.75 mm, beaked; ovule solitary, erect.

Rare, in semi-evergreen forests, ca 700 m.

Fl.: March.

Specimen examined: XCH 18193 (Courtallam).

Distribution: Western Ghats of Peninsular India.

2. PIPER L.

Ca 2100 species distributed pantropically; about 21 species have been reported from India.

Key to the species

1a. Erecct shrubs or sub-shrubs .... 8. **P. umbellatum**
1b. Herbs or climbing shrubs 2

2a. Flowers with connate bracts .... 6. **P. trichostachyon**
2b. Flowers with free bracts 3

3a. Petiole, peduncle with spreading hairs .... 2. **P. hymenophyllum**
3b. Petiole, peduncle glabrous 4

4a. Spikes erect 5
4b. Spikes pendulous 6

5a. Fruiting spike subglobose, short .... 4. **P. mullesua**
5b. Fruiting spike cylindric, elongate .... 3. **P. longum**

6a. Stout woody climbers, also cultivated .... 5. **P. nigrum**
6b. Slender climbers, wild 7

7a. Berries globose .... 1. **P. argyrophyllum**
7b. Berries ellipsoid .... 7. **P. trioicum**

1. Piper argyrophyllum Miq., Syst. Piperac. 330. 1844; Wight, Ic. t. 1941. 1853; Hook.f., Fl. Brit. India 5. 93. 1886; Gamble, Fl. Pres. Madras 2: 1207.1925; Kumari in A.N. Henry *et al.*, Fl. Tamil Nadu 2: 203. 1987; Matthew, Fl. Palni Hills 2: 1057. 1999. **(Fig. 43).**

Slender, terrestrial vines; stem ribbed, thickened at nodes. Leaves broad, elliptic to lanceolate, 6-10 x 2.5-4.5 cm, membranous, glabrous above, with scattered silver blotches beneath, base acute, often oblique, margin entire, apex acuminate, 3 nerved from the base; petioles to 1 cm, glabrous, ribbed. Spikes 3-5 cm, pendulous; peduncle to 1 cm, often glabrous. Flowers greenish, unisexual. Bracts obovate-orbicular, ca 2 mm, closely imbricate, medially adnate to spike axis, with free margins, adaxially crispate hairy. Stamens 2-4; filaments short and wide; anthers 4 celled, distinct, glabrous, ca 0.5 x 0.6 mm. Ovary ovoid-ellipsoid, glabrous, ca 0.8 x 0.6 mm; styles 3 or 4, reflexed, caducous. Berries globose, ca 4.5 x 3.8 mm, glabrous, smooth.

Occasional, in evergreen forests between 900-1500 m.

Fl. & Fr.: August – September.

Specimens examined: XCH 13592, 16645 (Kalakad); 18795 (Mahendragiri).

Distribution: Peninsula, Sri Lanka, introduced into Mascarena islands.

2. **Piper hymenophyllum** Miq. in London J. Bot. 5. 554. 1846; Wight, Ic. t. 1942. 1853; Hook.f., Fl. Brit. India 5: 93. 1886; Gamble, Fl. Pres. Madras 2: 1207. 1925; Kumari in A.N. Henry *et al.*, Fl. Tamil Nadu 2: 203. 1987; Matthew, Fl. Palni Hills 2: 1058. 1999. *Piper nilghirianum* C.DC. in DC., Prodr. 16(1): 364. 1869. **(Fig. 44).**

Terrestrial vines; stem pubescent, thickened at nodes, 3-5 ribbed. Leaves broadly elliptic-ovate, 7-10 x 2.5-4 cm, thinly coriaceous, pellucid when dry, glabrous above, crispate hairy beneath, base rounded, more or less oblique when young, margin entire, apex shortly acuminate; petioles small, pubescent with crisped hairs, to 1.5 cm. Spikes long, glabrous, 6.5-12 x 0.3-0.4 cm; peduncle to 1 cm, longer than petiole of opposed leaf. Male: Bracts orbiculate-ovate, closely imbricate, margin thin. Stamens 3 or 4; filaments stout; anthers reniform, rounded, decurrent to the filament. Female: Bracts orbicular-ovate; bracteoles confluent with bract. Ovary ovate-ellipsoid, ca 1.6 x 1.7 mm, styles 3 or 4, stout, recurved. Berries ellipsoid-globose.

Rare, in evergreen forests between 1200-1500 m.

Fl.: September – October.

Fr.: November – April.

Specimens examined: XCH 10353, 13605, 14304, 17206, 17208, 19690 (Kalakad).

Distribution: Peninsular India.

3. **Piper longum** L., Sp. Pl. 29. 1753; Hook f., Fl. Brit. India 5: 83. 1886; Gamble, Fl. Pres. Madras 2: 1205. 1925; Kumari in A.N. Henry *et al.*, Fl. Tamil Nadu 2:204. 1987. *Chavica roxburghii* Miq., Syst. Piperac. 239. 1843; Wight, Ic. t. 1928. 1853.

Perennial vines; stems herbaceous, slender, prostrate or ascending. Leaves ovate or cordate, 4-4.5 x 1.3-1.5 cm, thinly membranous, pellucid when dry, glabrous at maturity except minutely puberulous midrib and veins, base unequally cordate, amplexicaul, margin entire, apex acute; petiole to 0.2 cm. Spikes 1-2 x 0.4 cm; peduncle to 2 cm, minutely puberulous, longer than petiole of opposed leaf; Female flower yellowish white, crowded; bracts peltate, stalked, membranous, inconspicuously puberulous on margin. Ovary obovoid, ca 0.7 x 0.7 cm; stigma 4-5 lobed, short. Berries concrescent forming fleshy spadix, small, obovoid ca 0.2 cm across.

Cultivated

Fl. & Fr.: November – March.

Specimen examined: XCH 14688 (Kallar).

Distribution: India, Sri Lanka and Malay Peninsula.

4. **Piper mullesua** Buch.-Ham. ex D.Don, Prodr. Fl. Nepal. 20. 1825; Kumari in A.N. Henry *et al.*, Fl. Tamil Nadu 2: 204 1987; Matthew, Fl. Palni Hills 2: 1058. 1999. *Chavica sphaerostachya* Miq., Syst. Piperac. 279. 1843; Wight, Ic.t. Pl. Ind. Orient. t. 1931. 1853. *Piper brachystachyum* Wallich ex Hook.f., Fl. Brit. India 5: 87. 1886; Gamble, Fl. Pres. Madras 2: 1206. 1925; Fyson, Fl. S. Ind. Hill Stat. 499. t. 433. 1932. **(Fig. 45).**

Slender, branched, extensive climbers; branchlets slightly pubescent. Leaves alternate, elliptic-ovate or lanceolate, 3.5-8 x 1.5-3 cm, glabrous on both surfaces, base acute, margin entire, apex caudate-acuminate; petioles small, glabrous, to 0.5-1 cm. Spikes erect, globose-oblong, 0.5-1 x 0.3-0.6 cm; peduncle to 6 cm, glabrous. Anthers rounded. Female spike subglobose, yellowish green. Bracts orbicular, ca 0.5 mm across, imbricate, crispate hairy within. Ovary ovoid, glabrous, 1.5 x 1 mm; stigma 3 or 4 lobed, reflexed, caducous. Berries small, spherical-ovoid, 2 x 2 mm, smooth.

Rare in evergreen forest between 1200-1500 m.

Fl.: April – July.

Fr.: Throughout the year.

Specimens examined: XCH 10444, 10462, 12125, 12879, 14902, 15039, 15390, 19826 (Kalakad); 18879 (Papanasam); 19950, 19826 (Sivagiri).

Distribution: Peninsular and Northeast India.

5. **Piper nigrum** L., Sp. Pl. 28. 1753; Hook.f., Fl. Brit. India 5: 90. 1886; Gamble, Fl. Pres. Madras 2: 1206. 1925; Matthew, Fl. Tamil. Carnatic 3: 1352. 1983; Kumari in A.N. Henry *et al.*, Fl. Tamil Nadu 2: 204. 1987; Matthew, Fl. Palni Hills 2: 1059. 1999. **(Fig. 46).**

Tall, glabrous climbers; stem terete, thickened at nodes. Leaves opposite, broadly ovate-elliptic, coriaceous, glabrous, 5-6 ribbed, 1-2 pairs of ribs basal, others higher up, impressed above, base obtuse to broadly acute, or truncate, slightly symmetrical, margin entire, apex shortly acuminate; petioles to 3.5 cm; stipules deciduous, adnate to petiole. Spikes 4-7 cm; peduncle to 1.5 cm, glabrous. Flowers bisexual; greenish. Bracts ovate-orbicular. Stamens 2, on either side of pistil; filaments flat, 0.75 mm; anther lobes widely separated, partly embedded in fleshy filaments, crispate-hairy within the cavity on either side of the ovary. Ovary globose-ovoid, ca 1 mm, 3 angled; stigma 3 lobed. Berries spherical, black, ca 0.3 x 0.3 cm.

Rare, in evergreen forests between 800-1200 m. Also cultivated largely

Fl. & Fr.: September – February.

Specimens examined: XCH 10173, 10313 (Kothayar); 14431 (Kalakad); 14668 (Kulasekaram).

Distribution: India and Sri Lanka.

6. **Piper trichostachyon** (Miq.). C.DC. in DC., Prodr. 16(1): 242. 1869; Hook.f., Fl. Brit. India 5: 80. 1886; Gamble, Fl. Pres. Madras 2: 1205. 1925; Kumari in A.N. Henry *et al.*, Fl. Tamil Nadu 2: 204. 1987. *Muldera trichostachya* Miq. in London J. Bot. 5: 556. 1846; Wight, Ic. t. 1944. 1853. **(Fig. 47).**

Stout-stemmed climbers; nodes thickened. Leaves alternate, elliptic-ovate or lanceolate, 3-7.5 x 1-3.5 cm, coriaceous, glabrous, green, glaucous above, light green beneath, base slightly oblique, margin entire, apex acute; lateral nerves 2-3 pairs, prominent, anteriormost pair emerges from midrib about 1-1.5 cm above leaf base; nerves strong beneath; petiole to 1 cm, minutely pubescent. Spikes ca 3 cm, minutely pubescent; peduncle ca 3 mm, glabrous. Flowers greenish, unisexual. Female: Bracts decurrent at base, upper portion transformed into a fleshy conical cup with a narrow slit like mouth, puberulous. Ovary obovoid, ca 0.3 x 1 mm; style 0; stigma 3-4 lobed, short. Young berries oblong, ca 1 x 1.4 mm.

Rare, in evergreen forest between 1000-1200 m.

Fl. & Fr.: April – June.

Specimen examined: XCH 12568 (Kothayar).

Distribution: **Endemic** to India.

7. **Piper trioicum** Roxb., Fl. Ind. 1: 151. 1820; Wight, Ic. t. 1935 & 1936. 1853; Kumari in A.N. Henry *et al.*, Fl. Tamil Nadu 2: 204. 1987. *Piper attenuatum* Buch.-Ham. ex Miq., Syst. Piperac. 306. 1844; Hook.f., Fl. Brit. India 5: 92. 1886; Gamble, Fl. Pres. Madras 2: 1207. 1925; Matthew, Fl. Tamil. Carnatic 3: 1351. 1983. *Piper sylvestre* Wight, Ic. Pl. Ind. Orient. t. 1937. 1853; Hook.f., Fl. Brit. India 5: 93. 1886. **(Fig. 48).**

Slender climbers; branches zig-zag, glabrous, ribbed. Leaves alternate, ovate or broadly elliptic-ovate, or cordate, 2.5-12.5 x 1.5 x 6.5 cm, membranous, glabrous on both surfaces, with minute white dots on dorsal surface, base rounded, rarely cordate, apex abruptly acuminate, 3-7 nerved from the base; nerves connected by transverse reticulations, becoming dark green to black with age, bullate above; petiole to 1.5 cm, glabrous. Spikes axillary, 5-8 cm. Male: bracts ovate, membranous, ca 1.2 mm. Stamens 3; filaments ca 0.3 mm; anthers glabrous, globose, ca 0.2 mm. Female: bracts linearly obovate to elliptic, sessile, decurrent, with raised membranous margin, apex subobtuse; bracteoles connate, ca 1.5 x 1.5 mm. Ovary globose, ovoid, ca 3 x 2 mm; styles 3-4, recurved or closely pressed over ovary; stigma 4 lobed. Berries smooth, ellipsoid, ca 0.3 mm across.

Occasional, in moist deciduous and semi evergreen forests between 300-600 m.

Fl.: August – October; *Fr.:* April – May.

Specimens examined: XCH 10469, 11137, 11595, 16715, 19537 (Kalakad); 14715, 18628, 18631; 19365 (Balamore); 17038 (Papanasam); 13939, 15834, 19154 (Courtallam).

Distribution: India, east to Java.

8. **Piper umbellatum** L., Sp. Pl. 30. 1753. *Piper subpeltatum* Willd. Sp. P1. 1: 166. 1798; Hook.f. Fl Brit. India 5: 95. 1886. *Lepianthes umbellata* (L.) Raf. in Sylva Tellur. 84 1838; Kumari in A.N. Henry *et al*., Fl Tamil Nadu. 2: 202. 1987; Huber in Dass. & Fosberg, Rev. Handb. Fl. Ceylon 6: 289. 1987; Matthew, Fl. Palni Hills 2: 1053. 1999. *Pothomorphe subpeltata* (Willd.) Miq. Comm. Phytogr. 37.1840; Wight Icon. Pl. Ind Orient. t. 1925. 1853. **(Fig. 49).**

Subshurbs to 1 m. Leaves alternate, orbicular, 8-22 x 10-28 cm, membranous, slightly hairy, palmately nerved, 10-14 from base, apex obtuse, base deeply cordate, margin entire; petioles 8-22 cm, pubescent, base sheathing. Cylindrical spikes axillary, 3-10 x 0.2-0.3 cm, 1-6 in umbels. Flowers bisexual; peduncles 1-3.5 cm, glabrous; bracts 2, partly covering the flower, peltate, ca 0.5 mm across, margin ciliate. Perinath 0. Stamens 2; anthers minute, 2 celled, 4 lobed, distinct. Ovary obovoid, ca. 0.5 x 0.2 mm, glabrous.

Found occasionally in the shady places in moist deciduous and evergreen forests between 500-700 m.

Fl. & Fr.: July - September

Specimens examined: XCH 10545, 12827, 14433, 15166 (Papanasam and Upper Kothayar)

Distribution: Africa, Sri Lanka, India, South East Asia and Tropical America.

8. CHLORANTHACEAE R.Br. ex Sims.

Herbs, shrubs or trees, usually aromatic. Leaves opposite, usually toothed; petioles often connate into a sheath; stipules small, subulate, usually on the margin of the sheath. Flowers in terminal or pseudo-axillary spikes, heads or panicles, unisexual or hermaphrodite; male flowers without perianth; female also or with a three toothed limb adnate to the ovary. Stamens 1 or 3, connate; filaments very short, thick; anthers 2 celled or if 3, side ones 1 celled. Ovary 1 celled; ovule1, orthotropous, pendulous; style very short. Fruit a small, ovate or globose drupe; exocarp fleshy; endocarp crustaceous; seeds pendulous; testa membranous; albumen copious, fleshy; embryo minute; cotyledons divaricate or scarcely distinct; radicle inferior.

4 genera and 67 species worldwide; 2 genera and 7 species in India.

SARCANDRA Gard.

Sarcandra chloranthoides Gard. in Calcutta J. Nat. Hist. 6: 348. 1846; Wight, Ic t. 1946. 1853; Kumari in A.N. Henry *et al*., Fl. Tamil Nadu 2: 205. 1987; Matthew, Fl. Palni Hills 2: 1061. 1999; Mohanan & Sivadasan, Fl. Agasthyamala 557. 2002. *Chloranthus grandifolius* Miq , Fl. Ind. Bat. 1(1): 802. 1856. *Chloranthus brachystachys* sensu Hook.f., Fl. Brit. India 5: 100. 1886 p.p., non

Blume 1829; Gamble, Fl. Pres. Madras 2:1211. 1925. *Sarcandra irvingbaileyi* Swamy in Proc. Natl. Inst. Sci. India. Pl. B. Biol. Sci. 19: 303. ff. 1-7.t. 11/8-11 & t. 12/12-13. 1953. *S. grandiflora* (Miq.) Subram. & Henry in Bull. Bot. Surv. India 12: 5. 1970 (1972). **(Fig. 50).**

Subshrubs to 1 m; tender parts pubescent. Leaves broadly elliptic-lanceolate, chartaceous, after drying dark brown to pale brown in colour, glabrous, base acute to cuneate, margin coarsely spinous-serrate, apex acuminate, 7-17.5 x 3.5-7 cm; lateral nerves 5-7 pairs, often ending in serration, faint to prominent; petioles ca 1 cm, basally connate into a sheath, with small subulate stipular margins. Spikes terminal or pseudo-axillary, 2-3 branched, ca 8 cm. Flowers bisexual. Stamens and ovary subtended by bract. Perianth 0. Stamen 1 or 3, ca 5 mm, connate; filaments short, thick; anthers 2 celled. Ovary 1 locular; ovule 1, orthotropous, pendulous; style very short; stigma truncate. Drupes ovoid or globose, ca 5 x 4 mm; exocarp fleshy; endocarp crustaceous; seed pendulous; testa membranous; albumen copious, fleshy.

Frequent, in evergreen forests between 600-1000 m.

Fl.: peak in January – March. *Fr.:* Throughout the year.

Specimens examined: XCH 16664 (Sivasilam); 17891, 12792, 16754 (Papanasam); 15868, 16197, 18143 (Courtallam); 11871, 11654, 14316 (Kalakad); 16067 (Naraikadu); 13577, 12466, 11827, 10354, 10096 (Kakachi); 14965 (Kothayar); 17407, 18588, 18822 (Mahendragiri).

Distribution: Peninsular India and Sri Lanka.

9. MYRISTICACEAE R.Br.

Evergreen trees, often stellately tomentose. Leaves alternate, entire, often pellucid-punctate; stipules 0. Flowers small, dioecious, regular, in axillary or lateral panicles, cymes, umbels or fascicles, rarely in racemes. Bracteoles persistent or caducous. Perianth gamophyllous, inferior, with 3, sometimes 2 or 4 lobes, valvate in bud. Androecium in male flowers of 3 or more, extrorse anthers connate in a sessile or stipitate column, ring or disc; anthers 2 celled, linear or ovate. Ovary in female flowers superior, free, 1-locular; ovule 1, basal, erect, anatropous; stigma capitate, discoid or lobed. Fruit more or less fleshy, splitting into 2, rarely 4 valves; seed erect, enclosed in thin or fleshy, entire or lacerate, coloured, often aromatic arillus; testa usually thick; albumen copious, hard, ruminate; embryo very small, basal; cotyledons divaricate, flat or crumpled; radicle short or inferior.

Tropical family of some 19 genera containing about 440 species, usually from lowland rain forests; ca 5 genera and 18 species in India.

Key to the genera

1a. Filaments and connectives connate in a column, anthers around the column, fully attached .... 2. **Myristica**

1b. Filaments and connectives connate in a peltate disc, anthers stellately attached to the margin of the disc, free .... 1. **Knema**

1. KNEMA Lour.

Ca 93 species distributed in the Indomalayan region; 8 in India.

Knema attenuata (Wallich ex Hook.f. & Thomson) Warb., Monogr. Myrist. 590. 1897; Gamble, Fl. Pres. Madras 2: 1215.1925; Kumari in A.N. Henry *et al.*, Fl. Tamil Nadu 2: 205. 1987. *Myristica attenuata* Wallich ex Hook.f. & Thomson, Fl. Ind. 157. 1855; Hook.f., Fl. Brit. India 5: 110. 1886. *M. corticosa* Bedd., Fl. Sylv. T. 278. 1872. **(Fig. 51).**

Tamil: Chora pathiri.

Trees to 15 m high. Leaves oblong, broadly lanceolate or broadly elliptic-ovate, 12-17 x 4-6 cm, coriaceous, glaucous and rusty-pubescent beneath, base attenuate, apex acuminate; lateral nerves ca 17 pairs, prominent and regularly raised beneath, less so above, arched; petioles ca 1.5 cm, rusty tomentose. Flowers small, unisexual; male in fascicles on thick peduncles from the axils of leaves or on scars of fallen leaves; female similar, fewer. Perianth 3 lobed; lobes broadly ovate, ca 5 x 4 mm, prominently rusty-tomentose without, prominently concave. Androecium stalked; filaments and connectives connate in a peltate disc; anthers 8-20, free, attached stellately to the margin of the disc, dehiscing downwards. Ovary ovoid, 1 locular; ovule 1; style short, thick; stigmas 2, laciniate on the margins. Drupes oblong-ovoid, ca 5 x 4 cm, prominently tomentose; aril laciniate, short, crimson in colour at the apex only.

Occasional, in semi-evergreen forests between 700-1600 m.

Fl. & Fr.: March – August.

Specimens examined: XCH 15408, 19531 (Kalakad); 12995, 17853, 19017 (Papanasam); 15710, 15747, 16174 (Courtallam); 10437 (Kakachi); 15130 (Mancholai); 18558 (Mahendragiri).

Distribution: **Endemic** to Western Ghats.

2. MYRISTICA L.

Genus of about 80 tropical species from Asia to Malaysia; ca 7 species have been reported from India.

Key to the species

1a. Flowers in panicled cymes, peduncles slender .... 2. **M. fragrans**

1b. Flowers in short cymes of fascicles on short thick woody peduncle 2

2a. Leaves densely reddish tomentose beneath .... 3. **M. magnifica**

2b. Leaves glabrous and usually glaucous beneath .... 1. **M. dactyloides**

1. Myristica dactyloides Gaertn., Fruct. 1: 195. t. 41. f. 2 a-d. 1788; Sinclair in Gard. Bull. Straits Settlem. 23: 445. 1968; Matthew & Rani in Matthew, Fl. Tamil. Carnatic 3: 1353. 1983; Kumari in A.N. Henry *et al.*, Fl. Tamil Nadu 2: 205. 1987; Philcox in Dassan. & Clayton, Rev. Handb. Fl. Ceylon 11: 364. 1997; Matthew, Fl. Palni Hills 2: 1062. 1999; Mohanan & Sivadasan, Fl. Agasthyamala 559. 2002. *M. laurifolia* Hook.f. & Thomson, Fl. Ind. 163. 1855; Hook.f., Fl. Brit. India 5: 103. 1886 incl. var. *lanceolata*. *M. beddomei* King in Ann. Roy. Bot. Gard. Calc. 3: 291. t. 118. ff. 1-8. 1891; Gamble, Fl. Pres. Madras 2: 1214. 1925. **(Fig. 52).**

Trees to 15 m high, dioecious; branchlets glabrous except for velvety terminal bud. Leaves oblong or elliptic-lanceolate, 9.5-18 x 4-8 cm, coriaceous, glabrous, shining, base acute or obtuse, apex acute; lateral nerves ca 15 pairs, impressed above, obscure to faint; petioles thick, hard, to 1-3 cm. Male: flowers ca 10 in compact cymes on thick shoots, axillary, ca 6 mm; pedicels ca 5 mm, tomentose. Perianth 3 lobed, connate into tube; lobes ovate, ca 1.8 x 1.2 mm, obtuse. Staminal column ca 3 mm; filaments connate; anthers free, 2 celled, attached by their back, extrorse, ca 1.5 mm. Female: subsessile. Perianth 3 lobed, like male. Ovary ovoid, ca 3 x 1.5 mm, appressed pubescent; ovule 1, basal, erect; style 2 fid. Drupes elliptic or ovoid, ca 4 x 3 cm, densely orange or rusty tomentose; pericarp hard; stalk stout, ca 1.5 cm; seed ovoid, ca 2 x 1.2 cm, smooth, brown to black; testa hard; aril fleshy, laciniate.

Occasional, in moist-deciduous to semi-evergreen forests between 400-1000 m.

Fl. & Fr.: March – May.

Specimens examined: XCH 19875 (Sivagiri); 14640 (Kallar); 18786, 18830, 18831 (Mahendragiri).

Distribution: India and Sri Lanka.

2. Myristica fragrans Houtt., Nat. Hist. 3: 333. 1774; Hook.f., Fl. Brit. India 5: 102. 1886; Gamble, Fl. Pres. Madras 2: 1214. 1925; Kumari in A.N. Henry *et al.*, Fl. Tamil Nadu 2: 205. 1987; Philcox in Dassan. & Clayton, Rev. Handb. Fl. Ceylon 11: 366. 1997. **(Fig. 53).**

Trees to 6 m, glabrous. Leaves oblong-ovate, or elliptic-lanceolate, 4-8 x 1.6-2.5 cm, chartaceous, glabrous, minutely pubescent when young, base acute, apex acute to shortly acuminate; lateral nerves ca 8 pairs, obscure; petioles to 1 cm. Drupes ovoid-globose, ca 4 x 3 cm, yellow, glabrous.

Cultivated.

Fr.: May.

Specimen examined: XCH 15872 (Courtallam).

Distribution: Native to the Moluccas, but cultivated throughout the world.

3. **Myristica magnifica** Bedd., Fl. Sylv. t. 262. 1872; Hook.f., Fl. Brit. India 5: 104. 1886; Gamble, Fl. Pres. Madras 2: 1214. 1925. *M. fatua* Houtt. var. *magnifica* (Bedd.) Sinclair in Gard. Bull. Straits Settlem. 23: 282. 1968; Kumari in A.N. Henry *et al.* Fl. Tamil Nadu 2: 205. 1987. **(Fig. 54).**

Trees to 7 m; branchlets clothed with golden pubescence when young. Leaves oblong or elliptic-lanceolate, 14.5-19.5 x 5-7 cm, coriaceous, densely reddish tomentose beneath, smooth and glabrous or midrib tomentose above, base rounded, apex acuminate; lateral nerves ca 20 pairs; petioles to 1 cm, golden tomentose. Flowers ovoid, in short cymes or fascicles on thick woody peduncle, small, densely rusty tomentose. Male: Perianth 3 lobed; lobes ovate, ca 3 x 1.5 mm, obtuse, tomentose. Staminal column to 0.8 mm long; anthers extrorse, ca 1.5 mm. Female: Perianth like male. Ovary ovoid; ovule 1, basal; style 2-fid.

Rare on river banks in moist-deciduous to evergreen forests between 400-1000 m.

Fl.: April.

Specimens examined: XCH 18911 (Papanasam); 14777 (Korimoni).

Distribution: Western Ghats of Peninsular India.

10. LAURACEAE Juss.

Erect, aromatic trees or shrubs, or leafless twining parasites (*Cassytha*), usually evergreen. Leaves alternate, rarely opposite or subopposite, entire; stipules 0. Flowers regular, small, hermaphrodite or dioecious, in axillary or lateral cymes, racemes or panicles, sometimes in umbellules. Bracts deciduous, often involucrate, sometimes 0. Perianth usually inferior; tube short, often enlarging in fruit: lobes usually 6, sometimes more, sometimes only 4, in 2 series, equal or unequal, deciduous or persistent in fruit, sometimes enlarged, as are the pedicels. Stamens usually a multiple of the perianth lobes, in 2-4 series on the tube; filaments of the inner rows usually, of the outer sometimes with a pair of glands attached or alongside; anthers erect, 2-4 celled; the cells dehiscing by upcurved, at length deciduous, vascular lids, the 4th row of stamens often reduced to glands; in female flowers the stamens replaced by staminodes, usually linear or clavate; the inner biglandular as in male. Ovary sessile at the base of the perianth tube, 1 celled; ovule solitary, anatropous, pendulous from the top of the cell. Fruit a dry or a fleshy berry or drupe, naked or enclosed in the perianth tube or supported by more or less enlarged perianth tube and lobes; pedicel frequently thickened; seeds pendulous; testa membranous; albumen 0; cotyledons plano-convex, fleshy; radicle minute, superior.

About 45 genera in tropical and warm regions, especially south-east Asia and Brazil; ca. 16 genera and 154 species have been reported from India.

Key to the genera

1a. Leaves sub-verticillate .... 1. **Actinodophne**
1b. Leaves not sub-verticillate 2

2a. Leaves 3 veined at base 3
2b. Leaves penninerved 4

3a. Flowers in umbellules, arranged in clusters or short racemes in the axils of upper leaves or on the branchlets between the nodes .... 8. **Neolitsea**
3b. Flowers in axillary, lateral or sub-terminal cymes or panicles ..... 5. **Cinnamomum**

4a. Anthers 2 celled 5
4b. Anthers 4 celled 7

5a. Fruits enclosed by an accrescent perianth tube.. 6. **Cryptocarya**
5b. Fruits not enclosed by an accrescent perianth tube 6

6a. Fruiting perianth persistant; leaves chiefly near the ends of the branchlets .... 3. **Apollonias**
6b. Fruiting perianth deciduous; leaves not at the ends of the branchlets.... 4. **Beilschmiedia**

7a. Involucral bracts deciduous or 0 8
7b. Involucral bracts persistent .... 7. **Litsea**

8a. Fruiting pedicels enlarged, thick .... 2. **Alseodaphne**
8b. Fruiting pedicels hardly enlarged 9

9a. Fruiting perianth lobes reflexed; leaves not fascicled at the ends of the branchlets .... 10. **Persea**
9b. Fruiting perianth lobes erect, cup like; leaves fascicled at the ends of the branchlets .... 9. **Ocotea**

1. ACTINODAPHNE Nees

Genus of 150 species from Sri Lanka, tropical Asia and Malaysia; 15 species have been reported from India.

Key to the species

1a. Lateral veins of the leaves 9-10 pairs; berries ovoid .... 1. **A. bourdillonii**
1b. Lateral veins of leaves 4-8 pairs; berries globose........................ 2. **A. campanulata**

1. Actinodaphne bourdillonii Gamble in Kew Bull. 1925: 129. 1925 et Fl. Pres. Madras 2: 1231. 1925; Chandrasek. in A.N. Henry *et al.*, Fl. Tamil Nadu 2: 206. 1987. **(Fig. 55).**

Trees to 5 m; branchlets pubescent. Leaves oblong-lanceolate, 15-17 x 3.5-4 cm, chartaceous, base attenuate, margin entire, glabrescent above except veins, tomentose beneath on nerves, margin hirsute, densely above; lateral veins 9-10 pairs, prominent; petioles thick, to 1 cm, densely hirsute. Inflorescence umbels, lateral, hirsute; peduncle to 0.5 cm, hirsute; pedicels to 1.5 cm, hirsute. Young berries ovoid, ca 3 x 2 mm. Fruiting perianth tube ca 0.5 x 0.4 cm, entire; lobes deciduous.

Rare, in shola forests between 800-1200 m.

Fr.: April – June.

Specimens examined: XCH 12115 (Kakachi); 12552 (Kothayar).

Distribution: **Endemic** to India.

2. **Actinodaphne campanulata** Hook.f., Fl. Brit. India 5: 148. 1886; Gamble, Fl. Pres. Madras 2: 1230. 1925; Chandrasek. in A.N. Henry *et al.*, Fl. Tamil Nadu, 2: 206. 1987. *A. salicina* sensu Bedd., Fl. Sylv. t. 195. 1873, non Meisner 1864. **(Fig. 56).**

Trees to 4 m high; branches terete, puberulous; branchlets fulvous villous, turgid at nodes. Leaves 4-8 together, elliptic-oblong, 4.5-8.5 x 1.5-3 cm, thinly coriaceous, glabrescent on both surfaces, glaucous beneath, base narrow-cuneate, margin entire, apex obtuse; lateral nerves 4-8 pairs, prominent beneath, obscure above, curved upwards, joined and looped towards margin, reticulation transverse, fine; petioles to 1.5 cm, glabrous, grooved adaxially. Inflorescence axillary. Flowers 5 in an umbellule, ca 4.5 x 3 mm, trimerous, unisexual; pedicels ca 4 mm; involucral bracts 6, ovate, ca 5 x 4.5 mm, concave, imbricate, caducous, villose without, glabrous within. Perianth lobes 6 (3 + 3), subequal, 2.5-3 x 1.5-2 mm, rounded, villous without, glabrous within, acute, minutely reticulate. Stamens in male flowers 9, in 3 whorls; filaments up to 1.5 mm, tawny tomentose; anthers oblong, 1.25-2 mm; pistillodes 0 or 2 mm. Staminode in female flowers 1.5-2.5 mm. Ovary ovoid, ca 1.5 x 1 mm, attenuate into style; style ca 1.5 mm; stigma dilated. Berries 1.5 x 1 cm, red, globose. Fruiting perianth tube ca 2 x 3.5 mm.

Rare, in evergreen forests between 1000-1500 m.

Fl. & Fr.: March – September.

Specimens examined: XCH 18775 (Kalakad); 10170 (Kothayar); 10304 (Kakachi); 18762, 18751, 18764 (Mahendragiri).

Distribution: **Endemic** to India.

2. ALSEODAPHNE Nees

Genus of about 50 species from Sri Lanka, India and Western Malaysia; 9 species have been reported from India.

Alseodaphne semecarpifolia Nees in Wallich, Pl. Asiat. Rar. 2: 72 1831; Gamble, Fl. Pres. Madras 2: 1226. 1925. *A. semecarpifolia* Nees var. *parvifolia* Hook.f., Fl. Brit. India 5: 144. 1886; Chandrasek. in A.N. Henry *et al.*, Fl. Tamil Nadu 2: 207. 1987. *A. semecarpifolia* sensu Wight, Ic. t. 1827. 1852, non Nees 1831. **(Fig. 57).**

Trees to 15 m high. Leaves alternate, crowded at the ends of the branchlets, elliptic, oblanceolate or obovate, 4-7.5 x 1.3-3.2 cm, chartaceous, glabrous, reddish beneath when dry, shining, base acute to cuneate, apex obtuse or emarginate or obtusely acute; lateral nerves obscure, venose-reticulate; petioles to 1 cm long. Flowers small, bisexual, in axillary or subterminal panicles; peduncle ca 5 cm; pedicels filiform, to 5 mm. Perianth tube short, 6 lobed; lobes oblong, ca 3 x 1 mm, pubescent. Stamens 9 perfect, 1st and 2nd rows opposite to perianth lobes, introrse, 3rd row opposite to 1st row, with two large glands on the inner base of filaments; staminodes of 4th row opposite to 2nd row, cordate, stipitate; filaments hairy; anthers 4 celled. Fruit an ellipsoid berry, ca 3 x 1 cm, smooth, seated on the top of the enlarged, thick, fleshy, often warted pedicel; perianth lobes scarcely enlarged, persistent.

Occasional, near stream banks or falls in deciduous forests between 150-500 m.

Fl.: March; *Fr.:* July – October.

Specimens examined: 11465 (Courtallam); 12292 (Papanasam); 13317, 16400, 16443, 17638 (Thirukurungudi); 15325 (Kalakad); 16022 (Naraikadu).

Distribution: Peninsular India, Sri Lanka.

3. APOLLONIAS Nees

2 species worldwide; 1 from Macronesia and 1 from India.

Apollonias arnottii Nees, Syst. Laurin. 670. 1836; Hook.f., Fl. Brit. India 5: 121. 1886; Gamble Fl. Pres. Madras 2: 1219. 1925; Chandrasek. in A.N. Henry *et al.*, Fl. Tamil Nadu 2: 206. 1987. **(Fig. 58).**

Trees to 3 m high; branchlets sparsely hairy. Leaves scattered near the ends of the branchlets, narrowly lanceolate to elliptic, ca 2.8 x 1.2 cm, thinly coriaceous, quite glabrous on both surfaces, penninerved, base acute, margin entire, apex acuminate; lateral nerves 6-8 pairs; petioles to 1 cm. Inflorescence racemes, axillary or subterminal; peduncle to 3 cm. Flowers bisexual; pedicels to 1 mm. Perianth tube ca 5 mm; lobes 6, in two rows, subequal, ovate, 2.25-2.75 x 1-1.5 mm, hairy within. Stamens 9, perfect, those of the 1st and 2nd rows introrse, opposite the perianth lobes; of the 3rd row opposite the first, extrorse with pair of stalked glands at their base, to 1.5-2 mm, those of the 4th row reduced to cordate staminodes, ca 1.25 mm; filaments 1-2 mm, hairy; anthers 2 celled, 0.5-1 mm. Ovary sessile, obconic, ca 1.5 mm; ovule ca 1 mm; style thick, ca 0.5 mm; stigma depressed capitate, 0.25 mm.

Rare, in evergreen forests between 1000-1200 m.

Fl.: April.

Specimen examined: XCH 12425 (Kakachi).

Distribution: **Endemic** to India.

4. BEILSCHMIEDIA Nees

Ca 200 species distributed in the tropical regions; around 9 species have been reported from India.

Beilschmiedia tirunelvelica V.S.Manickam, C.Murugan, G.J.Jothi & V.Sundaresan, Nord. J. Botany 24(4): 407. 2C05. **(Fig. 59).**

Trees to 3-5 m; bark smooth, dark brown; terminal bud flattened, enclosed by 2 lanceolate (5-8 mm long) scales, glabrous; internodes 1-8 cm, terete; branchlets flat; glabrous. Leaves opposite, simple, oblong-lanceolate, 7-12 x 3-5 cm, thinly coriaceous, glabrous, base cuneate, margin entire, apex acuminate; nerves 5-7 pairs, prominent on both surfaces, arcuate, reticulation prominent; petiole 1-1.5 cm, channeled above. Racemes, supra-axillary, 2-5 cm, glabrous; peduncles upto 1 cm. Flowers pale green, 5-6 mm in diam., bisexual, regular; pedicels slender, 5-7 mm, black when dry. Perianth tube 2-3 mm, obconical; lobes 4, shallowly lobed, ovate, ca 2.5 mm, deciduous, margin ciliate, apex obtuse, apiculate. Stamens 4, in 2 whorls; filaments ca 1.5 mm, flat; anthers 2 celled, introrse; staminodes 2, ovoid-cordate, hairy, ca 1.2 mm. Ovary ovoid, ca 0.5 mm, glabrous, 1 celled, free from perianth tube; ovule 1, pendulous; style ca 0.75 mm, slender, articulated, caducous; stigma simple, acute. Drupes oblong, ca 4 x 2 cm, glabrous, pink-purple; cotyledons 2, plano-convex.

In montane forests at 700-1400 m.

Fl. & Fr.: March-April

Specimens examined: MH 19103 (Poongulam), XCH 18652 (Mahandragiri, on the way to Kailasaparvatham)

Distribution: **Endemic** to Tirunelveli and Kanyakumari Hills

5. CINNAMOMUM Blume

Genus of about 350 species from South, East and South-East Asia to Australia; in India, it is represented by 46 species.

Key to the species

1a. Panicles with few flowers 2

1b. Panicles with many flowers 3

2a. Perianth lobes ca 1.5 mm long...1. **C. filipedicellatum**
2b. Perianth lobes ca 5 mm long..5. **C. travancoricum**

3a. Leaves glabrous 4
3b. Leaves hairy...6

4a. Lateral nerves running almost to apex, nervules horizontal and parallel ....2. **C. malabatrum**
4b. Lateral nerves running about three fourth up, nervules irregularly reticulate..........5

5a. Leaves lanceolate-ovate or narrowly elliptic, lateral nerves prominent; peduncles shorter than leaves, ca 5 cm long ....6. **C. verum**
5b. Leaves broadly ovate or oblong or ellipitic, lateral nerves obscure; peduncles usually longer than leaves, ca 11 cm long ....7. **C. wightii**

6a. Flowers ca 8 x 3 mm; leaves 3-5 ribbed; arising from the base.......3. **C. perrottetii**
6b. Flowers ca 5 x 2.5 mm; leaves 3 ribbed; arising from or a little above the base ....4. **C. sulphuratum**

1. **Cinnamomum filipedicellatum** Kosterm. in Bull. Bot. Surv. Ind. 25: 92. 1983 (1985); Chandrasek. in A.N. Henry *et al*., Fl. Tamil Nadu 2: 207. 1987. *C. gracile* Hook.f., Fl. Brit. India 5: 133. 1886, non Miq. 1864; Gamble, Fl. Pres. Madras 2: 1223.1925. **(Fig. 60).**

Trees to 4 m; branchlets slender, glabrous; terminal buds small, minutely sericeous. Leaves opposite and alternate, broadly ovate or elliptic, 4-9 x 2.5-4.5 cm, thinly coriaceous, smooth, glossy above, glabrous and pale beneath, base cuneate or rounded, margin entire, apex broadly acuminate; nerves slender, prominent; secondary nerves slender, obscure, parallel, subhorizonal; petioles slender, glabrous, to 1.5 cm, concave above. Inflorescence axillary or pseudo-terminal, glabrous except at nodes, slender, few flowered, branched, with filiform, 1-3 cm long branches; peduncle slender, glabrous. Flowers ca 3 x 2.5 mm, minutely sparsely sericeous; pedicels filiform, to 1 cm. Perianth tube ca 1 mm; lobes 6, ovate, ca 1.5 x 1.25 mm, acute, densely appressed pilose within. Stamens 9 perfect, ca 2 mm; filaments short, ca 0.5 mm, pilose; anthers ca 1.5 mm. Staminodes acicular, ca 0.5 mm. Ovary ellipsoid, ca 1 x 0.5 mm; style short, with minute, peltate stigma. Fruits ellipsoid, ca 5 x 0.8 cm; cup shallow, ca 3 x 6 mm bearing broad, obtuse, glabrous bases of perianth lobes, 1 mm.

Occasional, in evergreen forests between 1000-1500 m.

Fl. & Fr.: February – April.

Specimens examined: 12135, 12502, 12901, 14903, 15022, 15044 (Kalakad); 18778 (Mahendragiri).

Distribution: **Endemic** to India.

2. **Cinnamomum malabatrum** (Burm.f.) Blume, Bijdr. 568. 1826; Chandrasek. in A.N. Henry *et al*., Fl. Tamil Nadu 2: 208. 1987. *Laurus malabatrum* Burm.f., Fl. Ind. 92. 1867; *Cinnamomum iners* sensu Gamble, Fl. Pres. Madras 2: 1225. 1925. **(Fig. 61).**

Trees to 4 m high. Leaves opposite-alternate, oblong-lanceolate, 15-20.5 x 2-4 cm, coriaceous, glabrous on both surfaces, glaucous above, base acute, margin entire, apex acute; nerves arising little above base, side ribs running almost to apex; transverse nervules horizontal, parallel; petioles to 1.5 cm. Perianth tube ca 1.5 x 2 mm, 6 lobed; lobes subequal, ovate, ca 2 x 1.5 mm, acute, pubescent. Stamens 9, in 3 whorls, perfect; filaments ca 1.5 mm; glands ca 1 mm; anthers ca 1 mm; staminodes slightly shorter than fertile stamens, cordate, ca 1.75 mm. Ovary ovoid-ellipsoid, ca 2 x 1.5 mm, sessile; ovule ca 1.5 mm; style ca 1 mm; stigma obscurely lobed.

Occasional, in moist deciduous forests between 400-600 m.

Fl.: January – March.

Specimen examined: XCH 14652 (Pechiparai).

Distribution: **Endemic** to India.

3. **Cinnamomum perrottetii** Meisner in DC., Prodr. 15(1): 22, 504. 1864; Hook.f., Fl. Brit. India. 5: 134. 1886; Gamble, Fl. Pres. Madras 2: 1225. 1925; Chandrasek. in A.N. Henry *et al.*, Fl. Tamil Nadu 2: 208. 1987. **(Fig. 62).**

Trees to 8 m high; branchlets densely minutely tomentose. Leaves opposite and subopposite, ovate-elliptic, 6-11 x 2.5 cm, coriaceous, young leaves dense fulvous tomentose, adult leaves glabrous above, subglaucous, sparsely tomentose beneath, base cuneate, margin entire, apex acute or obtuse; lateral nerves almost reaching the tip; petioles to 1 cm. Inflorescence in the axils of terminal leaves, 4.5-10 cm, densely, minutely fulvous tomentose. Flowers pale yellow, ca 8 x 3 mm; pedicels to ca 1.5 mm. Perianth tube 1.5-3 mm; lobes fleshy, oblong-ovate, 2.5-3 x 1.5-2 mm, acute, densely appressed pilose. Stamens 9, 2-3 mm; filaments slender, 0.75-1.5 mm, pilose; anthers 0.75-1.5 mm; staminodes 3, ca 1.5 mm. Ovary ellipsoid, ca 1.5 x 0.75 mm, sessile; style thick, ca 1.5 mm; stigma peltate.

Rare, in evergreen and shola forests between 500-1500 m.

Fl. & Fr.: March – June.

Specimens examined: XCH 12496, 14915 (Kothayar).

Distribution: **Endemic** to India.

4. **Cinnamomum sulphuratum** Nees in Wallich, Pl. Asiat. Rar. 2: 74. 1831; Hook.f., Fl. Brit. India 5: 132. 1886; Gamble, Fl. Pres. Madras 2: 1225. 1925; Chandrasek. in A.N. Henry *et al.*, Fl. Tamil Nadu 2: 208. 1987. **(Fig. 63).**

Trees to 7 m high; branchlets slender, angular, densely, minutely subappressed fulvous-tomentose. Leaves opposite or subopposite, elliptic-oblong or ovate-elliptic, 2-15 x 1.5-4 cm, coriaceous, fulvous-tomentose beneath when young, glabrescent later, glabrous above, base attenuate, margin entire, apex acuminate; nerves slender, prominent beneath, lateral nerves arising at or little above the base; side ribs nearly to apex; petioles to 1 cm, flat or slightly concave above. Inflorescence axillary or terminal, ca 15 cm, lax, densely, minutely fulvous-tomentose. Flowers pale yellow, ca 5 x 2.5 mm. Perianth tube ca 1.5 mm; lobes ovate, ca 3.5 x 2 mm, subequal. Stamens 9, 1.5-2.5 mm; staminodes 3, ca 1 mm. Ovary ellipsoid, ca 2 x 0.75 mm, sessile; style thick, ca 0.75 mm; stigma discoid or obscurely 3 lobed.

Rare, in evergreen forests between 1000-1500 m.

Fl.: April – May.

Specimens examined: XCH 12442, 12546, 18875 (Kalakad).

Distribution: **Endemic** to India.

5. **Cinnamomum travancoricum** Gamble in Kew Bull. 1925: 128. 1925 et Fl. Pres. Madras. 2: 1224. 1925; Chandrasek. in A.N. Henry *et al*., Fl. Tamil Nadu 2: 208. 1987. **(Fig. 64).**

Trees to 9 m high; branchlets slender, angular, sparsely appressed pilose. Leaves opposite, elliptic or ovate-elliptic, 6-9.5 x 2-4 cm, thinly coriaceous, base acute, margin entire, apex obtusely acuminate; midrib slender, prominent; laterals slender. Inflorescence 1-6 cm, pseudoterminal and axillary, pubescent. Flowers white. Perianth tube ca 1 mm; lobes oblong, ca 5 mm, acute, densely subappressed pilose. Stamens 2-3 mm; filaments 1-1.25 mm; glands ca 0.5 mm; anthers ca 1-2 mm; staminodes 1-1.5 mm. Ovary globose, ca 1.5 x 0.7 mm; style short, ca 1 mm; stigma peltate.

Rare, in evergreen forests between 800-1500 m.

Fl.: March.

Specimens examined: XCH 18632, 12930 (Mahendragiri).

Distribution: Endemic to India.

6. **Cinnamomum verum** J.Presl, Prir. Rostl. 2: 36. t. 7. 1825; Chandrasek. in A.N. Henry *et al*. Fl. Tamil Nadu 2: 208. 1987. *C. zeylanicum* Blume, Bijdr. 568. 1826; Wight, Ic.t. 123. 1829; Hook.f., Fl. Brit. India 5: 131. 1886; Gamble, Fl. Pres. Madras 2: 1224. 1925. **(Fig. 65).**

Trees to 4 m high; branchlets slender. Leaves opposite or subopposite, lanceolate-ovate or narrowly elliptic, 4-9 x 2-4.5 cm, coriaceous, smooth, shining above, glabrous beneath, base attenuate, apex acuminate, margin entire, apex acute to acuminate; nerves prominent; lateral veins about three-fourth up; petioles to 1.5 cm, slightly concave above. Inflorescence panicles, to 7 cm, main peduncle with few stiff short branches, axillary and pseudo-terminal. Flowers pale yellow, ca 9.5 x 3 mm; pedicels to 1.5 mm. Perianth tube ca 7 mm, silky; lobes oblong-lanceolate, ca 2 x 1.5 mm, acute. Stamens 9, ca1.25-1.5 mm; filaments thick, short, ca 0.75 mm; anthers 0.5-0.75 mm; staminodes ca 1 mm. Ovary ellipsoid.

Rare, in evergreen forests between 1200-1500 m.

Fl.: March.

Specimens examined: XCH 13511, 13642, 12546 (Kalakad).

Distribution: India, Sri Lanka, Myanmar, Malaysia.

7. **Cinnamomum wightii** Meisner in DC., Prodr. 15(1): 11. 1864; Hook.f., Fl. Brit. India 5: 132. 1886; Gamble, Fl. Pres. Madras 2: 1224. 1925; Fyson, Fl. South Indian Hill st. 2: t. 437. 1932; Chandrasek. in A.N. Henry *et al.*, Fl. Tamil Nadu 2: 209. 1987. **(Fig. 66).**

Trees to 7 m high; branches and branchlets thick, black when dry. Leaves broadly ovate or oblong or elliptic, 3-6 x 1.5-4 cm, thick-coriaceous, glaucous beneath, shining above, base rounded to truncate, margin entire, often undulate, apex obtuse or subacute, 3-5 ribbed, finely reticulate beneath; petioles to 2 cm, thick, black when dry. Berries oblong-globose, ca 2 x 1 cm, pale-brown, supported by a truncate cup.

Very rare, in evergreen forests between 800-1200 m.

Fr.: January.

Specimens examined: XCH 14312 (Kalakad); 16761 (Papanasam).

Distribution: Western Ghats.

6. CRYPTOCARYA R.Br.

Genus of about 300 pantropic species distributed in warm and tropical regions of the world, especially in Asia; in India, it is represented by 15 species.

Cryptocarya beddomei Gamble in Kew Bull. 1925: 127. 1925 et Fl. Pres. Madras 2: 1218. 1925; Chandrasek. in A.N. Henry *et al.*, Fl. Tamil Nadu 2: 209. 1987. **(Fig. 67).**

Trees to 7 m high; branchlets woody, lenticellate. Leaves oblong-ovate or elliptic, 6.5-9 x 3-4 cm, coriaceous, shining, when dry chocolate-brown above, glaucous beneath, base rounded, apex subacute; lateral nerves ca 6 pairs, obscure above, prominent beneath, reticulation prominent beneath; petioles to 1.5 cm. Flowers in axillary or terminal pedunculate panicles.

Rare, in moist deciduous-semievergreen forests.

Fl.: September onwards.

Specimens examined: XCH 10525 (Mancholai); 18715 (Mahendragiri).

Distribution: Peninsular India.

7. LITSEA Lamk.

Genus of about 300 species, tropical, but not in Africa; a few species are extra-tropical; 45 in India.

Key to the species

1a. Perianth lobes incomplete or 0 2
1b. Perianth lobes complete 3

2a. Branchlets, inflorescence and leaves beneath densely fulvous-tomentose; nerves 10-15 pairs; flower buds ca 5 mm across .... 8. **L. quinqueflora**
2b. Branchlets, inflorescence and leaves beneath glabrescent, glossy; nerves 6-8 pairs; flower buds ca 1 mm across .... 6. **L. ligustrina**

3a. Leaves glabrous beneath or nearly so 4
3b. Leaves pubescent or tomentose beneath 8

4a. Umbels peduncled, umbel bearing male flowers solitary .... 9. **L. venulosa**
4b. Umbels subsessile, umbel bearing male flowers in clusters or racemes 5

5a. Umbels in axillary clusters 6
5b. Umbels in racemes 7

6a. Leaves elliptic-obovate, obtuse-acute at apex, berries subglobose .... 3. **L. coriacea**
6b. Leaves oblong-lanceolate, acuminate at apex, berries oblong .... 5. **L. laevigata**

7a. Leaves 9.5-13.5 cm long, very coriaceous .... 7. **L. oleoides**
7b. Leaves 4-7.5 cm long, subcoriaceous .... 1. **L. beddomei**

8a. Umbels sessile or subsessile, in axillary or lateral clusters .... 2. **L. bourdillonii**
8b. Umbels in racemes 9

9a. Leaves 5-11 cm long; lateral nerves 7-8 pairs, facing towards apex, transverse nervules prominent beneath, obscure above, not parallel .... 10. **L. wightiana**
9b. Leaves ca 19 cm long; lateral nerves 12 pairs, facing towards margin, transverse nervules prominent on both surfaces, parallel .... 4. **L. insignis**

1. Litsea beddomei Hook.f., Fl. Brit. India 5: 177. 1886; Gamble, Fl. Pres. Madras 2: 1237. 1925; Chandrasek. in A.N. Henry *et al.*, Fl. Tamil Nadu 2: 209. 1987. **(Fig. 68).**

Trees to 4 m; branchlets terete, pubescent. Leaves oblong, linear-lanceolate; 4-7.5 x 1-2 cm, thinly coriaceous, glabrous above, glaucous beneath, base acute, margin entire, apex acute; lateral nerves 8-10 pairs; petioles to 6 mm. Umbels in racemes, ca 2 cm. Flowers unisexual, 4 in each umbel, 3.5 mm. Bracts 4, 3 x 2.5-3 mm. Perianth tube ca 2.5 mm, pubescent; lobes 6 (3 + 3), 2.5 x 1.25 mm, membranous, pubescent without. Male flowers: Stamens 9; filaments ca 1.2 mm; anthers 1 mm, villose at base; staminodes 9, ca 1 mm. Female flowers: Ovary oblong, ca 1.5 mm; style ca 1 mm; stigma capitate, ca 0.75 mm.

Rare, in evergreen forests between 1200-1500 m.

Fr.: September – December.

Specimens examined: XCH 13578, 13585, 17189 (Kakachi); 12533 (Kothayar); 17525 (Mahendragiri).

Distribution: **Endemic** to India.

2. Litsea bourdillonii Gamble in Kew Bull. 1925: 131. 1925 et Fl. Pres. Madras 2: 1237. 1925; Chandrasek. in A.N. Henry *et al.*, Fl. Tamil Nadu 2: 209. 1987. **(Fig. 69).**

Trees to 7 m high; branchlets woody, densely fulvous-pubescent, blackish-brown when dry. Leaves alternate, opposite or subopposite at the ends of the branchlets, broadly oblanceolate or obovate, 12-18 x 5.5-6.5 cm, chartaceous, reddish-brown beneath when dry, fulvous-pubescent on nerves, more so on midrib, sometimes glabrous, base acute, apex shortly abruptly acute; lateral nerves 10-15 pairs, prominent, closer towards the base. Berries (immature) many in dense tufts on branchlets, subglobose, on a cup shaped, 2-4 cleft, fulvous-pubescent, subsessile perianth tube.

Rare, in semi-evergreen forests, ca 700 m.

Fr.: June.

Specimen examined: XCH 12935 (Papanasam).

Distribution: Peninsular India.

3. Litsea coriacea (Heyne ex Meisner) Hook.f. Fl Brit. India, 5: 166. 1886; Gamble, Fl. Pres. Madras 2: 1236. 1925; Chandrasek. in A.N. Henry *et al.*, Fl. Tamil Nadu 2: 209. 1987.

Small trees to 4 m high; branchlets slender, bark pale. Leaves elliptic-obovate or oblanceolate, 4-5 x 2-2.5 cm, apex acute-obtuse, coriaceous, red-brown beneath; lateral nerves ca 7 pairs, transverse nerves obscure; petioles 0.5-0.7 cm, fulvous pubescent. Inflorescence an umbel, axillary, creamy in colour. Berries subglobose, on a cup shaped, enlarged perianth tube.

Occasional in evergreen forests up to 1500 m.

Fl. & Fr.: July-October.

Specimens examined: XCH 10952, 13478, 17372, 19814 (Kalakad, Sivagiri, Upper Kothayar).

Distribution: **Endemic** to South India.

4. Litsea insignis Gamble in Kew Bull. 1925: 130. 1925 et Fl. Pres. Madras 2: 1237. 1925; Chandrasek. in A.N. Henry *et al.*, Fl. Tamil Nadu 2: 210. 1987; Matthew, Fl. Palni Hills 2: 1073. 1999. **(Fig. 70).**

Trees to 10 m high; branchlets woody, corky, pubescent, grayish when dry. Leaves broadly elliptic-oblong or obovate, ca 19 x 10.5 cm, coriaceous, roughly fulvous-pubescent beneath; lateral nerves ca 12 pairs, prominent; transverse nervules parallel, conspicuous; petioles up to 3 cm, fulvous pubescent. Inflorescence is an umbel, many, yellowish, on old wood, leaving tubercles.

Rare, in evergreen forests, ca 1200 m.

Fl.: August.

Specimen examined: XCH 13637 (Kothayar).

Distribution: Peninsular India.

5. **Litsea laevigata** (Nees) Gamble, Fl. Pres. Madras 2: 1236. 1925; Chandrasek. in A.N. Henry *et al.*, Fl. Tamil Nadu 2: 210. 1987. *Tetranthera attenuata* Nees var. *laevigata* Nees, Syst. Laurin. 677. 1836. *Tetranthera salicifolia* Roxb. ex Nees in Wallich, Pl. Asiat. Rar. 2: 66. 1831. *Litsea salicifolia* (Roxb. ex Nees) Hook.f., Fl. Brit. India 6: 167. 1886 p.p. **(Fig. 71).**

Trees to 6 m high. Leaves alternate, oblong-lanceolate, 8.5-14 x 3.5-5.5 cm, chartaceous, greenish above, reddish-brown beneath when dry, glabrous, acute to narrowed at base, acuminate at apex; lateral nerves ca 8 pairs, prominent with reticulation beneath, less so above; petioles ca 1 cm. Berries oblong, ca 1 x 0.5 cm, seated on a small obconic perianth tube; perianth tube ca 8 mm, puberulous; stalk to 1.5 cm, puberulous.

Rare, in semi-evergreen forests, ca 700 m.

Fr.: May.

Specimen examined: XCH 12735 (Mahendragiri).

Distribution: Southern parts of Peninsular India.

6. **Litsea ligustrina** (Nees) Hook.f., Fl. Brit. India, 5: 158. 1886; Gamble, Fl. Pres. Madras 2: 1235. 1925; Chandrasek. in A.N. Henry *et al.*, Fl. Tamil Nadu 2: 210. 1987. *Tetranthera ligustrina* Nees in Wallich, Pl. Asiat. Rar. 2: 65. 1831; Wight, Ic. t. 1835. 1852. **(Fig. 72).**

Small trees to 2 m high; young branchlets pubescent, lenticellate, black when dry. Leaves subopposite, crowded at the end of the branchlets, elliptic-obovate, 2.5-6 x 1-2 cm, chartaceous, glossy, areolate above, glabrescent beneath, base acute, margin entire, reflexed, apex acute-obtuse; lateral nerves 6-8 pairs; petioles 2-5 mm, glabrescent. Flowers 5, in solitary umbels; peduncle 6-7 mm. Involucral bracts 4, ca 3.5 x 3 mm; margin ciliate; pedicels to 3 mm, pubescent. Male flowers: Perianth 0. Filaments 0.75 mm; anthers 0.75 mm. Pistillode ovoid. Ovary ca 1 x 1 mm; style 1 mm; stigma 1 mm. Fruits globose, ca 8 mm across, subsucculent; fruiting pedicel thick, ca 5 x 2 mm.

Occasional, in semievergreen to evergreen forests between 600-1200 m, mostly along stream banks.

Fl. & Fr.: October – March.

Specimens examined: XCH 19071 (Papanasam), 13814 (Courtallam), 18443, 19814, 19844, 19974 (Sivagiri), 10514, 11997 (Mancholai), 10952, 11754, 14280 (Kalakad), 18689 (Mahendragiri).

Distribution: India and China.

7. **Litsea oleoides** (Meisner) Hook.f., Fl. Brit. India 5: 175. 1886; Gamble, Fl. Pres. Madras 2: 1236. 1925; Chandrasek. in A.N. Henry *et al.*, Fl. Tamil Nadu 2: 211. 1987; Matthew, Fl. Palni Hills 2: 1673. 1999. *Tetranthera oleoides* Meisner in DC., Prodr. 15(1): 195. 1864.

(Fig. 73).

Trees to 6 m high; branchlets lenticellate, glabrous or minutely puberulous, black when dry. Leaves elliptic-oblong, 9.5-13.5 x 3-4 cm, thick coriaceous, glabrous, subacute or retuse at apex, rounded at base; lateral nerves ca 12 pairs, distantly arranged, faint; petioles thick, up to 1.5 cm, blackish when dry. Berries depressed globose, ca 1.8 cm across, seated on a short, flattened undulate, perianth tube ca 2 cm across, puberulous or glabrous; fruiting pedicel thick, ca 2.5 x 0.2 cm, reddish.

Rare, in evergreen forests, ca 1000 m.

Fr.: September.

Specimens examined: XCH 13096, 10536 (Mancholai).

Distribution: Peninsular India.

8. **Litsea quinqueflora** (Dennst.) C.R. Suresh in Nicolson *et al.*, Interpret. van Rheede's Hort. Mal. 158. 1988; Kosterm. in Dassan. *et al.*, Rev. Handb. Fl. Ceylon 9: 146. 1995. *Darwinia quinqueflora* Dennst., Schlussel Hort. Mal. in Allg. Deutsche Garten Mag. 2(2): 12, 20, 31. 1818. *Litsea tomentosa* Nees in Wallich, Pl. As. Rar. 2: 65. 1831, non Blume. *Tetranthera japonica* Wight ex Wallich in Wallich, Pl. As. Rar. 2: 65. 1831, non Spreng. *T. japonica* forma *foeminea* Wight ex Steud., Nomencl. ed. 2, 2: 672. 1841. *T. velutina* Blume, Ann. Mus. Bot. Lugduno Batavum 1(24): 376. 1851; Hook.f., Fl. Brit. India 5: 180. 1886. *T. tomentosa* Roxb. ex Wallich in Wight, Ic. Pl. Ind. Or. 5: 14. t. 1834. 1852; Hook.f., Fl. Brit. India 5: 157. 1886. *Litsea deccanensis* Gamble, Fl. Pres. Madras 2: 1235. 1925; Matthew & Rani in Matthew, Fl. Tamil. Carnatic 3: 1360. 1983; Chandrasek. in A.N. Henry *et al.*, Fl. Tamil Nadu 2: 210. 1987. **(Fig. 74).**

Tamil: Perumbandali.

Trees to 10 m high; branchlets fulvous-tomentose. Leaves spirally arranged, crowded at the ends of the branchlets, elliptic-ovate or oblong-ovate or elliptic-oblanceolate, 7.5-24.75 x 3-7.5 cm, chartaceous, densely fulvous-tomentose beneath, only on nerves above, suborbicular to acute at base, apex acute or shortly acuminate; lateral nerves ca 12 pairs, distantly arranged, prominent beneath; petioles up to 2 cm. Flowers in umbels, umbels large, solitary or few together, axillary or in axils of fallen leaves; peduncles to 1.5 cm, tomentose; involucral bracts 4 lobed; lobes suborbicular, ca 7 x 5.6 cm, hooded at apex, puberulous without. Male many; tepals 2-4 or more, linear, ca 3.5 x 0.5 mm, deciduous, pubescent without and within at base. Stamens 20 or more; filaments slender, villous, ca 0.5 mm; anthers ca 1 mm; staminodes filiform, ca 0.5 mm. Female few; tepals caducous. Ovary oblong-globose, ca 1.5 x 0.7 mm; style to 1.5 mm; stigma irregularly lobed.

Rare, in moist-deciduous to semi-evergreen forests between 400-600 m.

Fl.: September – December.

Specimens examined: XCH 10868, 11690 (Kalakad).

Distribution: Southern parts of Peninsular India, Sri Lanka.

9. Litsea venulosa (Meisner) Hook.f., Fl. Brit. India 5: 161. 1886; Gamble, Fl. Pres. Madras 2: 1235. 1925; Chandrasek. in A.N. Henry *et al.*, Fl. Tamil Nadu 2: 211. 1987. *Tetranthera venulosa* Meisner in DC., Prodr. 15(1): 187. 1864. **(Fig. 75).**

Straggling shrubs to 2 m high; branches and branchlets slender. Leaves alternate, oblong, elliptic-oblong or lanceolate, 7.5-10.5 x 2.8-3.4 cm, chartaceous, membranous when young, glabrous, glaucous beneath, pale grayish-green or brown when dry, obtusely acuminate at apex, truncate to acute at base; lateral nerves 6-10 pairs, very slender, faint beneath, obscure above; petioles to 1 cm, slender. Inflorescence an umbel; male solitary; female often racemed, pedunculate, peduncles to 1 cm. Flowers 3-5 in umbels, white; involucral bracts 4 lobed; lobes oblong-orbicular, hooded, ca 5 x 3.5 mm. Perianth tubular at base, 5 lobed; lobes oblong, ca 3 x 2.4 mm, glossy. Stamens in male 10; filaments ca 2.5 mm, 6 biglandular, 4 eglandular; with subspathulate heads; inner short, with large glands. Ovary ovoid, ca 0.75 x 0.5 mm, tapering into slender style; style to 1.25 mm; stigma discoid.

Rare, in evergreen forests between 900-1200 m, mostly along forest edges.

Fl.: Probably throughout the year.

Specimens examined: XCH 12826, 16696, 18994 (Papanasam); 20698 (Kothayar).

Distribution: Southern parts of Peninsular India.

10. Litsea wightiana (Nees) Hook.f. in Benth. & Hook.f., Gen. Pl. 3: 132. 1880; Hook.f., Fl. Brit. India 5: 177. 1886 p.p.; Gamble, Fl. Pres. Madras 2: 1238. 1925; Chandrasek. in A.N. Henry *et al.*, Fl. Tamil Nadu 2: 211. 1987; Matthew, Fl. Palni Hills 2: 1074. 1999. *Cylicodaphne wightiana* Nees in Wallich, Pl. Asiat. Rar. 2: 68. 1831; Wight, Ic. t. 1833. 1852. **(Fig. 76).**

Trees to 4 m high; branchlets rusty-tomentose, terete; bark light-brown, smooth. Leaves oblong-obovate, 5-11 x 3-5 cm, coriaceous, glabrous on both surfaces, greenish above with impressed veins, light-brown and tomentose beneath; lateral nerves 7-8 pairs, prominent beneath, obscure above; petioles to 2 cm, rusty tomentose. Inflorescence of racemes arranged in umbels, axillary, peduncles to 1.5 mm, rusty-pubescent, bracts suborbicular, ca 4 x 3.6 mm, concave, rusty pubescent without, glabrous within, each umbel with 5 flowers. Perianth 6 lobed; lobes lanceolate, unequal, ca 2-3 x 1 mm, pubescent without. Stamens ca 12, glandular; filaments bearded. Ovary oblong-globose, ca 2 x 1 mm; style to 2.5 mm; stigma discoid. Berries ovoid, ca 1.5 x 0.7 cm, seated on cup shaped perianth tube, fruiting pedicel thickened, reddish.

Frequent, in evergreen forests between 800-1200 m.

Fl.: August – December; *Fr.:* January – April.

Specimens examined: XCH 19472 (Kalakad); 19131 (Courtallam); 16054 (Naraikadu); 12123, 20074, 20092, 10066, 10330, 10546, 10581, 13507, 13599 (Kakachi); 12515, 12549, 13626, 14960, 17233, 19294, 19396 (Kothayar).

Distribution: **Endemic** to India.

8. NEOLITSEA Merr.

Species ca 60 in the tropical and subtropical mainland of Asia, Malaysia, New Guinea and tropical Australia; ca 7 species are found in India.

Key to the species

1a. Fruiting perianth very small as compared to the berry, perianth tube crenate .... 1. **N. cassia**

1b. Fruiting perianth rather large as compared to the berry, perianth tube entire .... 2. **N. scrobiculata**

1. Neolitsea cassia (L.) Kosterm. in J. Sci. Res. (Jakarta) 1: 85. 1952; Chandrasek. in A.N. Henry *et al.*, Fl. Tamil Nadu 2: 211. 1987; Kosterm. in Dassan. *et al.*, Rev. Handb. Fl. Ceylon 9: 164. 1995. *Laurus cassia* L., Sp. Pl. 1: 369. 1753, et ed. 2: 528. 1762. *L. involucrata* Vahl ex Lam., Encycl. 3: 445. 1793. *Litsea zeylanica* CAG. & Fr. Nees in Amoen. Bot. Bonn., Fasc. 1: 58. t. 5. 1823; Hook.f., Fl. Brit. India 5: 178. 1886 p.p. *Tetradenia zeylanica* (CAG. & Fr. Nees) Nees in Wallich, Pl. As. Rar. 2: 64. 1831. *Neolitsea zeylanica* (CAG. & Fr. Nees) Merr. in Philipp. J. Sci. 1, Suppl. 57. 1906; Gamble, Fl. Pres. Madras 2: 1239. 1925. *N. involucrata* (Vahl ex Lam.) Alston in Trimen, Handb. Fl. Ceylon 6: 141. 1931. **(Fig. 77).**

Small trees to 4 m high; branchlets slender. Leaves spirally arranged, crowded at the end of the branchlets, lanceolate or elliptic or oblong-lanceolate, 6-12 x 1.5-3 cm, subcoriaceous, smooth, glossy, glabrous, reticulate-areolate, base acute to cuneate, apex gradually tapering; 3 ribbed above base; lateral nerves 2-4 pairs, obscure; petioles slender, to 1 cm. Flowers in sessile umbels, pale yellow, sericeous. Involucral bracts 4, suborbicular, ca 4 x 3.8 mm, concave, pubescent without, deciduous. Female flowers: Perianth 4 lobed; lobes oblong-spathulate, ca 2 x 0.8 mm, pubescent without. Staminodes 6; outer eglandular; inner 2 with stipitate, orbicular, basal glands; filaments thread-like, sparsely sericeous. Ovary ovoid-globose, ca 1 mm across.

Occasional, in moist-deciduous to semi-evergreen forests between 400-800 m.

Fl.: March

Specimens examined: XCH 19246 (Courtallam); 17871 (Papanasam); 14371, 19490, 19602 (Kalakad), 18606, 18701, 18765 (Mahendragiri); 10508, 10663, 10659 (Mancholai).

Distribution: India, Sri Lanka.

2. **Neolitsea scrobiculata** (Meisner) Gamble, Fl. Pres. Madras, 2: 1240. 1925; Chandrasek. in A.N. Henry *et al.*, Fl. Tamil Nadu 2: 212. 1987; Matthew Fl. Palni Hills 2: 1076. 1999. *Litsea scrobiculata* Meisner in DC., Prodr. 15(1): 223. 1864. *L. zeylanica* sensu Hook. f, Fl. Brit. India 5: 178. 1886 p.p. non Nees 1823. **(Fig. 78).**

Trees to 10 m high. Leaves alternate, elliptic to ovate, 4-9 x 1.5-3.5 cm, coriaceous, glabrous, green above, glaucous beneath, base acute-cuneate, margin entire, apex caudate-acuminate, 3 nerved from base, pitted on either side of nerves; petioles to 2.5 cm long. Flowers 3-8, in axillary, subsessile umbels. Involucral bracts concave, reticulate, 3-5 mm across. Male flowers: Pale green, ca 2.5 x 1 cm, densely appressed pilose; pedicels 1 mm long. Perianth tube ca 1 mm; lobes narrowly oblong; pistillode ca 1 mm. Female flowers: Yellow, ca 3 x 1.5 mm; perianth tube ca 1.5 x 1.25 mm; lobes lanceolate, free, 1.5 x 0.75 mm. Staminode ca 1 mm. Ovary ovoid, 0.75 x 0.5 mm; style 0.75 mm, thick; stigma ca 0.5 mm, peltate. Berries globular, ca 1 x 1 cm, on perianth tube.

Occasional, in evergreen and semievergreen forests between 800-1500 m.

Fl. & Fr.: February – April.

Specimens examined: XCH 10724, 11951, 15164 (Mancholai); 12374 (Papanasam); 10887 (Kalakad); 10316, 10342, 13478 (Kakachi); 18156 (Courtallam); 10285, 12480 (Kothayar); 17511, 18753 (Mahendragiri).

Distribution: **Endemic** to India.

9. OCOTEA Aubl.

Genus of about 425 species distributed mainly in tropical and subtropical part of the America; 1 in India.

Ocotea lancifolia (Schott) Mez. in Jahrb. Konigl. Bot. Gart. Berlin 5: 254 1889. *Phoebe lanceolata* Nees, Syst. Laurin. 109. 1836; Wight, Ic. t. 1821. 1852; Hook.f., Fl. Brit. India 5: 141. 1886; Gamble, Fl. Pres. Madras 2: 1228. 1925; Chandrasek. in A.N. Henry *et al*., Fl. Tamil Nadu 2: 212. 1987. **(Fig. 79).**

Trees to 3 high; branchlets slender, puberulous. Leaves alternate, elliptic-lanceolate to oblanceolate, 3-9 x 1.5-2.5 cm, thinly coriaceous, quite glabrous on both surfaces, penninerved, base cuneate, margin entire, apex acuminate; lateral nerves 8-10 pairs, prominent beneath; petiole to 1 cm. Inflorescence panicles, axillary, ca 4.5 cm. Flowers bisexual, ca 5 x 2.25 mm. Pedicels slender, ca 3 mm. Perianth tube short; lobes 6 (3 + 3); outer ovate, ca 2 x 1.5 mm; inner ca 1.8 x 0.7 mm, acute, glabrous without, pubescent within. Stamens 9, perfect, in 3 rows; of 1st and 2nd rows introrse, eglandular; of 3rd row extrorse, glandular; gland 1-2 mm; filaments ca 0.5 mm; anthers 4 celled, ca 0.6 mm; of 4th row staminodes, cordate, ca 0.9 mm. Ovary globose, ca 1 x 0.75 mm; style thick, ca 0.5 mm; stigma depressed-capitate, ca 0.25 mm.

Fl.: March.

Specimens examined: XCH 10970, 15048 (Kalakad); 18150, 19246 (Courtallam); 14942 (Kothayar); 18863 (Papanasam).

Distribution: India, Myanmar.

10. PERSEA L.

Genus of about 150 pantropic and extra-tropical species; not in Africa; 2 in India.

Persea macrantha (Nees) Kosterm. in Reinwardtia 6: 193. 1962; Chandrasek. in A.N. Henry *et al*., Fl. Tamil Nadu 2: 212. 1987. *Machilus macrantha* Nees in Wallich, Pl. Asiat. Rar. 2: 70. 1831; Hook.f., Fl. Brit. India 5: 140. 1886; Gamble, Fl. Pres. Madras 2: 1227. 1925. **(Fig. 80).**

Trees to 6 m. Leaves simple, alternate, elliptic to oblong, 3.5-13 x 1.5-6.5 cm, coriaceous, glabrous on both surfaces, glaucous beneath, base acute to obliquely truncate, margin entire, apex obtuse to acute; lateral nerves 6-10 pairs; petioles to 3.5 cm. Inflorescence panicles, diffusely branched, terminal and in upper axils of leaves, ca 12 cm. Flowers pale green, trimerous, bisexual, ca 8 x 5 mm; pedicels to 4 mm. Perianth tube short, ca 1 mm; lobes 6 (3+3); outer whorl smaller, obovate, ca 4.5 x 1.8 mm, pubescent, ciliate, acute; inner oblanceolate, ca 6 x 2 mm, pubescent, ciliate, acute. Stamens 9, in 3 whorls; of first two whorls introrse, eglandular, of 3rd whorl extrorse, glandular; glands ca 1 mm; filaments 2-3 mm, pubescent; anthers 4 celled, ca 1 mm; staminodes 3, ca 2.5 mm, cordate, stipitate.

Ovary globose, ca 2 x 1 mm; style ca 2 mm; ovule ca 1 x 0.5 mm, pendulous; stigma capitate.

Occasional, in evergreen forests between 800-1000 m.

Fl.: December.

Specimens examined: 20199 (Kothayar); 18284 (Courtallam).

Distribution: India, Sri Lanka.

11. HERNANDIACEAE Blume

Trees or climbing shrubs. Leaves alternate, simple or pinnate, entire or lobed; stipules 0. Flowers hermaphodite, monoecious or polygamous, regular, usually small, white or greenish. Perianth superior; lobes free or shortly united in two, valvate 3-5 merous whorls or in one, imbricate, 4-8 merous whorl. Stamens 3-5, in one whorl; anthers 2 celled, dehiscing introrsely or laterally by 2 valves; staminodes glandular, in 1 or 2 whorls outside the stamens or absent. Ovary inferior, 1 celled; ovule solitary, pendulous, anatropous. Fruit dry, often longitudinally ribbed, sometimes with 2-4 longitudinal wings or with 2 terminal wings formed by accrescent periath lobes or else wingless and enclosed in an inflated cupule; seed solitary; albumen 0; cotyledons large, more or less lobed, twisted.

5 genera and 62 species in tropical and subtropical parts of the world; ca 3 genera and 6 species in India.

GYROCARPUS Jacq.

4 species distributed in warm and tropical regions; 1 in India.

Gyrocarpus americanus Jacq., Select. Stirp. Amer. Hist. 282. 1763; Gamble, Fl. Pres. Madras 2: 1242. 1925; Matthew & Rani in Matthew, Fl. Tamil. Carnatic 3: 1366. 1983; Sumithr. in Dassan. & Fosberg, Rev. Handb. Fl. Ceylon 6: 119.1987. *G. asiaticus* Willd., Sp. Pl. 4: 982. 1806; Verdc. in Kew Bull. 21: 252. 1967; Chandrasek. in A.N. Henry *et al.*, Fl. Tamil Nadu 2: 213. 1987; Matthew, Fl. Palni Hills 2: 1080.1999. *G. jacquinii* Roxb., Pl. Cor. t. 1.1795 p.p.; Bedd., Fl. Sylv. t. 196. 1872; Hook.f., Fl. Brit. India 5: 461. 1878. **(Fig. 81).**

Tamil: Thannukku.

Trees to 6 m; branchlets stout, brittle. Leaves alternate, 3-5 lobed when young, later broadly ovate, 4-13 x 3-17 cm, subcoriaceous, dark green above, paler and tawny pubescent beneath, base truncate, margin entire, apex acute to acuminate, 3-5 nerved at base; petioles to 8 cm, pubescent. Cymes corymbose, subterminal, much branched, compact; peduncle ca 9 cm, with single bisexual flower with two male flowers on either side. Male flower: Pedicel short, to 1 mm; perianth 4-7 lobed; lobes yellowish, oblong-ovate, ca 2 x 1 mm, prominently connate at base, cucullate at apex. Stamens 4-7, much exserted; filaments to 2 mm, pubescent; anthers oblong, to 1 mm, connectives apiculate; staminodes clavate, ca 1 mm, pubescent, alternate with stamens. Bisexual flower: Tepals 2, basally adnate to the ovary, appressed-pubescent. Drupes obovoid-globose, longitudinally ribbed, crowned by 2 spathulate wings.

Occasional in the foot hills.

Fl.: March.

Specimens examined: XCH 12305, 12357 (Papanasam); 15234 (Kalakad); 18327 (Sivagiri).

Distribution: Throughout tropics.

12. PROTEACEAE Juss.

Trees or shrubs, rarely herbs. Leaves alternate, rarely opposite or verticillate, entire, toothed or pinnatisect; stipules 0. Flowers capitate, spicate, racemose or rarely solitary, hermaphrodite, regular or irregular. Bracts small, caducous or nearly obsolete, sometimes imbricate, hardened in fruit forming a cone; bracteoles 0-2. Perianth inferior, 4 lobed, at first valvately connate in a tube with gibbous base, at length more or less separating with recurved tips. Stamens 4, on and opposite to the perianth lobes; anthers erect; connectives continuous with the filaments; cells 2, introrse, dehiscing longitudinally, Disc of 4 hypogynous scales, sometimes 0. Ovary superior, sessile or stipitate, 1 celled, often oblique; style terminal, thickened at tip; stigma terminal or lateral; ovules solitary or in pairs or many, biseriate. Fruit a nut or a drupe, indehiscent or ultimately opening on the inner edge or at the tip, or a follicle or capsule with often coriaceous valves; seeds solitary or in pairs or few; testa membranous or coriaceous, sometimes winged; albumen 0; cotyledons compressed or fleshy, often unequal; radicle short, inferior or lateral.

75 genera and 1710 species worldwide; 2 genera and 7 species in India.

HELICIA Lour.

This genus consists of 87 species in S.E. Asia, Malaysia and Eastern Australia; 4 reported from India.

Helicia nilagirica Bedd., in Madras J. Lit. Sci. Ser. 3. 1: 56. t. 11. 1864; Hook.f., Fl. Brit. India 5: 190. 1886; Gamble, Fl. Pres. Madras 2: 1243. 1925; Kumari in A.N. Henry *et al.*, Fl. Tamil Nadu 2: 213. 1987. **(Fig. 82).**

Trees to 5 m high; branchlets woody, ca 1 cm broad. Leaves alternate, broadly elliptic-oblanceolate, 5-13 x 2.5-6 cm, coriaceous, base acute to cuneate, margin entire or distantly dentate, apex acute; lateral nerves ca 6 pairs, prominent beneath, reticulation prominent; petioles to 4 cm. Flowers in terminal and axillary racemes to 25 cm, in bud ca 2.5 cm; pedicels to 2 mm, sparsely puberulous. Perianth tube slender, 4 lobed; lobes linear-oblong to ca 8 mm, sparsely pubescent without, revolute. Stamens 4, didynamous, inserted on the revolute base of perianth lobes, subsessile; anthers linear, to 3 mm; connectives produced in an apiculus, dehiscing longitudinally. Disc of 4 hypogynous glands, fused into basal cup. Ovary superior, oblong, to 5 mm, 1 locular; ovules 2 per locule; style slender, to 5 mm; stigma clavate, to ca 2.5 mm. Nut broadly oblong, ca 10 x 6.5 mm, with persistent style.

Very rare, on the edges of evergreen forests between 800-1200 m

Fl. & Fr.: August onwards.

Specimens examined: XCH 12464, 10309, 10188, 14884.

Distribution: Peninsular India.

13. THYMELAEACEAE Juss.

Shrubs or trees with tough, fibrous bark, rarely herbs. Leaves alternate or opposite, simple, entire. Flowers usually hermaphrodite, in axillary of terminal heads, umbels, clusters or racemes. Bracts various, often forming an involucre. Perianth regular, tubular or campanulate, often corolline. 4-5. Lobed; lobes imbricate in bud, often with scales at their bases within. Stamens as many or twice as many as the lobes of the perianth, those opposite the lobes usually attached above the alternate ones; filaments short; anthers erect, 2-celled, dehiscing longiludinally. Disc hypogynous, annular, cupular or of scales, sometimes 0. Ovary superior, 1-2 celled; ovule 1 in each cell, anatropous, pendulous from the apex of the cell; style short or long, terminal or excentric; stigma usually capitate. Fruit a small berry, drupe or capsule; seed solitary or one in each cell, pendulous or lateral; testa crustaceous, rarely membranous; albumen fleshy or 0; cotyledons fleshy, usually thick; radicle short, superior.

About 45 genera and 500 species, almost cosmopolitan in both temperate and tropical regions; especially well represented in Africa; ca 15 genera and 31 species in India.

GNIDIA L.

Ca 160 species distributed in the tropics; 1 in India.

Gnidia glauca (Fresen.) Gilg in Bot. Jahrb. Syst. 19: 265. 1894 in obs. var. **glauca**; Towns. in Dassan. & Fosberg, Rev. Handb. Fl. Ceylon 2: 505. 1981; Kumari in A.N. Henry *et al.*, Fl. Tamil Nadu 2: 214. 1987; Matthew, Fl. Palni Hills 2: 1083. 1999. *Lasiosiphon glaucus* Fresen. in Flora 21: 603. 1838. *Gnidia eriocephala* Meisner in Denksehr. Koenigl. Baier. Bot. Ges. Regensburg 3: 292. 1841; Wight, Ic. t. 1859. 1852. *Lasiosiphon eriocephalus* (Meisner) Decne. in Jacq., Voy. 148. 1844; Hook.f., Fl. Brit., India 5: 197. 1886; Gamble, Fl. Pres. Madras 2: 1244. 1925. **(Fig. 83).**

A much branched, bushy shrub or small tree to 3 m high; branchlets terete, with prominent leaf-scars. Leaves crowded towards summit of branchlets, linear-oblong or oblong-oblanceolate, 3-7.5 x 1-1.5 cm, subcoriaceous, minutely appressed-pubescent beneath, base obtuse, margin entire, apex acute; lateral nerves ca 6 pairs, looping towards apex, obscure; petioles to 2 mm long, minutely appressed-pubescent. Heads ca 2 cm across. Flowers many, yellow, when dry brownish, dense, subtended by oblong-ovate, white bracts; pedicels thick, ca 2 mm, densely pilose. Perianth tube to 1 cm, glabrous within, densely pilose without; lobes equal, oblong, ca 3.5 x 2 mm, glabrous within, appressed pilose without. Stamens 10; filaments to 0.5 mm; anthers oblong, to 2.5 mm. Ovary oblong, ca 4 x 1.2 mm, stipitate; style short, to 4 mm.

On exposed, bare slopes in high altitude grassland.

Fl.: January – March.

Specimen examined: XCH 15254 (Kalakad).

Distribution: South and West India, Tropical Africa from Nigeria to Sudan and Ethiopia, South to Malawi, Zambia and Mozambique.

14. ELAEAGNACEAE Juss.

Trees or shrubs, often armed, with silvery or coppery scales on the branches, inflorescence and leaves beneath. Leaves alternate or opposite, entire, estipulate, Flowers hermaphrodite or dioecious, solitary, in axillary fascicles, spikes or racemes. Perianth in hermaphrodite or male flowers tubular, constricted above ovary; in female flowers reduced to a 2-4 lobed limb. Filaments free; anthers oblong, 2 celled. Ovary free, 1 celled; style filliform; stigma lateral; ovule 1. Fruit a nut covered by the thickened perianth base.

3 genera and ca 100 species mainly in N. Hemisphere; 2 genera and ca 19 species in India.

ELAEAGNUS L.

Ca 90 species in the world; 14 in India.

Elaeagnus conferta Roxb., Fl. Ind. 1: 460 1820. *E. latifolia* L., Sp. Pl. 121. 1753 p.p.; Wight, Ic.t. 1856. 1852; Hook.f., Fl. Brit. India 2: 202. 1886 p.p. *E. kologa* Schlecht. in DC., Prodr. 14: 611. 1857; Gamble, Fl. Pres. Madras 2: 1246. 1925; Matthew & Rani in Matthew, Fl. Tamil. Carnatic 3: 1370. 1983; Kumari in A.N. Henry *et al.*, Fl. Tamil Nadu 2: 214. 1987; Matthew, Fl. Palni Hills 2: 1084. 1999; Mohanan & Sivadasan, Fl. Agasthyamala 576. 2002. **(Fig. 84).**

Woody stragglers; branchlets lepidote. Leaves alternate, elliptic or ovate-oblong, 3-7.5 x 1.6-3.6 cm, coriaceous, densely scaly beneath, less so above when young coppery-brown when young, margin entire, apex shortly, obtusely acuminate;

lateral nerves obscure, toothed irregularly; petioles to 5 mm long, coppery-brown after drying. Flowers 3-6 or more, in axillary fascicles, bisexual, actinomorphic. Pedicels ca 5 mm, scaly. Perianth tube urceolate, to 5 mm, scaly without, 4 lobed; lobes yellowish, ovate, ca 2.5 x 2 mm, scaly on both surfaces. Stamens 4, included, alternate with perianth lobes; filaments ca 0.5 mm; anthers oblong, to 1 mm. Ovary oblong-ovoid, ca 1.8 mm, scaly, 1-locular; ovule 1, erect; style to 6 mm, scaly, in-rolled. Nuts oblong, ca 2.5 x 1.2 cm, densely scaly, orange-red on ripening.

Rare, on forest edges in evergreen forests above 1000 m.

Fl. & Fr.: March – May, almost throughout the year.

Specimens examined: XCH 14947, 14998, 12514 (Kothayar); 18642 (Mahendragiri).

Distribution: Peninsular India.

15. LORANTHACEAE Juss.

Evergreen shrubs, rarely herbs, usually parasitic on the branches of trees; stems much branched, often jointed. Leaves opposite or sometimes alternate, coriaceous, entire, sometimes reduced to scales or 0; stipules 0. Flowers actinomorphic or zygomorphic, bisexual or unisexual, sometimes large and brightly coloured, sometimes inconspicuous, in racemes or spikes or heads or fascicles, usually bracteate and often with 2 or more bracteoles. Calyculus adnate to ovary; limb annular or cupular, entire or shortly toothed, sometimes 0. Corolla 3-8 lobed, valvate, free or more or less connate in a tube. Stamens as many as and opposite to petals and more or less adnate to petals; filaments slender or 0; anthers basifixed or dorsifixed, opening by longitudinal slits or by rows of cells or by pores. Ovary inferior, 1 locular, usually without distinct placenta; style short or long; stigma simple. Fruit a 1 seeded berry or drupe with fleshy pericarp and often viscid mesocarp; seed solitary, without distinct testa; albumen fleshy or none; embryo straight, usually fleshy; radicle superior.

A family of ca 84 genera and about 950 species, found in tropical and temperate regions; 8 genera and 49 species in India.

Key to the genera

1a. Leaves opposite, decussate; corolla lobes spirally twisted when opening ... 3. **Helicanthes**
1b. Leaves alternate to subopposite; corolla lobes not twisted 2

2b. Petals 6; calyculus supported by a bract and two bracteoles 3
2a. Petals 4-5; calyculus supported by a bract, no bracteoles 4

3a. Inflorescence axis decussately flattened ....2. **Elytranthe**
3b. Inflorescence axis terete.. 5. **Macrosolen**

4a. Petals 5.. 5
4b. Petals 4.. 6

5a. Corolla actinomorphic; leaves subopposite. ....1. **Dendrophthoe**
5b. Corolla zygomorphic; leaves alternate ....7. **Taxillus**
6a. Corolla zygomorphic; calyculus attenuate at base...................................6. **Scurrula**
6b. Corolla actinomorphic; calyculus rounded at base..........................4. **Helixanthera**

1. DENDROPHTHOE Mart.

A genus of 30-35 species centred in southeast Asia, widespread in the tropical regions; 4 in India.

Key to the species

1a. Flowers in racemes ....1. **D. falcata**
1b. Flowers in axillary fascicles or cymes...2
2a. Flowers in umbellate cymes; corolla very straight, erect .... 2. **D. memecylifolia**
2b. Flowers solitary or other than umbellate cymes; corolla cleft to the middle, lobes recurved ....3. **D. neelgherrensis**

1. **Dendrophthoe falcata** (L.f.) Blume in Schult.f., Syst. Veg. 7: 1730. 1830; Matthew & Rani in Matthew, Fl. Tamil. Carnatic 3: 1372. 1983; Chandrasek. in A.N. Henry *et al.*, Fl. Tamil Nadu 2: 215. 1987; Wiens in Dassan. & Fosberg, Rev. Handb. Fl. Ceylon 6: 134. 1987; Matthew, Fl. Palni Hills 2: 1087. 1999; Mohanan & Sivadasan, Fl. Agasthyamala 579. 2002; K. Rajasekaran in N.P. Balakr. *et al.,* Fl. India. 23: 3. 2012. *Loranthus falcatus* L.f., Suppl. Pl. 211. 1781. *L. longiflorus* Desr. in Lam., Encycl. 3: 598. 1792; Hook.f., Fl. Brit. India 5: 214. 1886; Gamble, Fl. Pres. Madras 2: 1253. 1925. *L. longiflorus* Desr. var. *falcatus* (L.f.) Kurz, For. Fl. Burma 2: 321. 1877; Hook.f., Fl. Brit. India 5: 215. 1886; Gamble, Fl. Pres. Madras 2: 1253. 1925. **(Fig. 85).**

Branchlets woody, terete. Leaves subopposite, very varaiable, usually oblong-ovate, 7.5-11.5 x 2-4 cm or broadly ovate, 7-12.5 x 5-7 cm or linear oblong, 4.5-10 x 0.5-0.7 cm, coriaceous, glabrous, base obtuse to acute, margin entire, apex obtuse; lateral nerves ca 6 pairs; petioles to 1 cm. Racemes axillary, shortly peduncled, ca 4 cm, with ca 30 flowers. Calyculus ca 1 mm. Corolla creamy-white at base, red towards apex and greenish on lobes; tube ca 4 cm, minutely tuberculate within, 5 lobed; lobes oblong, ca 1 cm, twisted and curved. Stamens 5; filaments reddish, erect, free from corolla lobes, ca 3 mm; anthers yellow, linear-oblong, ca 5 mm. Ovary ca 3 mm long; style to 5.5 cm.

Frequent, in the foot hills of deciduous forests as parasites on other plants.

Fl.: Throughout the year.

Specimens examined: XCH 11749, 19719 (Kalakad); 12368, 12387, 14488, 16852, 16921, 13180, 17857 (Papanasam); 11408, 15531, 13921, 14000 (Courtallam); 10753 (Mancholai); 16220, 16269 (Thirukurungudi); 12093 (Mancholai).

Distribution: Sri Lanka, India, Thailand, Indo-China, Australia.

2. **Dendrophthoe memecylifolia** (Wight & Arn.) Danser in Bull. Jard. Bot. Buitenzorg Ser. 3. 10: 310. 1929; Chandrasek. in A.N. Henry *et al.*, Fl. Tamil Nadu 2: 215. 1987; Matthew, Fl. Palni Hills 2: 1088. 1999; K. Rajasekaran in N.P. Balakr. *et al.,* Fl. India. 23: 5. 2012. *Loranthus memecylifolius* Wight & Arn., Prodr. 383. 1834; Hook.f., Fl. Brit. India 5: 217. 1886; Gamble, Fl. Pres. Madras 2: 1254. 1925. **(Fig. 86).**

Woody shrubs; branchlets terete, swollen at mature nodal regions. Leaves oblong-ovate, 2.5-8.5 x 1-2.7 cm, coriaceous, blackish-green when dry, cuneate at base, obtuse at apex; lateral nerves obscure, midrib prominent. Flowers ca 8, in axillary or lateral peduncled, umbellate cymes; peduncles ca 1.5 cm; pedicels ca 5 mm. Calyculus tubular, to 2.5 cm, minutely 5 toothed. Corolla yellowish or orange in colour; tube to 1 cm, ribbed and puberulous within, 5 lobed; lobes linear-oblong, ca 1 cm, reflexed. Stamens 5; filaments ca 5 mm, puberulous at base; anthers oblong, to 2 mm. Ovary obovoid, ca 5 mm; style to 2 cm; stigma capitate. Berries obovoid, ca 1.5 cm.

Rare, in dense evergreen forests between 1200-1500 m.

Fl.: January – March; *Fr.:* March – August.

Specimens examined: XCH 10064, 12261 (Kakachi); 12518, 14887, 14995 (Kothayar); 18775 (Mahendragiri).

Distribution: Peninsular India.

3. **Dendrophthoe neelgherrensis** (Wight & Arn.) Tiegh. in Bull. Soc. Bot. France 42: 252. 1895; Chandrasek. in A.N. Henry *et al.*, Fl. Tamil Nadu 2: 215. 1987; Wiens in Dassan. & Fosberg, Rev. Handb. Fl. Ceylon 6: 130. 1987; Matthew, Fl. Palni Hills 2: 1088. 1999; Mohanan & Sivadasan, Fl. Agasthyamala 579. 2002; K. Rajasekaran in N.P. Balakr. *et al.,* Fl. India. 23: 5. 2012. *Loranthus neelgherrensis* Wight & Arn., Prodr. 382. 1834; Wight, Ic. t. 1020. 1845; Hook.f., Fl. Brit. India 5: 216. 1886; Gamble, Fl. Pres. Madras 2: 1254. 1925.

Branchlets terete. Leaves ovate-lanceolate, 8-9.5 x 1.3-1.6 cm, coriaceous, base cuneate to attenuate, apex acute; lateral nerves ca 5 pairs, faint to obscure; midrib prominent; petioles ca 5 mm. Racemes crowded in axillary clusters or solitary, slightly pedunculate. Bracts ovate, acute, ca 3 mm. Calyculus cupular, ca 5 mm, obscurely 5 toothed. Corolla dark reddish, white at middle; tube ca 2 cm, 5 lobed, lobes oblong, ca 1 cm. Stamens 5; filaments ca 5 mm, reddish; anthers oblong, to 2 mm. Ovary oblong, ca 7 mm; style to 2.5 cm; stigma capitate.

Occasional, in deciduous forests.

Fl.: May.

Specimens examined: XCH 12716, 14765 (Korimoni).

Distribution: Sri Lanka, Peninsular India.

2. ELYTRANTHE (Blume) Blume

Ca 11 species in the Indomalayan realm, 5 in India.

Key to the species

1a. Leaves oblong-ovate or oblanceolate, less than 5.5 cm long; bracts and bracteoles less conspicuous .... 1. **E. capitellata**

1b. Leaves broadly elliptic-ovate, or oblong or broadly lanceolate, more than 5.5 cm long; bracts and bracteoles much conspicuous .... 2. **E. parasitica**

1. Elytranthe capitellata (Wight & Arn.) Engl. in Engl. & Prantl, Nat. Pflanzenfam. 3(1): 189 1889; Gamble, Fl. Pres. Madras 2: 1255. 1925; K. Rajasekaran in N.P. Balakr. *et al.,* Fl. India. 23: 8. 2012. *Loranthus capitellatus* Wight & Arn., Prodr. 1: 382. 1834; Hook.f., Fl. Brit. India 5: 221. 1886. *Macrosolen capitellatus* (Wight & Arn.) Danser in Blumea 2: 36. 1936; Wiens in Dassan. & Fosberg, Rev. Handb. Fl. Ceylon 6: 126. 1987; Chandrasek. in A.N. Henry *et al.*, Fl. Tamil Nadu 2: 217. 1987; Matthew, Fl. Palni Hills 2: 1093. 1999. **(Fig. 87).**

Branchlets woody, terete. Leaves oblong-ovate or oblanceolate, 4-5.5 x 1-2 cm, chartaceous, base attenuate, obtuse or acute at apex; lateral nerves obscure; midrib raised beneath; petioles ca 3 mm. Spike usually with two pairs of decussate, apically crowded flowers. Bracts and bracteoles rounded to obtuse. Calyculus cupular, ca 3 mm. Corolla (buds) reddish yellow at base, greenish at middle, red at upper portion; tube ca 4.5 mm, 6 lobed; lobes linear-spathulate, ca 6 x 0.6 mm. Stamens 6; filaments ca 1.5 mm; anthers oblong, to 1 mm. Style to 6 mm; stigma globose.

Rare, in deciduous forests.

Fl: May.

Distribution: Peninsular India, Sri Lanka.

Specimen examined: XCH 12759 (Mahendragiri).

2. Elytranthe parasitica (L.) Danser, Bull. Jard. Bot. Buitenzorg III, 10: 315 1929; K. Rajasekaran in N.P. Balakr. *et al.,* Fl. India. 23: 9. 2012. *Lonicera parasitica* L., Sp. Pl. 175. 1753. *Loranthus loniceroides* L., Sp. Pl. ed. 2. 1: 473. 1762; Wight, Ic. t. 203. 1839; Hook.f., Fl. Brit. India 5: 221. 1886. *Elytranthe loniceroides* (L.) G.Don, Gen. Syst. 3: 427. 1834; Gamble, Fl. Pres. Madras 2: 1255. 1925. *Macrosolen parasiticus* (L.) Danser in Blumea 2: 36. 1936; Chandrasek. in A.N. Henry *et al.*, Fl. Tamil Nadu 2: 217. 1987; Wiens in Dassan. & Fosberg, Rev. Handb. Fl. Ceylon 6: 128. 1987; Matthew, Fl. Palni Hills 2: 1093. 1999; Mohanan & Sivadasan, Fl. Agasthyamala 582. 2002. **(Fig. 88).**

Branchlets terete, woody, pale blackish or brownish when dry, lenticellate. Leaves oblong or elliptic-ovate, or lanceolate, 5.5-9 x 1.8-3.2 cm, coriaceous, base truncate or cuneate, apex obtuse, sessile or very shortly petioled. Spike with 3-4 pairs of decussate, apically crowded flowers. Calyculus cupular, ca 5 mm. Corolla pinkish; tube 2.5 cm, 6 lobed; lobes linear-oblong, ca 1.5 x 0.4 cm, fully reflexed. Stamens 6; filaments flattened, very shortly free from lobe; anthers linear-oblong, ca 8 mm, yellowish. Style ca 4 cm; stigma globose, minutely 4 lobed.

Rare, in evergreen forests above 1200 m.

Fl.: September.

Specimens examined: XCH 10634, 10310 (Kakachi).

Distribution: Peninsular India, Sri Lanka.

3. HELICANTHES Danser

Monotypic and endemic to Deccan peninsula.

Helicanthes elasticus (Desv.) Danser in Verh. Kon. Ned. Akad. Wetensch., Afd. Natuurk. Tweede Sect. 2. 29(6): 55. 1933; Matthew & Rani in Matthew, Fl. Tamil. Carnatic 3: 1373. 1983; Chandrasek. in A.N. Henry *et al.*, Fl. Tamil Nadu 2: 216. 1987; Matthew, Fl. Palni Hills 2: 1090. 1999; Mohanan & Sivadasan, Fl. Agasthyamala 580. 2002; K. Rajasekaran in N.P. Balakr. *et al.,* Fl. India. 23: 10. 2012. *Loranthus elasticus* Desv. in Lam., Encycl. 3: 599. 1789; Wight, Ic.t. 343. 1840; Hook.f., Fl. Brit. India 5: 216. 1886; Gamble, Fl. Pres. Madras 2: 1254. 1925. *L. euphorbiae* Wight, Ic. t. 1063. 1846. **(Fig. 89).**

Branchlets woody, dichotomous, slightly zig-zag. Leaves opposite, decussate, thick coriaceous, when dry straw in colour, ovate, 5-6.5 x 3-4 cm or oblong-ovate, 3-4.5 x 1-2 cm or elliptic-ovate or oblanceolate, 4-6.5 x 1.2-1.6 cm, 3 ribbed, obscure, sessile. Flowers sessile, fascicled at the nodes of branches. Calyculus cupular, ca 4 x 2.5 mm, entire. Corolla red, with green ribs; tube ca 1.5 cm, broad towards apex, ribbed within; ribs minutely dotted, 5 lobed; lobes linear, ca 1.5 cm, reflexed, dotted at margins within. Stamens 5; filaments ca 1 cm; anthers narrowly oblong, to 4 mm. Ovary subglobose, ca 2 x 2 mm; ovule solitary; style filiform, to 3.25 cm; stigma obovoid. Berries oblong-globose, ca 8 mm.

Occasional, in deciduous forests.

Fl. & Fr.: December – April; July – September.

Specimens examined: XCH 17625 (Thirukurungudi); 18318 (Sivagiri); 14783 (Korimoni).

Distribution: Peninsular India.

4. HELIXANTHERA Lour.

A genus of about 25 species occurring throughout southeast Asia, but also with species in Africa; 10 in India.

Key to the species

1a. Flowers sessile, in spikes, in cavities of a thickened rachis 2
1b. Flowers pedicelled, in racemes, solitary or fascicled on the nodes of old branchlets 3

2a. Leaves elliptic-ovate, obtuse to acute at apex, glabrous .... 1. **H. hookeriana**
2b. Leaves ovate or orbicular, obtuse at apex, stellate pubescent ...... 3. **H. lepidophylla**

3a. Flowers lax, pedicels straight; drupes ovoid-ellipsoid .... 4. **H. wallichiana**
3b. Flowers condensed, pedicels curved; drupes subglobose .... 2. **H. intermedia**

1. **Helixanthera hookeriana** (Wight & Arn.) Danser in Bull. Jard. Bot. Buitenzorg Ser. 3. 10: 317. 1929; Chandrasek. in A.N. Henry *et al.*, Fl. Tamil Nadu 2: 216. 1987; Wiens in Dassan. & Fosberg, Rev. Handb. Fl. Ceylon 6: 148. 1987; Matthew, Fl. Palni Hills 2: 1090. 1999; K. Rajasekaran in N.P. Balakr. *et al.*, Fl. India. 23: 13. 2012. *Loranthus hookerianus* Wight & Arn., Prodr. 1: 381. 1834; Hook.f., Fl. Brit. India 5: 207. 1886; Gamble, Fl. Pres. Madras 2: 1251. 1925. **(Fig. 90).**

Branchlets woody, terete, lenticellate. Leaves alternate, usually opposite when young, elliptic-ovate, 4-8.5 x 1.5-3.5 cm, chartaceous, base cuneate, obtuse to acute at apex; lateral nerves ca 5 pairs, distantly arranged, looping; midrib raised on both surfaces; petioles up to 1 cm. Spikes to 6 cm, rusty tomentose when young. Calyculus flask shaped, ca 1 mm. Corolla red; tube to 3 mm, 4 lobed; lobes oblong-spathulate, ca 4 x 1 mm. Stamens 4, introrse; filaments ca 2.5 mm, rusty red; anthers oblong, to 1.5 mm, reddish. Ovary globose, ca 1 x 1 mm; ovule solitary. Drupes ovoid-globose, ca 5 x 3 mm, brownish-purple.

Occasional, in moist deciduous to semi-evergreen forests between 300-500 m.

Fl. & Fr.: May – August.

Specimens examined: XCH 16494 (Thirukurungudi); 15908 (Courtallam).

Distribution: Peninsular India, Sri Lanka.

2. **Helixanthera intermedia** (Wight) Danser in Bull. Jard. Bot. Buitenzorg Ser. 3. 10: 317. 1929; Chandrasek. in A.N. Henry *et al.*, Fl. Tamil Nadu 2: 216. 1987; Matthew, Fl. Palni Hills 2: 1091. 1999; Mohanan & Sivadasan, Fl. Agasthyamala 581. 2002; K. Rajasekaran in N.P. Balakr. *et al.*, Fl. India. 23: 13. 2012. *Loranthus intermedius* Wight in Calcutta J. Nat. Hist. 6: 361. 1846; Hook.f., Fl. Brit. India 5: 205. 1886; Gamble, Fl. Pres. Madras 2: 1251. 1925. **(Fig. 91).**

Branchlets woody, terete, pale coppery-brown when dry, lenticellate. Leaves opposite or subopposite, subcoriaceous, ovate, or elliptic, 4-9.3 x 1.6-3.8 cm, coppery brown beneath when dry, paler above, base cuneate, apex obtuse or obtusely acute; lateral nerves ca 5 pairs, obscure; pedicels ca 5 mm. Racemes axillary and lateral, ca 6 cm. Bracts ovate, ca 2 mm, pubescent without. Calyx cupular, ca 3 mm, pubescent without. Corolla red; tube to 1.8 cm, 4 lobed; lobes linear-spathulate, ca 6 x 1.5 mm, reflexed. Stamens 4; filaments ca 6 mm; anthers linear-oblong, to 4 mm. Ovary globose, ca 2.5 mm across; style to 1.75 cm; stigma ovoid. Drupes subglobose, ca 4 mm across.

Rare, in evergreen forests above 1000 m.

Fl. & Fr.: July – October.

Specimens examined: XCH 10422, 10583 (Kakachi); 17248 (Kothayar); 19436 (Alagiapandiapuram).

Distribution: Peninsular India.

3. **Helixanthera lepidophylla** (Walp.) Danser in Bull. Jard. Bot. Buitenzorg Ser. 3. 10: 317. 1929; Chandrasek. in A.N. Henry *et al.*, Fl. Tamil Nadu 2: 216. 1987; K. Rajasekaran in N.P. Balakr. *et al.,* Fl. India. 23: 15. 2012. *Loranthus lepidophyllus* Walp., Ann. 2: 730. 1852. *L. wightii* Hook.f., Fl. Brit. India 5: 207. 1886; Gamble, Fl. Pres. Madras 2: 1251. 1925. **(Fig. 92).**

Branchlets woody, terete, pale, minutely lenticellate. Leaves opposite and alternate, ovate or orbicular, 1-2.5 x 0.8-1.8 cm, subcoriaceous, stellate-pubescent beneath, later often glabrous, brownish or blackish-green when dry, paler beneath, obtusely acute at base, apex obtuse; lateral nerves obscure; midrib raised beneath; petioles 1-3 mm. Spikes to 3.5 cm, axillary and lateral. Flowers two in each axis or solitary. Calyculus cupular, ca 1.65 mm, stellate-pubescent without, minutely 4 toothed. Corolla pale yellowish; tube to 5 mm, 4 lobed; lobes linear-obovate, ca 6 mm, stellate-pubescent without, reflexed, apex recurved. Stamens 4, introrse; filaments ca 3 mm; anthers obovate, ca 1 x 8 mm. Ovary globose; ovule solitary.

Occasional, in semi-evergreen forests, above 400 m.

Fl.: June.

Specimens examined: XCH 13109 (Papanasam); 19869 (Sivagiri).

Distribution: Peninsular India.

4. **Helixanthera wallichiana** (Schult.f.) Danser in Bull. Jard. Bot. Buitenzorg Ser. 3. 10: 319. 1929; Chandrasek. in A.N. Henry *et al.*, Fl. Tamil Nadu 2: 216. 1987; Matthew, Fl. Palni Hills 2: 1092. 1999; Mohanan & Sivadasan, Fl. Agasthyamala 581. 2002; K. Rajasekaran in N.P. Balakr. *et al.,* Fl. India. 23: 18. 2012. *Loranthus wallichianus* Schult.f. in Roem. & Schult, Syst. Veg. 7: 100. 1829; Wight., Ic. t. 143. 1839; Hook.f., Fl. Brit. India 5: 205. 1886; Gamble, Fl. Pres. Madras 2: 1251. 1925. **(Fig. 93).**

Branchlets woody, terete, lenticellate, grayish. Leaves elliptic-ovate, 4-7.5 x 1.8-4 cm, subcoriaceous, when dry brownish, base attenuate, apex obtuse or subacute; lateral nerves ca 7 pairs, obscure; midrib prominently raised beneath; petioles thick, up to 1.5 cm. Racemes to 6.5 cm. Flowers (buds) laxly arranged; bracteole ovate, concave, ca 1 mm, ciliate; pedicels ca 5 mm. Calyx cupular, ca 2 mm, minutely 4 toothed. Corolla pinkish; tube to 6.5 mm, 4 lobed; lobes linear-spathulate, ca 3 x 1 mm. Stamens 4; filaments ca 2 mm; anthers linear-oblong, to 1.2 mm. Ovary globose, ca 0.8 mm across; style to 9 mm; stigma connate. Drupes ovoid or ellipsoid, ca 5 mm.

Occasional, in moist-deciduous to evergreen forests between 300-1200 m.

Fl. & Fr.: June – September.

Specimens examined: XCH 11048 (Kalakad); 13964 (Courtallam); 12512 (Kothayar).

Distribution: Peninsular India.

5. MACROSOLEN (Blume) Rchb.

A genus of 35-40 species throughout southeastern Asia and India; 3 in India.

1. **Macrosolen albicaulis** Wiens in Sri Lanka J. Sci. Biol. Sci. 9(2): 45. 1971 et in Dassan. & Fosberg, Rev. Handb. Fl. Ceylon 6: 126. 1987. **(Fig. 94).**

Branchlets pale yellow in colour, woody, terete, lenticellate. Leaves lanceolate or elliptic-lanceolate, 6-11 x 1.5-2.8 cm, pale in colour, base truncate, obtusely acute at apex or shortly acuminate; lateral nerves faint, irregular; petioles up to 5 mm. Spikes shortly pedunculate. Bracts and bracteoles usually rounded. Flowers and fruits not seen.

Occasional, in deciduous forests at ca 300 m.

Specimen examined: XCH 15526 (Courtallam).

Distribution: Sri Lanka, India.

Note: Doubtful in identity, parasitic on *Syzygium cumini.*

6. SCURRULA L.

A genus of 23 species, widely distributed in Southeast Asia and India; 10 in India.

Scurrula parasitica L., Sp. Pl. 110. 1753; Matthew & Rani in Matthew, Fl. Tamil. Carnatic 3: 1374. 1983; Chandrasek. in A.N. Henry *et al.*, Fl. Tamil Nadu 2: 217. 1987; Wiens in Dassan. & Fosberg, Rev. Handb. Fl. Ceylon 6: 146. 1987; Matthew, Fl. Palni Hills 2: 1094. 1999; Mohanan & Sivadasan, Fl. Agasthyamala 583. 2002; K. Rajasekaran in N.P. Balakr. *et al.*, Fl. India. 23: 27. 2012. *L. scurrula* L., Sp. Pl. ed. 2. 1: 472. 1762; Hook.f., Fl. Brit. India 5: 208. 1886. *Loranthus* *buddleioides* Desv. in Lam., Encycl. 3: 600. 1792; Gamble, Fl. Pres. Madras 2: 1251. 1925. **(Fig. 95).**

Branchlets woody, terete, dark brownish. Leaves subopposite, oblong-orbicular or elliptic, broadly obovate, 6-8 x 3.5-4.2 cm, coppery-brown above when dry, pale green beneath, rusty stellate tomentose, more so beneath, base rounded or subtruncate, apex obtuse; lateral nerves ca 6 pairs, forked and looping, faint; petioles ca 1 cm. Flowers in axillary fascicles, densely covered with mealy brownish tomentum. Calyculus cupular, ca 4 mm, brown-tomentose without. Corolla pale yellowish; tube to 1.8 cm, brown tomentose without, 4 lobed; lobes spathulate, ca 6 x 1.4 mm, slightly concave, brown tomentose without. Stamens 4; filaments ca 2.1 mm; anthers oblong, to 1.4 mm. Ovary obconic, ca 6 x 4 mm, 1 locular; ovule 1, solitary; style adnate to corolla tube.

Rare, in deciduous forests.

Fl.: September.

Specimens examined: XCH 17856 (Papanasam); 10732 (Mancholai).

Distribution: India, Sri Lanka, Indo-Malaysia.

7. TAXILLUS Tiegh.

A genus of 12 to 16 species occurring in southern Asia through the Malay Peninsula southward to Borneo and the Philippines; 13 species in India.

Key to the species

1a. Corolla densely tomentose 2
1b. Corolla minutely stellate hairy or glabrous .... 1. **T. cuneatus**
2a. Leaves black above when dry, densely rusty-floccose beneath ..... 3. **T. tomentosus**
2b. Leaves dark greenish above when dry, thinly grey or tawny pubescent beneath .... 2. **T. recurvus**

1. Taxillus cuneatus (B. Heyne ex Roth) Danser in Bull. Jard. Bot. Buitenzorg Ser. 3. 10: 354. 1929; Matthew & Rani in Matthew, Fl. Tamil. Carnatic 3: 1376. 1983; Chandrasek. in A.N. Henry *et al.*, Fl. Tamil Nadu 2: 217. 1987; Wiens in Dassan. & Fosberg, Rev. Handb. Fl. Ceylon 6: 144. 1987; Matthew, Fl. Palni Hills 2: 1095. 1999; Mohanan & Sivadasan, Fl. Agasthyamala 584. 2002; K. Rajasekaran in N.P. Balakr. *et al.,* Fl. India. 23: 32. 2012. *Loranthus cuneatus* B. Heyne ex Roth, Nov. Pl. Sp. 193. 1821; Hook.f., Fl. Brit. India 5: 214. 1886; Gamble, Fl. Pres. Madras 2: 1253. 1925. **(Fig. 96).**

Branchlets woody, grayish, with prominent swollen nodes. Leaves alternate, obovate-spathulate, 1.5-3 x 0.3-2.5 cm, minutely stellate hairy when young, cuneate at base, apex obtuse or obscurely retuse; lateral nerves obscure; petioles 3-10 mm. Cymes axillary, 3 flowered, shortly pedunculate. Pedicels ca 3 mm.

Calyculus cupular, ca 3.5 mm, margin minutely toothed. Corolla pale greenish; tube to 1.65 cm, sparsely stellate-pubescent, 5 lobed; lobes linear-oblong, ca 6.6 mm, fully reflexed, acute. Stamens 5, included; filaments ca 2 mm; anthers linear-oblong, to 2 mm, Ovary cylindric, ca 2.5 mm; style to 2.25 cm; stigma capitate, reddish.

Occasional, in deciduous forests.

Fl. & Fr.: Throughout the year.

Specimens examined: XCH 16410, 17584 (Thirukurungudi); 14190 (Papanasam); 12023 (Mancholai); 17427 (Mahendragiri).

Distribution: Peninsular India, Sri Lanka.

2. **Taxillus recurvus** (Wallich ex DC.) Tiegh. in Bull. Soc. Bot. France 42: 256. 1895; Chandrasek. in A.N. Henry *et al.*, Fl. Tamil Nadu 2: 218. 1987; Matthew, Fl. Palni Hills 2: 1096. 1999; K. Rajasekaran in N.P. Balakr. *et al.*, Fl. India. 23: 34. 2012. *Loranthus recurvus* Wallich ex DC., Prodr. 4: 299. 1830; Hook.f., Fl. Brit. India 5: 213. 1886; Gamble, Fl. Pres. Madras 2: 1252. 1925. *L. candolleanus* Wight & Arn., Prodr. 1: 385. 1834; Wight, Ic. t. 305. 1840. **(Fig. 97).**

Branchlets smooth, rusty tomentose. Leaves obovate, 3 nerved at base, ca 2 x 1 cm, base cuneate, apex obtuse; lateral nerves obscure; petioles ca 5 mm. Cymes ca 6 flowered, stellate hairy. Calyculus ca 2 mm, irregularly 5 lobed. Corolla yellowish-green; tube slightly curved, rusty-villous, ca 1.2 cm, 5 lobed.

Rare, in evergreen forests above 1000 m.

Fl.: September.

Specimen examined: XCH 10762 (Kalakad).

Distribution: Western Ghats of Peninsular India.

3. **Taxillus tomentosus** (B. Heyne ex Roth) Tiegh. in Bull. Soc. Bot. France 42: 256. 1895; Chandrasek. in A.N. Henry *et al.*, Fl. Tamil Nadu 2: 218. 1987; Wiens in Dassan. & Fosberg, Rev. Handb. Fl. Ceylon 6: 139. 1987; Matthew, Fl. Palni Hills 2: 1096. 1999; Mohanan & Sivadasan, Fl. Agasthyamala 584. 2002; K. Rajasekaran in N.P. Balakr. *et al.*, Fl. India. 23: 36. 2012. *Loranthus tomentosus* B. Heyne ex Roth, Nov. Pl. Sp. 191. 1821; Hook.f., Fl. Brit. India 5: 212. 1886; Gamble, Fl. Pres. Madras 2: 1252. 1925.

Branchlets woody, terete, dark brown, rusty tomentose when young. Leaves obovate or oblong-ovate, 3-6 x 1-2.5 cm, chartaceous, glabrous or sparsely pubescent above, densely, conspicuously rusty-floccose beneath, dark brown or black above when dry; lateral nerves 3-6 pairs, arching; petioles ca 5 mm, rusty tomentose. Cymes of 3-4 flowers. Pedicels ca 3 mm. Calyculus ca 2.5 mm, rusty tomentose, 5 toothed. Corolla yellowish; tube ca 1.75 cm, densely rusty-tomentose without, less so within on filaments, 5 lobed; lobes spathulate, ca 7.5 mm, rusty tomentose without, fully reflexed. Stamens 5, free; filaments to 2.5 mm, tomentose; anthers oblong, to 1.5 mm. Ovary subglobose; style to 1.8 cm; stigma capitate.

Occasional, in evergreen forests between 1100-1500 m.

Fl.: Throughout the year.

Specimens examined: XCH 10812, 10981, 11060, 19561 (Kalakad); 10314, 10455, 12429, 17211 (Kakachi); 17397 (Mahendragiri); 10115, 10273, 14937 (Kothayar); 14733 (Balamore).

Distribution: Peninsular India, Sri Lanka.

16. VISCACEAE Batsch

Chlorophyllous, parasitic shrubs or herbs, monoecious or dioecious. Stems usually forked, nodes articulated or swollen. Leaves present or absent, opposite, simple and entire, sometimes reduced to scales. Inflorescence axillary or terminal, solitary or aggregated. Flowers unisexual, monochlamydous; perianth lobes 2-4, valvate. Stamens as many as the peianth lobes or less. Ovary inferior, unilocular. Fruits fleshy, one-seeded, berry-like.

7 genera and 400 species distributed mainly in the tropical regions; 4 genera and 19 species in India.

1. VISCUM L.

A genus of ca 67 species occurring throughout the tropical and subtropical areas of the old world, but with the greatest concentration of species in Tropical Africa and the Malagasy. 15 species are from India, of which 4 are endemic.

Key to the species

1a. Leafy shrubs 2

1b. Leafless shrubs 3

2a. Berries globose, apically acute .... 4. **V. orientale**

2b. Berries oblong, apically truncate .... 3. **V. heyneanum**

3a. Berries glossy white; branchlets flattened .... 2. **V. articulatum**

3b. Berries greenish; branchlets not flattened .... 1. **V. angulatum**

1. Viscum angulatum B. Heyne ex DC., Prodr. 4: 283. 1830; Hook.f., Fl. Brit. India 5: 225. 1886; Gamble, Fl. Pres. Madras 2: 1259. 1925; Matthew, Fl. Palni Hills 2: 1100. 1999; Mohanan & Sivadasan, Fl. Agasthyamala 585. 2002; V.N. Sanjai & N.P. Balakr. in N.P. Balakr. *et al.,* Fl. India. 23: 49. 2012. *V. ramosissimum* Wight, Icon. Pl. Ind. Orient. 3: t. 1017. tantum, 1845. **(Fig. 98).**

Leafless, slender parasitic shrubs; main stem terete; branchlets 4 angled. Flowers unisexual, minute, in axillary triads, sessile, solitary or whorled at nodes. The male flowers usually lateral; female flowers central. Perrianth slightly 4 lobed, tubular in male flowers, united with ovary in female flowers. Ovary globose, minute, 1 locular; ovule solitary; stigma subsessile.

Rare, in deciduous forests.

Fl.: March.

Specimen examined: XCH 14134 (Sivagiri); XCH 11740 (Kalakad); 16319, 16247, 17621 (Thirukurungudi); 14134 (Papanasam).

Distribution: Asia and Australia

2. **Viscum articulatum** Burm.f., Fl. Ind. 211. 1768; Hook.f., Fl. Brit. India 5: 226. 1886 p.p.; Gamble, Fl. Pres. Madras 2: 1259. 1925; Wiens in Dassan. & Fosberg, Rev. Handb. Fl. Ceylon 6: 416. 1987; Chandrasek. in A.N. Henry *et al.*, Fl. Tamil Nadu 2: 218. 1987; Matthew, Fl. Palni Hills 2: 1100. 1999; Mohanan & Sivadasan, Fl. Agasthyamala 586. 2002; V.N. Sanjai & N.P. Balakr. in N.P. Balakr. *et al.*, Fl. India. 23: 50. 2012. **(Fig. 99).**

Leafless, pendulous, di-trichotomously branched, yellowish green shrubs, dark brown or greenish when dry; internodes very variable in length and breadth, flattened, broadening gradually from base to top, twisted near base. Flowers in axillary fascicles of 1-3 in triads; central flower usually female; bracts cup shaped, at the top of internodes. Male: Perinath 4 lobed; lobes ovate, minute, reflexed. Berries globose, ca 4 mm across, with persistent stigma; seeds globose, ca 3 x 3 mm.

Occasional, in disturbed sites in deciduous forests.

Fl. & Fr.: Throughout the year.

Specimens examined: XCH 11726, 14270 (Kalakad), 13165, 14119 (Karayar); 11178, 11406, 13747, 13910, 18038 (Courtallam).

Distribution: Sri Lanka, India, widely distributed in South Eastern Asia.

3. **Viscum heyneanum** DC., Prodr. 4: 278. 1830; Chandrasek. in A.N. Henry *et al.*, Fl. Tamil Nadu 2: 218. 1987; Wiens in Dassan. & Fosberg, Rev. Handb. Fl. Ceylon, 6: 412. 1987; Matthew, Fl. Palni Hills 2: 1100. 1999; Mohanan & Sivadasan, Fl. Agasthyamala 586. 2002; V.N. Sanjai & N.P. Balakr. in N.P. Balakr. *et al.*, Fl. India. 51: 36. 2012. *V. verruculosum* Wight & Arn., Prodr. 1: 379. 1834; Hook.f., Fl. Brit. India 5: 224. 1886; Gamble, Fl. Pres. Madras 2: 1258. 1925. *V. orbiculatum* Wight, Ic. t. 1016. 1845; Hook.f., Fl. Brit. India 5: 224. 1886; Gamble, Fl. Pres. Madras 2: 1258. 1925. *V. heyneanum* DC. var. *liocarpum* Danser in Blumea 4: 305. 1941. **(Fig. 100).**

Leafy, bushy shrubs; branchlets terete, slightly angled when young. Leaves elliptic-ovate or obovate, 2-3.5 x 0.8-1.8 cm, base acute to cuneate, margin falsely crenulate, apex acute or obtuse; lateral nerves 3-5 pairs, looping at apex, obscure; petioles ca 0.2 mm. Flowers 4-6 or more, in axillary fascicles, in peduncled triads, rarely all flowers female, pale yellow. Perinath 3-4 lobed in male, adnate to ovary in female. Stamens 3-4. Ovary 1-locular; ovules 1-2; style 0; stigma capitate. Berries oblong, ca 5 x 2 mm, prominently warty.

Occasional, in deciduous forests.

Fl. & Fr.: Throughout the year.

Specimens examined: XCH 10911 (Kalakad); 13105, 16881, 16928 (Papanasam); 15982 (Donavur); 14831, 17515, 18583 (Mahendragiri); 19441 (Alagiapandipuram).

Distribution: Peninsular India and Sri Lanka.

4. Viscum orientale Willd., Sp. Pl. 4: 737. 1805; Hook.f., Fl. Brit. India 5: 224. 1886 p.p.; Gamble, Fl. Pres. Madras 2: 1258. 1925; Matthew & Rani in Matthew, Fl. Tamil. Carnatic 3: 1378. 1983; Chandrasek. in A.N. Henry *et al.*, Fl. Tamil Nadu 2: 219. 1987; Wiens in Dassan. & Fosberg, Rev. Handb. Fl. Ceylon 6: 419. 1987; Matthew, Fl. Palni Hills 2: 1101. 1999; V.N. Sanjai & N.P. Balakr. in N.P. Balakr. *et al.*, Fl. India. 23: 54. 2012. **(Fig. 101).**

Large, much branched shrub, black or dark brown when dry; branchlets angular, dense and often whorled apically. Leaves elliptic or oblong-ovate or obovate, 3-6 x 1.2-2.3 cm, base attenuate, apex obtuse or acute, 3 ribbed at base, obscure. Flowers few or many, in sessile or peduncled clusters. Perinath 3 lobed, reflexed, free in male, adnate to ovary in female. Stamens 3. Ovary globose. Berries globose, smooth.

Occasional, in deciduous forests to evergreen forests between 1500-1200 m.

Fl. & Fr.: Throughout the year.

Specimens examined: XCH 11082, 11714 (Kalakad), 16405, 17667 (Thirukurungudi); 18030, 18195 (Courtallam); 14107 (Papanasam); 14653 (Pechiparai); 12438 (Kakachi).

Distribution: India, Bangladesh, Sri Lanka, China, Malay islands, New Guinea, Australia.

17. SANTALACEAE R.Br.

Trees, shrubs or herbs, many being parasitic or semiparasitic on the branches or roots of other plants. Leaves alternate or opposite, entire, sometimes scale like or 0; stipules 0. Flowers bisexual or unisexual, regular, small, solitary or in simple or compound cymes; bracts usually small; bracteoles 2, sometimes 1 or both obsolete. Perianth superior or partly inferior; 3-8 lobed or toothed; lobes often with a tuft of hairs behind the anthers. Stamens as many as and opposite to perianth lobes attached to or above the perianth base; anthers 2 celled. Disc epigynous or perigynous. Ovary 1 locular, inferior; ovules 2-3, adnate to or pendulous from central column; style short; stigma entire or 3-6-lobed. Fruit a nut or drupe; seeds globose or ovoid, smooth, rugose or sulcate; testa thin or wanting; albumen copious, fleshy; embryo usually terete; radicle superior.

48 genera and 1025 species worldwide; 10 genera and 18 species in India.

Key to the genera

1a. Trees with often axillary spines; anther cells divergent; drupes 1.5 cm across .... 3. **Scleropyrum**

1b. Trees or shrubs without axillary spines; anther cells distinct; drupes 0.5 – 0.7 cm across 2

2a. Leaves opposite; perianth usually 5 lobed; stamens with interposed processes .... 2. **Santalum**

2b. Leaves alternate; perianth usually 3 lobed; stamens without interposed processes .... 1. **Osyris**

1. OSYRIS L.

Ca 6 species in tropical and subtropical parts of Asia and Africa; 2 in India.

Osyris quadripartita Salzm. ex Decne. in Ann. Sci. Nat. Bot. ser. 2. 6: 65. 1836; Matthew & Rani in Matthew, Fl. Tamil. Carnatic 3: 1380. 1983; Kumari in A.N. Henry *et al.*, Fl. Tamil Nadu 2: 219. 1987; V.N. Sanjai & N.P. Balakr. in N.P. Balakr. *et al.*, Fl. India. 23: 61. 2012. *O. wightiana* Wallich ex Wight, Ic. t. 1853. 1852; Matthew, Fl. Palni Hills 2: 1103. 1999. *O. arborea* Wallich ex DC. in DC., Prodr. 14: 633. 1857; Hook.f., Fl. Brit. India 5: 232. 1886; Gamble, Fl. Pres. Madras 2: 1262. 1925. *O. quadripartita* var. *puberula* (Hook.f.) Kumari in Biol. Mem. 2(1 & 2): 126. 1977 *et* in A.N. Henry *et al.*, Fl. Tamil Nadu 2: 219. 1987. **(Fig. 102).**

Monoecious or dioecious shrubs to 2 m; branchlets angular, when dry dark brown or blackish. Leaves alternate, elliptic or elliptic-ovate or obovate, 1-5 x 0.5-3 cm, subcoriaceous, blackish when dry, glabrous, glaucescent, base acute to cuneate, margin entire, apex prominently mucronate; lateral nerves ca 5 pairs, faint to obscure; petioles ca 0.3 mm long. Flowers 3-merous, solitary or in short, axillary cymes, polygamous; male flowers in clusters, bisexual flowers often solitary. Pedicels ca 3 mm. Male: Perinath 3 lobed; lobes triangular, ca 1 x 0.4 mm, thick, basally connate. Stamens 3, free, ca 7 mm, inserted at the base of perinath; anther cells distinct; disc angular between stamens. Female: Perianth thick; lobes decurrent and adnate to ovary, freely 3 lobed above; lobes ovate, thick, ca 2 x 1 mm. Ovary inferior, sunk into perianth tube, oblong-globose; ovules 3, pendulous; style very short; stigma 3 lobed. Drupes ovoid-globose, ca 1 x 0.75 cm; orange; seeds globose, ca 5 mm across.

Frequent, in moist-deciduous to evergreen forests between 500-1200 m.

Fl.: March – June; *Fr.:* Throughout the year.

Specimens examined: XCH 10826, 11037, 14367, 15192, 15276, 19502 (Kalakad); XCH 15128 (Mancholai).

Distribution: Sri Lanka, Peninsular India to subtropical Himalayas.

2. SANTALUM L.

Ca 17 species in Southeast Asia and Australia; 1 in India.

Santalum album L., Sp. Pl. 349. 1753; Hook.f., Fl. Brit. India 5: 231. 1886; Gamble, Fl. Pres. Madras 2: 1261. 1925; Matthew & Rani in Matthew, Fl. Tamil. Carnatic 3: 1381. 1983; Kumari in A.N. Henry *et al.*, Fl. Tamil Nadu 2: 219. 1987; Matthew, Fl. Palni Hills 2: 1103. 1999; V.N. Sanjai & N.P. Balakr. in N.P. Balakr. *et al.,* Fl. India. 23: 64. 2012. **(Fig. 103).**

Tamil: Sandanam.

Trees to 5 m; branchlets slender, glabrous. Leaves opposite, rarely alternate, elliptic-ovate to lanceolate; 5-7 x 2-3 cm, chartaceous, acute at both ends; lateral nerves ca 10 pairs, parallel, faint, looping near margin; petioles to 1.5 cm. Flowers bisexual, in paniculate cymes; cymes 3-chotomous, terminal and axillary. Pedicels filiform, to 5 mm. Perinath 5 lobed, reddish, connate at base, to 1.8 mm; lobes ovate, ca 2 x 1 mm, glaucescent without. Stamens 5; filaments to 1.5 mm; anthers obovate, to 1 mm, alternate with fleshy scales. Ovary half-inferior, oblong-ovate, ca 1.5 x 0.5 mm; ovules 2, pendulous from beneath the top of long acuminate central column; style elongate, to 1.2 mm; stigma 2 lobed. Drupes subglobose, ca 6 mm across, annulate on the top by the deciduous perianth, beaked with the basal part of style.

Frequent in deciduous forests.

Fl. & Frt.: Throughout the year.

Specimens examined: XCH 11303, 11430, 11436, 13728, 13876, 15737 (Courtallam); 12377, 14130, 14481, 16942 (Papanasam); 10656, 10680, 11941, 12224 (Mancholai); 17567 (Mahendragiri); 16331 (Thirukurungudi); 11757, 15228, 19720 (Kalakad).

Distribution: India, Malaysia, Indonesia

3. SCLEROPYRUM Arn.

4 species in tropical and subtropical Asia; 1 in India.

Scleropyrum pentandrum (Dennst.) Mabberley in Taxon 26: 533. 1977; Kumari in A.N. Henry *et al.*, Fl. Tamil Nadu 2: 219. 1987; Mohanan & Sivadasan, Fl. Agasthyamala 587. 2002; V.N. Sanjai & N.P. Balakr. in N.P. Balakr. *et al.,* Fl.

India. 23: 65. 2012. *Pothos pentandrus* Dennst., Schluess. Hort. Ind. Malab. 16, 24, 33. 1818. *Sphaerocarya wallichiana* Wight & Arn. in Edinburgh New Philos. J. 15: 180. 1833. *Scleropyrum wallichianum* (Wight & Arn.) Arn. in Mag. Zool. Bot. 2: 550. 1838; Hook.f., Fl. Brit. India 5: 234. 1886; Gamble, Fl. Pres. Madras 2: 1262. 1925. **(Fig. 104).**

Trees to 5 m, with thick, sharp, conical spines. Leaves alternate, elliptic to broadly ovate or oblong-ovate, 8-12.5 x 4.5-5 cm, coriaceous, base rounded, margin entire, apex obtuse; lateral nerves ca 3 pairs, distantly arranged, facing towards apex, with raised midrib beneath; petioles thick, to 1 cm. Flowers bisexual, in short catkin like inflorescence. Spikes at the leafless nodes of the branchlets. Perianth tube solid, 5 lobed; lobes ovate, ca 0.3 x 1.5 mm, pink to cream in colour, valvate or subimbricate, with a tuft of hairs behind the stamens. Stamens 5, inserted at the base of perianth lobes; filaments bifid, with 1 celled anthers; anther cells dehiscing transversely. Disc annular. Female: In drooping racemes. Perianth jointed with ovary wall. Ovary inferior; ovules 3, pendulous from a central column; style short; stigma large, peltate. Fruit a pyriform drupe, ovoid, ca 2 x 1.5 cm, on thickened pedicel, crowned by perianth lobes; seed 1, globose, ca 1 cm across.

Occasional, in disturbed deciduous forests.

Fl. & Fr.: Throughout the year.

Specimens examined: XCH 11674, 15200 (Kalakad); 17713 (Thirukurungudi); 13116, 14432, 15661 (Papanasam).

Distribution: Peninsular India, Sri Lanka.

18. BALANOPHORACEAE Rich.

Parasitic, fleshy, achlorophyllous herbs; rhizome warted with lenticels. Stems stout, cylindric or obovoid, waxy. Leaves imbricate. Flowers unisexual, actinomorphic, in stoutly peduncled heads. Male: pedicellate, bracteate. Tepals 4-9, valvate. Stamens as many as tepals, connate into a synandrium; anthers extrorse, dehiscing longitudinally. Female: sessile, minute, embedded, bracteate. Tepals 0. Ovary unilocular; ovule 1.

17 genera and 67 species worldwide; 3 genera and 8 species in India.

BALANOPHORA J.R.Forst. & G.Forst.

Ca 20 species distributed in tropical Africa, tropical Australia, tropical and subtropical Asia and Pacific regions; around 5 species in India.

Balanophora indica (Arn.) Griff. in Trans. Linn. Soca London 20: 95. 1846; Hook.f., Fl. Brit. India. 5: 237. 1886; Gamble, Fl. Pres. Madras. 2: 1263. 1925; N.P. Balakr. in N.P. Balakr. *et al.*, Fl. India. 23: 75. 2012. *Langsdorffia indica* Arn.

in Ann. Nat. Hist. 2: 37. 1838. *B. fungosa* var. *indica* (Arn.) B.Hansen in Dansk. Bot. Ark. 28: 100, t. 20. 1972; Matthew, Fl. Tamil Carnatic. 2: t. 623. 1982; 3: 1383. 1983 et Fl. Palni Hills. 2: 1105.1999; Kumari in Henry *et al.*, Fl. Tamil Nadu 2: 220. 1987; Folke. in Dassan. & Fosberg, Rev. Handb., Fl. Ceylon 6: 104. 1987. **(Fig. 105).**

Dioecious, glabrous herbs, to 14-20 cm; stems stout. Leaves ca 10, closely, spirally arranged, thick, 1.2-2.3 x 1.2-2 cm, pale brown, cucullate, apex acute-obtuse, base flat; stipules 0. Male: Heads ovoid-cylindric, pale pink-cream, ca 9 x 6 cm. Pedicels ca 1 cm. Bracts truncate. Tepals 4-9, oblong, apex subacute, fleshy, ca 5-7 x 3-4 mm. Synandrium oblong, ca 6 x 3 mm; anthers 4-9, horseshoe shaped. Female: Heads globose, ca 3 cm, rusty brown; spadicles ca 1 mm. Ovary globose, ca 0.15 mm; styles filiform, ca 0.5 mm; stigmas minute.

Rare on shaded forest floors between 600-1000 m.

Fl. & Fr.: January – March.

Specimens examined: XCH 12212 (Mancholai), XCH 17930 (Papanasam)

Distribution: Australia, Sri Lanka, India and Malaysia.

19. EUPHORBIACEAE Juss.

Monoecious or dioecious, herbs (erect or prostrate), shrubs (erect or straggling), climbers, twiners or trees, armed or unarmed; branchlets terete or angular, often grained, compressed or simple, often zig-zag, glabrous or pubescent or stellate tomentose; stems woody or succulent, often with acrid milky latex. Leaves simple or distichous, entire or palmately lobed, opposite or alternate or opposite and alternate (rarely whorled), chartaceous, coriaceous, membranous or fleshy, persistent or caducous, entire or serrate, crenate, dentate at margin, often glandular, penninerved, or palmi-nerved, glandular at base or eglandular; petioles present or absent, winged or not; stipules present or absent, interpetiolar or intrapetiolar, persistent or deciduous. Inflorescence axillary, pseudoaxillary, terminal, or subterminal, simple or branched spikes or racemes or umbellate, sometimes cauliflorous, often pedunculate. Flowers unisexual or bisexual, actinomorphic (pseudanthium often zygomorphic), solitary or in glomerules, often grouped into spiciform or capitate thyrses or pseudanthium (Cyathia); bracteate or ebracteate, often bracteolate; pedicellate. Perianth simple, usually calycine, rarely petaloid, sometimes with both sepals and petals; lobes free or connate, valvate or imbricate or both, rarely reduced or even 0, caducous or accrescent in fruit. Disc glands usually annular, entire or lobed or pulvinate, or of free scales or 0. Stamens 1 - many, free or connate, sometimes raised on a column, in 1 or 2 series, hypogynous; filaments equal or unequal; anthers mostly 2 celled, rarely 4 celled, parallel or divergent, basifixed or dorsifixed, simple or versatile; connectives present, rarely absent; broad, narrow, or truncate, longitudinally or transversely dehiscent; pistillode central or columnar or 0. Ovary superior or half-inferior, usually 3 or 4-12 locular; ovules 1 or 2 per locule; placentation axile or pendulous or laterally pendulous; styles 3 or as many as locules, free or connate or half-united, deeply or slightly bifid, rarely multifid, smooth or papillate, plumose or laciniate, erect or spreading. Fruit a capsular schizocarp, drupe or berry, globose or subglobose or ellipsoid-ovoid, smooth or verrucose or tuberculate, 1-3 lobed; seeds 1 or 2 per locule, with or without a caruncle or arillus, globose or trigonous, or ellipsoid-ovoid, glabrous or appressed hairy; testa smooth, crustaceous, pitted, foveolate or marbled.

Ca 322 genera and 8900 species, cosmopolitan in distribution; 70 genera and ca 410 species in India.

Key to the genera

1a. Each locule of the ovary with 2 ovules 2
1b. Each locule of the ovary with 1 ovule 17

2a. Leaves trifoliolate; inflorescence paniculate .... 8. **Bischofia**
2b. Leaves simple; inflorescence not paniculate 3

3a. Flowers in axillary or terminal spikes or racemes 4
3b. Flowers in axillary clusters 6

4a. Male bracts usually with 1 flower; disc present .... 5. **Antidesma**
4b. Male bracts usually with >3 or more flowers; disc absent 5

5a. Inflorescence catkin like racemes; usually in axils, male inflorescence upto 10 cm long; stamens 3 .... 6. **Aporosa**
5b. Inflorescence racemose, fasicled on trunks, rarely axillary, male inflorescence upto 25 cm long; stamens 6 .... 7. **Baccaurea**

6a. Leaves dentate; disc intrastaminal; fruit drupaceous 7
6b. Leaves entire; disc extrastaminal or absent; fruit capsular or baccate, rarely drupaceous 8

7a. Stamens >5; styles stigmatiform .... 17. **Drypetes**
7b. Stamens <5; styles dilated .... 34. **Putranjiva**

8a. Petals present; sepals imbricate .... 2. **Actephila**
8b. Petals absent; if present then sepals valvate 9

9a. Petals present; sepals valvate 10
9b. Petals absent; sepals imbricate 11

10a. Ovary with 2 carpels; fruit a drupe .... 11. **Bridelia**
10b. Ovary with 3 carpels; fruit a capsule .... 13. **Cleistanthus**

11a. Disc of male flowers annular; ovule anatropous .... 30. **Meineckia**
11b. Disc absent, if present dissected or annular; ovule hemitropous 12

12a. Leaf base symmetric; pistillode present in male flowers .... 22. **Flueggea**
12b. Leaf base asymmetric; pistillode absent in male flowers 13

13a. Dioecious; fruits dry, irregularly dehiscent .... 29. **Margaritaria**
13b. Monoecious; fruits dry or fleshy, when dry indehiscent or tardily or regularly dehiscent 14

14a. Disc usually present in male and female flowers; seeds not ventrally invaginated ....33. **Phyllanthus**
14b. Disc absent; seeds ventrally invaginated 15

15a. Fruits longitudinally 3-15 grooved, usually 3-6 locular .... 24. **Glochidion**
15b. Fruits not grooved, 3 locular.. 16

16a. Leaves black when dry; calyx of male flowers obconic, turbinate; styles united into a column on ovoid ovary ....10. **Breynia**
16b. Leaves brown when dry; calyx of male flowers discoid; styles 3, free 35. **Sauropus**

17a. Latex absent.. 18
17b. Latex usually present 26

18a. Petals present ....3. **Agrostistachys**
18b. Petals absent.. 19

19a. Climbers or twiners; styles connate into a distinct column 20
19b. Erect shrub or small tree; styles free or basally connate 21

20a. Plants with stinging hairs..37. **Tragia**
20b. Plants with glandular hairs...15. **Dalechampia**

21a. Indumentum stellate; stamens inflexed in bud (except in *Koilodepas*) 22
21b. Indumentum simple or stellate; stamens not inflexed in bud 23

22a. Inflorescence spiciform racemes, terminal; filaments slender, >2 mm long; styles bifid at the upper part ....18. **Epiprinus**
22b. Inflorescence spicate, axillary; filaments stout, <1 mm long; styles multifid at the upper part ....26. **Koilodepas**

23a. Female bracts large and foliaceous, concealing the flowers and fruits... 1. **Acalypha**
23b. Female bracts small, not foliaceous, not concealing the flowers and fruits 24

24a. Indumentum usually stellate; leaves often opposite, rarely alternate, glandular-granulose..28. **Mallotus**
24b. Indumentum simple, leaves alternate, not glandular-granulose 25

25a. Anther thecae parted by the connective; interstaminal disc present....12. **Claoxylon**
25b. Anther thecae not parted by the connective; interstaminal disc absent....27. **Macaranga**

26a. Latex clear or whitish; inflorescence dichasial paniculate to spicate 27
26b. Latex whitish; inflorescence cyathial, racemose or spicate......................................34

27a. Petals usually present...28
27b. Petals absent...36. **Suregada**

28a. Sepals of male flowers fused; completely covering the buds ....4. **Aleurites**
28b. Sepals of male flowers free; if connate not completely covering the buds 29

29a. Indumentum of stellate or lepidote hairs 30
29b. Indumentum simple 32

30a. Fruit capsular 31
30b. Fruit drupaceous .... 23. **Givotia**

31a. Stems not usually showing red exudates; inflorescences bisexual .... 14. **Croton**
31b. Stems showing red exudates; inflorescences unisexual .... 32. **Paracroton**

32a. Leaves not lobed 33
32b. Leaves palmately lobed .... 25. **Jatropha**

33a. Petals present in male and female flowers .... 16. **Dimorphocalyx**
33b. Petals present only in male flowers .... 9. **Blachia**

34a. Inflorescence cyathial, perianth absent .... 19. **Euphorbia**
34b. Inflorescence racemose or spicate, perianth present 35

35a. Woody herbs, ascending; <1 m high .... 31. **Microstachys**
35b. Shrubs or trees, >1 m high 36

36a. Male and female flowers in different thyrses, flowering when the tree is leafless .... 21. **Falconeria**
36b. Male and female flowers in same thyrses, flowering on leafy twigs .. 20. **Excoecaria**

1. ACALYPHA L.

Pantropical except in Hawaii and Pacific Islands, a few in temperate regions, ca 450 species; 10 species in India.

Key to the species

1a. Male flowers in axillary spikes; female flowers in terminal panicles .....4. **A. paniculata**
1b. Male and female flowers in axillary spikes 2

2a. Allomorphic flowers absent .... 2. **A. fruticosa**
2b. Allomorphic flowers present 3

3a. Bracts lacerate .... 1. **A. ciliata**
3b. Bracts dentate .... 3. **A. indica**

1. Acalypha ciliata Forssk., Fl. Aeg.-Arab. 162. 1775; Hook.f., Fl. Brit. India 5: 417. 1887; Gamble, Fl. Pres. Madras 2: 1330. 1925; Rani in Matthew, Fl. Tamil. Carnatic 3: 1406. 1983; Chandrab. in A.N. Henry *et al.*, Fl. Tamil Nadu 2: 220. 1987; Philcox in Dassan & Clayton, Rev. Handb. Fl. Ceylon 11: 131. 1997; N.P. Balakr. & T.Chakrab., Fam. Euphorb. India 125. 2007; S.R.M. Susila Rani & N.P. Balakr. in N.P. Balakr. *et al.,* Fl. India 23: 94. 2012. **(Fig. 106).**

Herbs to 30 cm, finely white-pilose. Leaves broadly ovate to elliptic-ovate, 5.3-7.3 x 3.3-4 cm, strigose-ciliate,broadly cuneate to rounded at base, acuminate or caudate acuminate at apex, serrate at margin, petiolate; petioles to 7.5 cm, pilose. Inflorescence androgynous, to 3 cm. Male flowers: Pedicels to 0.1 mm. Sepals elliptic-ovate, ca 0.28 x 0.1 mm, tomentose. Stamens to 0.25 mm; anthers linear-oblong, to 0.15 mm. Female flowers: Crowded at base, bracteate. Bracts campanulate, ca 3.5 x 3 mm, well folded, many nerved, aciculate at apex, hispid, hispid-glandular at base without, ciliate within. Sepals very minute. Ovary subglobose, ca 1 x 1 mm, obscurely 3 lobed, sparsely pubescent; styles 3, ca 1.3 mm, aciculate. Allomorphic flowers: ca 4 x 1.2 mm, with two fimbriate, hispid whorls of bracts. Capsules ovoid-globose, ca 2 x 2 mm, sparsely hispid-glandular, ciliate; seeds ovoid, ca 2 x 1.3 mm, smooth.

Weed of cultivated and waste lands, often found in deciduous forests.

Fl.: & Fr.: November - January.

Specimens examined: XCH 17283 (Thirukurungudi). MH 15188, 39459, 14636.

Distribution: India, Sri Lanka, Pakistan, Tropical Africa and Southern Arabia.

2. **Acalypha fruticosa** Forssk., Fl. Aeg.-Arab. 161. 1775; Hook.f., Fl. Brit. India 5: 415. 1887; Gamble, Fl. Pres. Madras 2: 1331. 1925; Rani in Matthew, Fl. Tamil. Carnatic 3: 1406. t. 103 d. 1983; Chandrab. in A.N. Henry *et al.*, Fl. Tamil Nadu 2: 220. 1987; Philcox in Dassan. & Clayton, Rev. Handb. Fl. Ceylon 11: 136. 1997; Matthew, Fl. Palni Hills 2: 1112. 1999; N.P. Balakr. & T.Chakrab., Fam. Euphorb. India 125. 2007; S.R.M. Susila Rani & N.P. Balakr. in N.P. Balakr. *et al.*, Fl. India 23: 96. 2012. **(Fig. 107).**

Tamil: Sinni, Aathathazhai.

Strong-smelling shrubs to 1 m high; stems puberulous, covered with minute yellowish-white glands. Leaves ovate, 3.5-5.5 x 2-3.5 cm, widely scattered orange yellowish glands beneath, acute to truncate at base, acute or acuminate at apex, crenate-serrate at margin; midrib with lateral nerves pubescent; petioles 1-2 cm, pubescent; stipules linear-lanceolate, to 2 mm. Inflorescence axillary, androgynous, 1-3 cm. Male flowers: Pedicels to 0.35 mm, tomentose. Sepals lanceolate, ca 0.5 x 0.25 mm, acute at apex, tomentose without. Stamens free, ca 0.15 mm; anthers distinct, oblong, ca 1.14 mm. Female flowers: Pedicels to 3.5 mm, tomentose. Bracts leafy, shallowly campanulate, ca 2.5 x 8.8 mm, folded, obtusely glandular-dentate, pubescent within, many nerved, tomentose without, glandular, enlarged in fruit. Calyx 3 lobed, ovate-lanceolate, ca 1 x 0.5 mm, tomentose. Ovary globose, ca 2 mm, tomentose, glandular; styles 3, 0.5-2 cm, laciniate. Capsules globose, ca 8 x 3 mm, tomentose, glandular.

Common from foothills to dry deciduous forests especially on dry riverbeds and rocky slopes.

Fl.: June - August. *Fr.:* Throughout the year.

Specimens examined: XCH 17596, 13232 (Thirukurungudi), 11190, 16282 (Courtallam), 13683 (kothayar), 18386 (Sivagiri), MH 2800, 3279, 13341.

Distribution: India, Sri Lanka, Tropical Africa, Indonesia (Java.)

3. **Acalypha indica** L., Sp. Pl. 1003. 1753; Wight, Ic. t. 877. 1844-1845; Hook.f., Fl. Brit. India 5: 416. 1887; Gamble, Fl. Pres. Madras 2: 1330. 1925; Rani in Matthew, Fl. Tamil. Carnatic 3: 1408. t. 103 f. 1983; Chandrab. in A.N. Henry *et al.,* Fl. Tamil Nadu 2: 220. 1987; Philcox in Dassan. & Clayton, Rev. Handb. Fl. Ceylon 11: 132. 1997; Matthew, Fl. Palni Hills 2: 1112. 1999; N.P. Balakr. & T.Chakrab., Fam. Euphorb. India 125. 2007; S.R.M. Susila Rani & N.P. Balakr. in N.P. Balakr. *et al.,* Fl. India 23: 96. 2012. **(Fig. 108).**

Tamil: Kuppaimeni

Erect herbs to 60 cm high; stems branched from or near the base, angular; branchlets striate, thinly pubescent. Leaves broadly ovate or rhombic-ovate, 2-3 x 1.7-2.7 cm or more, sparsely, minutely pubescent beneath, truncate at base, acute at apex, crenate-serrate at margin. Pedicels ca 3 cm, longer than leaf blade, finely pubescent, filiform. Inflorescence androgynous, to 3 cm. Male flowers: Minute, greenish. Pedicels to 0.25, pubescent. Sepals ovate, ca 0.25 x 0.2 mm, ciliate. Stamens to 0.2 mm; anthers oblong, to 0.2 mm. Female flowers: Solitary within bracts. Pedicels to 1 mm, pubescent. Bracts campanulate, 3-6 mm, dentate-ciliate at margin, pubescent without, folded. Ovary globose, ca 2 x 2 mm, hispid; styles 3, aciculate, to 7 mm. Allomorphic flowers: 1 x 0.8 mm, long stalked, ciliate, or glandular-ciliate. Capsules globose, hirsute.

Weed in wastelands, in moist and shady places up to 500 m.

Fl.: & Fr.: Throughout the year.

Specimens examined: XCH 17079, 17832 (Thirukurungudi).

Distribution: India, Sri Lanka, Southeast Asia, Philippines, Ryukyu Islands, Taiwan, West central and Eastern tropical Africa, Madagascar and Mascarenes.

4. **Acalypha paniculata** Miq., Fl. Ned. Ind Bat. 1(2): 406. 1859; Hook.f., Fl. Brit. India 5: 415.1887; Gamble, Fl. Pres. Madras 2: 1330. 1925; Rani in Matthew, Fl. Tamil. Carnatic 3: 1409. t. 103 g. 1983; Philcox in Dassan. & Clayton, Rev. Handb. Fl. Ceylon 11: 135. 1997; N.P. Balakr. & T.Chakrab., Fam. Euphorb. India 126. 2007; S.R.M. Susila Rani & N.P. Balakr. in N.P. Balakr. *et al.,* Fl. India 23: 100. 2012. *A. racemosa* Wallich ex Baill., Etude Gen. Euphorb. 443. 1858, nom. illegit., vide Muell.Arg. in DC., Prodr. 15(2.2): 802. 1866; Chandrab. in A.N. Henry *et al.*, Fl. Tamil Nadu 2: 221. 1987; Matthew, Fl. Palni Hills 2: 113. 1999. **(Fig. 109).**

Undershrubs, shrubs or woody herbs, to 1 m high, finely pubescent; branchlets spreading, often compressed. Leaves broadly ovate, 3.5-10 x 1.5-5 cm, hirsute on both surfaces, rounded to truncate at base, acute to acuminate at apex, crenate-serrate at margin; petioles 1-10 cm, finely pubescent. Male flowers: Minute, green to red, in slender, elongate spikes 1.5-17 cm axillary, filiform. Pedicels to 0.45 mm. Sepals ovate, ca 0.5 x 0.35 mm, sparsely glandular hairy without. Stamens to 0.24 mm; anthers ca 0.2 mm. Female flowers: Greenish, in elongate, panicled racemes to 12-19 cm, mostly terminal or axillary, finely pubescent, with minute bracts subtending 3 flowers. Pedicels to 0.45 mm, pubescent. Calyx 6 lobed, lanceolate, ca 0.5 x 0.2 mm, acute at apex, pubescent without, margin glandular-hairy, persistent. Ovary depressed globose, ca 1.5 mm, glandular-ciliate; styles 3-partite, each with 3-5 aciculate hairs to 1 mm. Capsules globose, ca 3 x 3 mm, hardly enlarged, densely glandular hairy, reddish.

Occasional, in shady places in moist deciduous forests.

Fl. & Fr.: Throughout the year.

Specimens examined: XCH 11258, 13993, 15785, 19227, 19233 (Courtallam), 10892 (Kalakad), 11937 (Kothayar), 16383, 17873 (Thirukurungudi), 16661 (Papanasam), 18358, 18460 (Sivagiri), 18552, 18553 (Mahendragiri).

Distribution: India, Sri Lanka, Tropical Africa and Indonesia.

2. ACTEPHILA Blume

An Asian/Australian genus of ca 20 described species, still poorly understood taxonomically (Webster, 1994); 1 in India.

Actephila excelsa (Dalz.) Muell.Arg. in Linnaea 32: 78. 1863; Wight, Ic. t. 1910. 1852; Hook.f., Fl. Brit. India 5: 282. 1887; Gamble, Fl. Pres. Madras 2: 1283. 1925; Chandrab. in A.N. Henry *et al.*, Fl. Tamil Nadu 2: 221. 1987; Matthew, Fl. Palni Hills 2: 1113. 1999; N.P. Balakr. & T.Chakrab., Fam. Euphorb. India 397. 2007; T.Chakrab. & N.P. Balakr. in N.P. Balakr. *et al.* Fl. India 23: 500. 2012. **(Fig. 110).**

Monoecious, glabrous shrubs, to 2 m; bark yellowish or grayish; branchlets terete, woody, sparsely grained, lenticellate. Leaves simple, alternate, often opposite, very variable, elliptic-lanceolate or oblanceolate, or oblanceolate-acuminate, obovate, 2-17 x 1-4.7 cm, acute to cuneate at base, acute, shortly acuminate, obtuse or rarely emarginate at apex, entire at margin, chartaceous to subcoriaceous, shining, greenish or yellowish or brownish on drying, penninerved; lateral nerves 10-12 pairs, arcuate, obscure above, raised beneath; petioles 0.5-1.5 cm, hard; stipules deciduous. Flowers axillary, fascicled or solitary on stem, pedicellate. Male flowers: Many, each ca 7.4 x 5.8 mm. Pedicels 2.4 mm. Calyx 5 lobed; lobes rounded, ca 2.2 x 1.4 mm, often apiculate, subequal, imbricate, greenish. Corolla 5 lobed; lobes oblanceolate, ca 1.1 x 0.5 mm, minutely sinuate-crenulate, inserted under disc; disc prominent, ca 1 mm across, lobed, yellow. Stamens 5, inserted on disc, extrorse; filaments shortly combined, to 1 x 1.5 mm; anthers parallel, ca 0.5 mm across, basifixed, longitudinally dehiscent; pistillode in central, ca 0.8 mm, 3-fid. Female flowers: Solitary or few together, ca 1.6 x 0.5 cm, pedicelled. Pedicels to 1 cm, slender. Calyx 5 lobed, broadly oblong, ca 2.6 x 1.5 mm, imbricate, persistent. Corolla 5 lobed; lobes oblanceolate, ca 1.1 x 0.6 mm, sinuate, alternate with sepals, inserted under disc; disc to 1 mm across, yellow. Ovary globose, ca 1.5 x 1.5 mm, sessile, 3 locular; ovules 2 per locule, placentation axile; styles 3, to 0.53 mm long, hardly bifid at apex, connate at base;

stigmas obtuse, persistent. Fruits capsular, depressed globose, ca 1.2 x 1.7 cm, 3 lobed with woody epicarp, scabrous; seeds obtusely trigonous, to 0.8 x 0.7 mm; testa loose, fragile, thin.

Common along the border of moist deciduous forests, very rare in evergreen forests upto 1500 m.

Fl. & Fr.: May - July.

Specimens examined: XCH 12370, 13026, 13186, 16551, 16588, 16722, 16952, 17053, 17088, 17132, 17874, 17882, 19058 (Papanasam), 19183 (Courtallam), 18755 (Mahendragiri).

Distribution: India, Sri Lanka and Malaysia.

3. AGROSTISTACHYS Dalzell

Ca 10 species in Asiatic region from India, Sri Lanka and Myanmar to New Guinea; 2 in India.

Key to the species

1a. Leaf margin serrate .... 1. **A. indica**

1b. Leaf margin entire .... 2. **A. meeboldii**

1. Agrostistachys indica Dalz. in Hooker's J. Bot. Kew Gard. Misc. 2: 41. 1850; Bedd. Ic. t. 241. 1868-1874; Hook.f., Fl. Brit. India 5: 406. 1887; Gamble, Fl. Pres. Madras 2: 1317. 1925; Chandrab. in A.N. Henry *et al.*, Fl. Tamil Nadu 2: 221. 1987; Philcox in Dassan. & Clayton, Rev. Handb. Fl. Ceylon 11: 123. 1997; N.P. Balakr. & T.Chakrab., Fam. Euphorb. India 157. 2007; N.P. Balakr. in N.P. Balakr. *et al.*, Fl. India 23: 172. 2012. **(Fig. 111).**

Shrubs to 3 m. Leaves elliptic-oblong or lanceolate, 12.5-28 x 3-8.5 (10) cm, attenuate at base, acute at apex, coarsely serrate at margin, spinulose; teeth facing upwards; lateral nerves 14-16 pairs, prominently raised beneath; transverse nervules prominent; petioles to 2-5 cm long, glabrous or rarely sparsely villous. Male flowers: In bracts, partly exserted. Bracts broadly ovate or linear-lanceolate, ca 2 x 1-2 mm, pubescent without, ciliate. Pedicels to 1.3 mm. Calyx ca 1.5 x 0.75 mm, ciliate, pubescent without. Petals oblong, ca 1.5 x 1 mm, unequal; disc glands orbicular. Stamens ca 10; filaments to 3 mm; anthers to 1 mm; pistillode to 0.4 mm. Female flowers: Bracts ca 3 mm, auriculate at base, ciliate. Pedicels ca 0.6 mm. Calyx caducous. Petals lanceolate, ca 3 x 1 mm, pubescent without; disc faintly 6 lobed; lobes orange. Ovary ca 3 x 3 mm, pubescent; styles 3, ca 2 mm, free, spreading, each faintly bifid at apex, thick. Capsules ca 7.5 mm across, smooth.

Common, in evergreen forests between 800-1500 m .

Fl.: January - March. *Fr.:* April onwards.

Specimens examined: XCH 12842, 14670 (Petchiparai), 15014 (Kothayar), 16027, 16156, 16159, 19059 (Papanasam), 19244 (Courtallam), MH 14471, 420, 19914, 76543, 90518.

Distribution: Peninsular India, Sri Lanka, Southwest Malaysia and Philippines.

2. Agrostistachys meeboldii Pax & K.Hoffm. in Engler, Pflanzenr. IV. 147: 100. 1912; Gamble, Fl. Press. Madras 2: 1318. 1925; N.P. Balakr. & T.Chakrab., Fam. Euphorb. India 157. 2007; N.P. Balakr. in in N.P. Balakr. *et al.,* Fl. India 23: 172. 2012. *Sarcoclinium longifolium* Wight, Ic. Pl. Ind. Ori. 5: 24. t. 1887-1888. 1852. *Agrostistachys longifolia* (Wight) Benth. ex Hook.f., Fl. Brit. India 5: 407. 1887. *A. borneensis* auct. non. Becc. 1904; sensu K.M. Matthew & Rani in Matthew, Fl. Tamil. Carnatic. 3: 1409. 1983. **(Fig. 112).**

Tamil: Veppala maram

Trees to 10 m; branchlets at ends blackish with resins. Leaves narrowly oblanceolate, ca 34 x 6.2 cm or more, deeply attenuate at base, shortly, obtusely acuminate or acute at apex, entire at margin, often sparsely pubescent at base; lateral nerves ca 20 pairs, raised beneath; subsessile. Male flowers: Pedicels to 2 mm, puberulous; calyx lobes ca 2 mm, unequal, sparsely puberulous without. Corolla 4 lobed; lobes obovate, ca 2 x 1 mm; disc glands globose. Stamens ca 10; filaments 1-2 mm; anthers to 1 mm; pistillode to 2 mm. Female flowers: Ovary ca 3 x 3 mm; disc 6 glandular; glands partly connate. Capsules ca 7.5 mm. sparsely pubescent. Pedicels to 1 mm.

Rare, in evergreen forests between 600-1000 m.

Fl.: November-January. *Fr.:* February onwards.

Specimens examined: XCH 10246, 12575 (Kothayar), 15366, 11658 (Kalakad), 12167 (Papanasam).

Distribution: Peninsular India, Malaya and Sri Lanka.

4. ALEURITES J.R.Forst. & G.Forst.

A genus of 2 species extending from India to the pacific islands (Webster, 1994); 1 in India.

Aleurites moluccana (L.) Willd., Sp. Pl. 4: 590 1805; Hook.f., Fl. Brit. India 5: 384. 1887; Rani in Matthew, Fl. Tamil. Carnatic 3: 1411. 1983; Chandrab. in A.N. Henry *et al.*, Fl. Tamil Nadu 2: 221. 1987; Philcox in Dassan. & Clayton, Rev. Handb. Fl. Ceylon 11: 117. 1997; N.P. Balakr. & T.Chakrab., Fam. Euphorb. India 192. 2007; T.Chakrab. & N.P. Balakr. in N.P. Balakr. *et al.,* Fl. India 23: 207. 2012. *Jatropha moluccana* L., Sp. Pl. 1006. 1753. **(Fig. 113).**

Tamil: Katakrote

Monoecious trees up to 6 m; branches spreading, pubescent when young. Leaves simple, alternate, crowded at ends of branches, broadly ovate or oblong,

6-15 x 4.15 cm, pubescent to glabrescent, dark-brown above on drying, paler beneath, truncate to subcordate at base, acute to shortly acuminate at apex, entire or shallowly 3 lobed at margin, 3-7 veined at base; petioles to 18 cm, pubescent, with pair of glands at point of insertion with lamina; stipules caducous. Inflorescence lax, terminal panicled cymes, thyrsiform, androgynous. Flowers bracteate, pedicellate, pedicel to 3 cm. Male flowers: Calyx opening into two unequal lobes; lobes lanceolate, or oblong-lanceolate, ca 2 cm, pubescent at apex. Petals 5, spathulate, ca 2.7 x 1.5 cm, pubescent at base within, white; disc 5 glandular, lanceolate, ca 0.45 mm. Stamens 10, 5 + 5, on a hairy receptacle; filaments ca 1.5 mm; outer free, alternate with disc glands; inner ¾ united, puberulous; anthers ca 0.3 mm, longitudinally dehiscent, exserted. Female flowers: Pedicels to 1 cm. Petals and disc glands similar to those of male. Ovary subglobose, ca 1 x 0.5 cm, lepidote, 3 locular; ovules 3, pendulous; styles 3, to 8 mm, erect, deeply bifid, subspreading. Fruits drupaceous, ellipsoid-globose, ca 5.2 x 3.7 cm, laterally compressed, with hard, 1-10 celled stones, lepidote; seeds subglobose, ca 2.7 x 2.1 cm, smooth, pale brown.

Rare, in evergreen forests at ca 1000 m, occasionally planted in the forest rest house.

Fl.: March - May. *Fr.:* July - September

Specimens examined: XCH 10367, 10622 (Kothayar), 12452, 12806 (Papanasam).

Distribution: Probably native of the Malay and Pacific islands, but doubtfully native in India and Sri Lanka.

5. ANTIDESMA Burm. ex L.

A genus of ca 200 species, only 10 of which occur in Africa and Madagascar. The remainder of the species are Asian except for a few species in Australia and the Pacific islands (Webster, 1994); 15 species are reported from India.

Key to the species

1a. Pistillode present .... 3. **A. bunius**
1b. Pistillode absent 2

2a. Stamens 2-3 .... 1. **A. acidum**
2b. Stamens 3-5 3

3a. Leaves obovate-rhomboid, 4-7 cm long .... 2. **A. alexiteria**
3b. Leaves broadly elliptic-oblong, 7-17.5 cm long .... 4. **A. montanum**

1. **Antidesma acidum** Retz., Obs. Bot. 5: 30. 1788; Fischer in Bull. Misc. Inform. 65. 1932; Rani in Matthew, Fl. Tamil. Carnatic 3: 1412. 1983; Chandrab. in A.N. Henry *et al.*, Fl. Tamil Nadu 2: 242. 1987; Matthew, Fl. Palni Hills 2: 1114. 1999; T.Chakrab. & M.Gangop. in N.P. Balakr. *et al.* Fl. India 23: 356. 2012. *Stilago diandra* Roxb , Pl. Cor. t. 166. 1802. *Antidesma*

diandrum (Roxb.) Roth, Nov. Pl. Sp. 369. 1821; Hook.f., Fl. Brit. India 5: 361. 1887; Gamble, Fl. Pres. Madras 2: 1298.1925. *Stilago lanceolaria* Roxb., Fl. Ind. 3: 760. 1832. *Antidesma lanceolaria* (Roxb.) Wight, Ic. t. 766. 1844. **(Fig. 114).**

Trees to 3 m; branchlets pubescent. Leaves obovate, 4-9 x 1.5-3.5 cm, chartaceous, puberulous beneath on nerves, coppery-brown on drying, acute at base, obtuse to acute at apex, entire at margin; lateral nerves 5-6 pairs, arcuate; petioles to 5 mm, puberulous. Inflorescence ca 8 cm. Bracts oblong, ca 0.6 mm. Male flowers: Pedicels ca 1.5. Calyx tubular, 5 lobed. Stamens 2; filaments to 1.5 mm. Female flowers: Pedicels ca 1 mm. Calyx subcampanulate, 5 lobed. Ovary ellipsoid, ca 1.5 x 1 mm; styles 3, recurved. Drupes ca 4 x 2.8 mm; seed ca 4 x 2.2 mm, crinkled.

Occasional, on slopes of the moist deciduous forets at ca 600 m.

Fl.: & Fr.: May – July

Specimens examined: XCH 11239, 11264, 12659, 12691, 19322, 19326 (Courtallam). MH 19159.

Distribution: India and Sri Lanka.

2. Antidesma alexiteria L., Sp. Pl. 1027. 1753 p.p.; Hook.f., Fl. Brit. India 5: 359. 1887; Chandrab. in A.N. Henry *et al.*, Fl. Tamil Nadu 2: 242. 1987; Philcox in Dassan. & Clayton, Rev. Handb. Fl. Ceylon 11: 279. 1997; N.P. Balakr. & T.Chakrab., Fam. Euphorb. India 304. 2007; T.Chakrab. & M.Gangop. in N.P. Balakr. *et al.* Fl. India 23: 357. 2012. *A. zeylanicum* Lam., Encycl. 1: 207. 1792; Gamble, Fl. Pres. Madras 2: 1297. 1925; Rani in Matthew, Fl. Tamil. Carnatic 3: 1414. 1983. **(Fig. 115).**

Tamil: Nalaidali

Shrubs to 2 m high, much branched; branchlets slender, woody, terete, compressed, closely-grained, puberulous to glabrous. Leaves obovate-rhomboid, 4-7 x 1.5-2.3 cm, coriaceous, puberulous to glabrescent beneath, glabrous above, acute to cuneate at base, acuminate at apex, entire at margin, penninerved; lateral nerves 6-8 pairs, obscure; petioles to 3 mm, puberulous. Male inflorescence to 2 cm. Male flowers: Epedicellate. Calyx 3 lobed; lobes half-ovate, ca 0.5 x 0.75 mm, concave, puberulous; disc triangular. Stamens 3, inserted in hollows of disc; filaments to 1.5 mm, slender; anther cells ca 0.5 mm across. Female inflorescence to 6 cm. Female flowers: Solitary. Pedicels to 0.5 mm, puberulous. Calyx 3-partite, shallowly cupular, minute, thinly puberulous; disc cushion like. Ovary gibbously globose, ca 1.25 x 0.85 mm; styles 4, persistent. Drupes gibbously globose, ca 6 x 5 mm, reddish to bluish red; stalks to 0.13 mm, puberulous.

Common, on the border of moist-deciduous forests, often in rocky places.

Fl.: January – March & July – September; *Fr.:* Throughout the year.

Specimens examined: XCH 10696, 10956, 11478, 11698, 11710, 12733, 12808, 13092, 13097, 13812, 15746, 15866, 15886, 15902, 16148, 16206, 16235,

16320, 16322, 16482, 16527, 16652, 16712, 16863, 16892, 16934, 16935, 19144, 19162, 19461, 19470, 19952, 19955. MH 488, 2821, 8387, 19277, 27864, 32272.

Distribution: Peninsular India and Sri Lanka.

3. **Antidesma bunius** (L.) Spreng., Syst. Veg. 1: 826. 1824 (1825); Hook.f., Fl. Brit. India 5: 364. 1887 p.p.; Gamble, Fl. Pres. Madras 2: 1298. 1925; Chandrab. in A.N. Henry *et al.*, Fl. Tamil Nadu 2: 242. 1987; Philcox in Dassan. & Clayton, Rev. Handb. Fl. Ceylon 11: 277. 1997; Matthew, Fl. Palni Hills 2: 1114. 1999; N.P. Balakr. & T.Chakrab., Fam. Euphorb. India 304. 2007; T.Chakrab. & M.Gangop. in N.P. Balakr. *et al.* Fl. India 23: 358. 2012. *Stilago bunius* L., Mant. Pl. 122. 1767. *Antidesma floribundum* Tul. in Ann. Sci. Nat. Bot. Ser. 3, 15: 189. 1851. *A. bunius* var. *floribundum* (Tul.) Muell. Arg. in DC., Prodr. 15, 2: 263. 1866. **(Fig. 116).**

Trees to 3 m, pubescent when young, soon glabrous. Leaves very variable, elliptic-oblong or obovate or oblanceolate, 5.5-12 x 2-3 cm, subcoriaceous, glabrous except at base, acute at base, acute to acuminate at apex, entire at margin, penninerved; lateral nerves 6-10 pairs, arcuate, prominent beneath, petioles 0.5-1 cm, stout; stipules linear-elliptic-lanceolate, to 2 25 mm. Male inflorescence to 3 cm, pubescent. Male flowers: Minute, subsessile. Bracts to 0.8 mm, pubescent. Calyx cupular, ca 1 mm, 4-6 lobed, tomentose without, yellowish white; disc basally cupular, above 3 lobed. Pistillode truncate. Female inflorescence to 9 cm, pubescent. Female flowers: Solitary, sessile. Bracts to 0.8 mm, pubescent. Calyx cupular, ca 1 mm, 6 lobed; lobes truncate, tomentose without, reddish yellow; disc cushion like, slightly cupular. Ovary ovoid-globose, ca 1.85 x 1 mm, tomentose; styles 3, ca 0.6 mm, deeply bifid, shallowly recurved. Drupes gibbously globose, ca 1 x 0.55 cm, pubescent, reddish.

Very rare, in evergreen forests, especially in rocky places along streams between 1000-1500 m.

Fl.: & Fr.: March - April.

Specimens examined: XCH 17948, 17971, 19089, 19094 (Papanasam). MH 16324, 99389.

Distribution: Peninsular India, Sri Lanka, Thailand, Indo-china, China, Indonesia, Philippines, Tahiti and Christmas islands.

4. **Antidesma montanum** Blume, Bijdr., 1124. 1826-27; T.Chakrab. & N.P. Balakr. in J. Econ. Taxon. Bot. Add. Ser. 9: 18. 1992; T.Chakrab. & M.Gangop. in J. Econ. Taxon. Bot. 24: 26, f. 11. 2000; N.P. Balakr. & T.Chakrab., Fam. Euphorb. India 305. 2007; T.Chakrab. & M.Gangop. in N.P. Balakr. *et al.* Fl. India 23: 361. 2012. *A. acuminatum* Wight, Ic. Pl. Ind. Or 6: 12 t. 1991. 1853; Hook.f., Fl. Brit. India 5: 363. 1887. *A. menasu* (Tul.) Muell.Arg., in DC., Prodr. 15(2): 257. 1886. pp., *tantum quoad* Hohenacker 104. *A. menasu* var. *linearifolia* Hook.f., l.c. 364. 1887. *A. wattii* Hook.f., l.c. 366. **(Fig. 117).**

Trees to 6 m; branches woody, terete; branchlets lenticellate, pubescent. Leaves broadly elliptic-oblong, 7-17.5 x 3-5.5 cm, chartaceous to subcoriaceous, glabrous above, puberulous beneath, truncate at base, slightly oblique, acuminate at apex, mucronate, entire at margin, penninerved; lateral nerves 6-9 pairs, arcuate, raised beneath, subsessile or petioles to 5 mm, thick, pubescent; stipules linear-lanceolate, to 3 mm. Inflorescence spiciform racemes or sparingly panicled racemes. Male inflorescence to 8 cm. Male flowers: Small. Pedicels to 5 mm, pubescent. Calyx concave, 3-partite, truncate, ca 0.8 x 0.8 mm, puberulous, pale yellow; disc slightly lobed. Stamens 3, inserted in hollows of the disc; filaments equal, ca 2.1 mm, slender; anthers to 0.4 mm. Female Inflorescence to 7 cm. Female flowers: Greenish or reddish green. Pedicels to 6 mm, pubescent. Calyx 3-partite, each truncate, ca, 0.8 mm, acute at apex; disc cushion like. Ovary ovoid-globose, ca 3 x 1 mm; stigmas 4, short, thick, persistent. Drupes globose-ellipsoid, ca 3 x 1.7 mm, hardly enlarged; stalks to 1.7 mm; seeds ovoid, ca 2 x 1.8 mm, albumen fleshy.

Occasional, in moist deciduous to evergreen forests, very common in Kothayar.

Fl.: & Fr.: May - July

Specimens examined: XCH 10008, 10172, 10255, 10333, 10394, 10760, 11648, 11927, 12117, 12531, 12840, 13510, 15984, 16008, 16092, 16209, 16214, 16804, 17400, 17746, 18214, 18275, 18601, 18809, 18946, 18948, 19289, 19293, 19834, 19927, 20005. MH 99164, 90122.

Distribution: peninsular India.

Note: Stem galls are very characteristic feature.

6. APOROSA Blume

Asian tropics, from India and Sri Lanka to Indonesia & Malaysia, ca 80 species; 6 species in India.

Key to the species

1a. Male catkins less than 1 cm long .... 3. **A. indo-acuminata**
1b. Male catkins more than 1 cm long 2

2a. Leaves ovate-oblong .... 1. **A. cardiosperma**
2b. Leaves obovate .... 2. **A. fusiformis**

1. **Aporosa cardiosperma** (Gaertn.) Merr. in J. Arnold Arb. 35: 139. 1954; N.P. Balakr. & T.Chakrab., Fam. Euphorb. India 310. 2007; T.Chakrab. & M.Gangop. in N.P. Balakr. *et al.* Fl. India 23: 368. 2012. *Croton cardiospermus* Gaertn., Fruct. 2: 120. t. 107. f. 11. 1791 (as *cardiospermum*). *Scepa lindleyana* wight, Ic. Pl. Ind. Or. 2: 5. t. 361. 1840. *Aporosa lindleyana* (Wight) Baill., Etude Gen. Euphorb. 645. 1858; Hook.f., Fl. Brit. India 5: 349. 1887. **(Fig. 118).**

Tamil: Vittil, Vettikan

Malayalam: Kotili, Vetti.

Trees to 7 m; branchlets grayish. Leaves ovate-oblong or elliptic, 7-11.5 x 3-4 cm, dark grayish-coppery on drying, paler beneath, shining above, obtuse to acute at base, acute to obtusely acuminate at apex; acumen to 1 cm, subentire or undulate at margin; lateral nerves 6-10 pairs; petioles to 1 cm long, thick. Male inflorescence to 4.5 cm. Male flowers: Sessile. Calyx obovate or ovate-oblong, ca 1 x 0.5 mm, margin ciliate. Stamens 3; filaments 1-1.5 mm; anthers ca 0.5 mm. Female flowers: Pedicels to 1.5 mm, tawny-puberulous. Calyx somewhat rhomboid, ca 1.5 x 1.8 mm, ciliate at margin. Ovary ellipsoid-globose, ca 5.3 x 2 mm, densely tawny puberulous; styles stout, ca 0.8 mm, apex recurved, hardly subplumose, persistent. Capsules subglobose or broadly ellipsoid, ca 1.5 x 1.5 cm, puberulous to glabrous; pedicels to 3 mm; seeds suborbicular plano-convex, ca 1 x 1 cm.

Common, in moist deciduous forests at ca 600 m.

Fl.: & Fr.: January - March.

Specimens examined: XCH 14576, 18572 (Mahendragiri), MH 52409, 61503.

Distribution: Peninsular India and Sri Lanka.

2. **Aporosa fusiformis** Thwaites, Enum. Pl. Zeyl. 288. 1861; Hook.f., Fl. Brit. India 5: 351. 1887; Gamble, Fl. Pres. Madras 2: 1309. 1925; Chandrab. in A.N. Henry *et al.,* Fl. Tamil Nadu 2: 221. 1987; N.P. Balakr. & T.Chakrab., Fam. Euphorb. India 310. 2007; T.Chakrab. & M.Gangop. in N.P. Balakr. *et al.* Fl. India 23: 368. 2012. **(Fig. 119).**

Trees to 3 m; branchlets grayish-brown. Leaves obovate, 6.5-9.5 x 3.5-4.5 cm, punctate on upper surface, pale coppery or green on drying, cuneate at base, obtusely acuminate at apex, acumen to 5 mm, entire at margin; lateral nerves ca 7 pairs, looping; petioles ca 1.5 cm. Male inflorescence to 1.5 cm. Bracts ovate, ca 2 x 1.5 mm, ciliate. Male flowers: Calyx 4 lobed; lobes triangular or ovate or elliptic, 1.5-2 x 1.5 mm, ciliate. Stamens 3; filaments to 2.5 mm; anthers to 0.5 mm. Female flowers and fruits not seen.

Rare, in evergreen forests at 900-1100 m.

Fl.: January - March.

Specimens examined: XCH 12262, 15052 (Kakachi), MH 60718, 70356.

Distribution: Peninsular India and Sri Lanka.

3. **Aporosa indo-acuminata** T.Chakrab. & N.P. Balakr. in Fl. India 23: 369. 2012 auct non. Thwaites, Enum. Pl. Zeyl. 288. 1861; Hook.f. Fl. Brit. India 5: 348. 1887; p.p.; Gamble, Fl. Pres. Madras 2: 1309. 1925; Chandrab. in A.N. Henry *et al.*, Fl. Tamil Nadu 2: 221. 1987; Philcox in Dassan. & Clayton, Rev.

Handb. Fl. Ceylon 11: 270. 1997; N.P. Balakr. & T.Chakrab., Fam. Euphorb. India 310. 2007; T.Chakrab. & M.Gangop. in N.P. Balakr. *et al.* Fl. India 23: 369. 2012. **(Fig. 120).**

Trees to 2 m; branchlets grayish to pale yellowish, hardly zig-zag. Leaves elliptic or oblong-elliptic, 6-9 x 2-3 cm, pale green on drying, attenuate at base, caudate-acuminate at apex, acumen to 1.5 cm long, sinuate to sub entire at margin; lateral nerves 5-9 pairs, arcuate, prominently raised; petioles to 1 cm, thin. Spikes to 7.5 mm, deciduous. Male flowers: Minute. Calyx 3 lobed; lobes oblong to oblanceolate, 1.5-1.7 x 0.6 mm, membranous, glaberulous. Stamens 3; filaments to 2 mm; anthers to 0.5 mm. Female flowers not seen. Capsules ovoid-globose, ca 1.2 x 1 cm, smooth; seeds plano-convex, ca 1.2 x 1 cm, orange in colour.

Occasional, in evergreen montane forests.

Fl.: January - March.

Specimens examined: XCH 17404, 17479 (Kothayar), 18085, 18823, 19410 (Mahendragiri), MH 77116, 99364.

Distribution: Peninsular India and Sri Lanka.

7. BACCAUREA Lour.

Asian tropics, from India to Philippines, New Guinea and Pacific island, ca 75 species; 3 species in India.

Baccaurea courtallensis (Wight) Muell.Arg. in DC., Prodr. 15(2): 459. 1866; Hook.f., Fl. Brit. India 5: 367. 1887; Gamble, Fl. Pres. Madras 2: 1310. 1925; Chandrab. in A.N. Henry *et al.*, F1. Tamil Nadu 2: 222. 1987; T.Chakrab. & M.Gangop. in J. Econ. Taxon. Bot. 21(3): 527. 1997; N.P. Balakr. & T.Chakrab., Fam. Euphorb. India 313. 2007; T.Chakrab. & M.Gangop. in N.P. Balakr. *et al.* Fl. India 23: 372. 2012. *Pierardia courtallensis* Wight, Ic. 5(2): 30. t. 1912. 1852. **(Fig. 121).**

Tamil: Mootupalam

Dioecious trees to 8 m high, with woody trunk; branchlets terete, sparsely grained, woody, often puberulous when young, dark to pale brown on drying. Leaves alternate, obovate-oblong or elliptic, 7-11.5 x 2-2.8 cm, chartaceous, sparsely furfuraceous, coppery brown on drying, cuneate or attenuate at base, obtusely acuminate at apex, entire or shallowly crenulate-denticulate towards apex at margin; lateral nerves 5-6 pairs, arching, puberulous beneath; petioles to 1 cm, puberulous; stipules lanceolate, to 3 mm. Inflorescence: cauliflorous, long racemose spikes; male to 8 cm; female to 21 cm. Flowers in fascicles, thick at base. Male flowers: Sessile. Calyx 4 lobed; lobes obovate or suborbicular or oblong, 1.3-1.5 x 0.9-1.5 mm, cupular, pubescent without, glabrous within, crimson red. Stamens 6, adnate to central convex column; filaments free, to 0.5 mm; anthers oblong, to 0.55 mm, basifixed, longitudinally dehiscent. Female flowers: Solitary. Pedicels to 3 mm. Calyx 5 lobed; lobes elliptic-lanceolate or linear-oblong or lanceolate, 2.6-3 x 0.8-1 mm, puberulous without. Ovary elliptic-ovoid or subglobose, to 3 x 2.8 mm, 2 locular, rusty-tomentose; ovules 4, 2 per locule, pendulous; styles 4, flabellate, each ca 0.7 mm, sessile; stigmas capitate. Capsules ellipsoid or ovoid, to 1.5 x 1 cm, coriaceous, pale brown; seeds globose-lanceolate, ca 0.5 x 0.25 mm, compressed.

Common, in moist deciduous to semi evergreen forests.

Fl.: & Fr.: March-June.

Specimens examined: XCH 12797, 14723, 15606, 16189, 17133, 18099, 19019. MH 14646, 19281, 76584, 88641.

Distribution: Peninsular India, extending to Orissa.

8. BISCHOFIA Blume

A genus of 2 species distributed from India and Sri Lanka to China, Indonesia, and the pacific islands; 1 in India.

Note: Airy Shaw places the genus in the monotypic family Bischofiaceae (Kew Bulletin 18: 253. 1965; Chandrab. in A.N. Henry *et al.*, Fl. Tamil Nadu 2: 241. 1987).

Bischofia javanica Blume, Bijdr. 1168. 1826; Hook.f., Fl. Brit. India 5: 345. 1887; Gamble, Fl. Pres. Madras 2: 1312. 1925; Rani in Matthew, Fl. Tamil. Carnatic 3: 1485. 1983; Matthew, Fl. Palni Hills 2: 1115. 1999; N.P. Balakr. & T.Chakrab., Fam. Euphorb. India 315. 2007; T.Chakrab. & M.Gangop. in N.P. Balakr. *et al.* Fl. India 23: 375. 2012. **(Fig. 122).**

Tamil: Thondi, Milachathayan

Malayalam: Cholavengai

Dioecious, glabrous, deciduous trees to 8(-15) m. Leaves alternate, 3-foliolate; lateral leaflets elliptic-ovate, 7.5-10 x 4.5-5.2 cm, subcoriaceous, pale brownish on drying, unequally rounded at base, acuminate at apex; acumen to 1 cm, coarsely crenate at margin, penninerved; lateral nerves ca 7 pairs, raised beneath; petiolules to 8 mm; terminal leaflet to 3 cm; petioles to 11 cm, angular. Inflorescence panicled raceme to 20 cm, axillary or lateral. Male flowers: Pedicels to ca 4 mm, slender. Calyx 5 lobed; lobes oblong, obtuse, ca 2 x 1 mm, prominently cucullate at apex, imbricate, concealing anthers. Stamens 5, exserted; filaments connate at base, ca 1 mm; anthers hemispheric or oblong, ca 2.6 mm long, parallel, longitudinally dehiscent. Female flowers: Pedicels to 6 mm, angular, ribbed. Calyx cupular, 5 lobed; lobes linear-lanceolate, ca 2 x 0.6 mm, caducous. Ovary ovoid, ca 3.4 mm, 3 locular; ovules 6, 2 per locule, pendulous; styles 3, 4-5 mm, reflexed, inner faces stigmatic. Fruit drupaceous, globose, ca 7 x 7 mm; pedicels to 1.5 cm; seeds triquetrous, ca 4 x 2 mm, smooth.

Occasional, in ravines in moist deciduous forests above 500 m.

Fl.: January - March. *Fr.:* June onwards.

Specimens examined: XCH 10898, 11049, 13984, 16097, 16133, 18065, 18066, 19907. MH 29210, 5653, 3129.

Distribution: India, Malay and Pacific islands.

9. BLACHIA Baill.

A genus of ca 12 species in tropical Asia from southern India to Hainan, the Philippines and perhaps Borneo (Webster, 1994); 3 in India.

Key to the species

1a. Calyx much enlarged in fruit; male flowers in racemes .... 1. **B. calycina**
1b. Calyx scarcely enlarged in fruit; male flowers in umbels .... 2. **B. umbellata**

1. Blachia calycina Benth. in J. Linn. Soc. Bot. 17: 226. 1880; Hook f., Fl. Brit. India 5: 402. 1887; Gamble, Fl. Pres. Madras 2: 1338. 1925; Chandrab. in A.N. Henry *et al.*, Fl. Tamil Nadu 2: 222. 1987; N.P. Balakr. & T.Chakrab., Fam. Euphorb. India 202. 2007; T.Chakrab. & N.P. Balakr. in N.P. Balakr. *et al.* Fl. India 23: 219. 2012. *Croton umbellatus* sensu Wight, Ic. t. 1874. 1852, non Willd. 1805. **(Fig. 123).**

Shrubs to 2 m. Leaves elliptic-rhomboid or elliptic-ovate, 4.5-11.5 x 1-3 cm, brown or blackish brown on drying, cuneate at base, acute or obtusely acuminate at apex; lateral nerves ca 10 pairs; petioles 0.3-1 cm, slender. Male flowers: Racemose. Peduncles to ca 3 cm. Pedicels to 1.5 cm, filiform. Sepals orbicular, ca 2.5 x 2 mm. Petals spathulate, ca 1.5 x 1.5 mm; disc ca 1.5 mm across. Stamens to 1.5 mm. Female flowers: Sepals oblong-ovate, ca 6 x 3 mm, persistent. Ovary ellipsoid, ca 2.5 x 2 mm; styles ca 3 mm. Capsules shallowly 4 lobed, ca 8 x 8 mm; seeds ca 7 x 3 mm.

Common, in open-deciduous forests between 100-400 m.

Fl.: & Fr.: June onwards.

Specimens examined: XCH 11378, 13878, 14533, 15971, 16553, 16932, 18058, 18059, 18060, 18344, 19712. MH 3385, 16226, 9942, 9529.

Distribution: Peninsular India.

2. Blachia umbellata (Willd.) Baill., Etud. Gen. Euphorb. 387. t. 19f; f. 18-20. 1858; Hook.f., Fl. Brit. India 5: 402. 1887; Gamble. Fl. Pres. Madras 2: 1338. 1925; Chandrab. in A.N. Henry *et al.,* Fl. Tamil Nadu 2: 222. 1987; Philcox in Dassan. & Clayton, Rev. Handb. Fl. Ceylon 11: 105. 1997; N.P. Balakr. & T.Chakrab., Fam. Euphorb. India 202. 2007; T.Chakrab. & N.P. Balakr. in

N.P. Balakr. *et al.* Fl. India 23: 220. 2012. *Croton umbellatus* Willd., Sp. Pl. 4: 545. 1805. *Codiaeum umbellatum* (Willd.) Muell.Arg. in DC., Prodr. 15(2): 1118. 1866. **(Fig. 124).**

Shrubs to 2 m. Leaves elliptic-rhombic or obovate, 4.5-10.5(15) x 2.3(5) cm, cuneate at base, obtusely acuminate or cuspidate acuminate at apex, entire or subentire at margin; lateral nerves ca 15 pairs, very slender; petioles to 1 cm. Male flowers: Umbellate or subumbellate. Peduncles to 4 cm. Pedicels to 1 cm. Calyx orbicular, ca 2 x 2 mm. Petals spathulate, ca 1.5 x 1.5 mm, puberulous; disc ca 1.2 x 1.5 mm; filaments to 2 mm; anthers to 1 mm. Female flowers: Pedicels to 0.2 cm. Sepals 5 lobed; lobes linear-lanceolate, ca 1 x 0.2 cm. Ovary ca 2 mm; styles 3, bifid. Capsules ca 1 cm, with scarcely enlarged persistent calyx; seeds ca 6.5 x 4.5 mm.

Occasional, in moist deciduous to semi evergreen forests between 300-700 m.

Fl.: & Fr.: June onwards.

Specimens examined: XCH 16307, 15794, 17123, 18344. MH 32928, 15203, 14126, 16557, 17123, 18453, 13356, 13361, 15921, 17676, 18395, 19779, 19786, 17653, 16902.

Distribution: Peninsular India and Sri Lanka.

10. BREYNIA J.R.Forst. & G.Forst.

Tropical Asia to Pacific Islands, ca 15 species: 3 species in India.

Key to the species

1a. Fruiting calyx usually larger than the fruit; fruits ca 5 x 3.2 mm ....2. **B. retusa**

1b. Fruiting calyx smaller than the fruit; fruits ca 8 x 8 mm .... 1. **B vitis-idaea**

1. **Breynia retusa** (Dennst.) Alston in Ann. Roy. Bot. Gard. Perad. 11: 204. 1929; Rani in Matthew, Fl. Tamil. Carnatic 3: 1415. tt. 99d & 101c 1983; Chandrab. in A.N. Henry *et al.,* Fl. Tamil Nadu 2: 222. 1987; T.Chakrab. & M.Gangop. in J. Econ. Taxon. Bot. 20: 505. f. 3. 1996; Philcox in Dassan. & Clayton, Rev. Handb. Fl. Ceylon 11: 240. 1997; Matthew, Fl. Palni Hills 2: 1116. 1999; N.P. Balakr. & T.Chakrab., Fam. Euphorb. India 340. 2007; T.Chakrab., M.Gangop. & N.P. Balakr. in N.P. Balakr. *et al.* Fl. India 23: 410. 2012. *Phyllanthus retusus* Dennst., Schluss. Hort. Ind. Malab. 24. 1818. *P. patens* Roxb., Fl. Ind. 3: 667. 1832. *P. turbinatus* Koenig ex. Roxb., Fl. Ind. 3: 666. 1832. *Melanthesa turbinata* (Koenig ex Roxb.) Wight, Ic. Pl. Ind. Or. 5, 2: 26. t. 1897. 1852; Thwaites, Enum. Pl. Zeyl. 285. 1861. *Melanthesopsis patens* Muell.Arg. in DC., Prodr. 15, 2: 437. 1866. *Breynia patens* (Muell. Arg.) Benth. in Benth. & Hook.f., Gen. Pl. 3: 277. 1880; Hook.f., Fl. Brit. India 5: 329. 1887; Gamble, Fl. Pres. Madras 2: 1304. 1925. **(Fig. 125).**

Tamil: Periseelai

Trees or shrubs to 2 m. Leaves elliptic to ovate-elliptic or elliptic-oblong, 1.3-2.5 x 0.9-1.3 cm, rounded to acute at base, obtuse or occasionally slightly emarginate at apex, entire at margin; lateral nerves 3-6 pairs, arched, raised; petioles to 3 mm, capillary. Male flowers: Arising from tuft of imbricate bracts. Bracts triangular, to 1.5 mm. Pedicels capillary, to 5 mm. Calyx turbinate, to 4 x 2.4 mm, faintly 6 lobed, rim of calyx tube thickened, crenate, yellow. Filaments united into a column, to 1.5 mm; anthers to 1 mm, with apical connectives. Female flowers: Solitary. Pedicels to 2 mm. Calyx campanulate, to 1.5 mm, 6 lobed (3+3); lobes obovate to suborbicular, 1.2-1.8 x 0.8-1.5 mm, thick, fleshy, apiculate. Ovary cupular, ca 0.8 mm; styles ¾ united, to 0.7 mm from base, above 3-fid, each bifid, arms recurved. Capsules depressed globose, ca 8 x 8 mm, dull red to orange red or light yellow.

Common in the plains and along the borders of moist deciduous forests at low elevations.

Fl.: & Fr.: April - July.

Specimens examined: XCH 10122, 10661, 10815, 10900, 11045, 11233, 11413, 11691, 11727, 12642, 12696, 12757, 12898, 13352, 13720, 13877, 14369, 15139, 15597, 15722, 15900, 16188, 16367, 16468, 16693, 17047, 18550, 19179, 19342, 19571. MH 99973, 19156, 27952, 3389.

Distribution: India, Sri Lanka, Bangladesh, Myanmar, Thailand, Malaya and Sumatra.

2. **Breynia vitis-idaea** (Burm.f.) C.E.C.Fisch. in Bull. Misc. Inform. Kew 1932: 65. 1932; Rani in Matthew, Fl. Tamil. Carnatic 3: 1416. 1983; Chandrab. in A.N. Henry *et al.,* Fl. Tamil Nadu 2: 222. 1987; Philcox in Dassan. & Clayton, Rev. Handb. Fl. Ceylon 11: 238. 1997; Matthew, Fl. Palni Hills 2: 1116. 1999; N.P. Balakr. & T.Chakrab., Fam. Euphorb. India 340. 2007; T.Chakrab., M.Gangop. & N.P. Balakr. in N.P. Balakr. *et al.* Fl. India 23: 411. 2012. *Rhamnus vitis-idaea* Burm.f., Fl. Ind. 561. 1768. p.p. *Phyllanthus rhamnoides* Retz., Obs. 5: 30. 1791. *Melanthesa rhamnoides* Blume, Bijdr. 591. 1825. *Phyllanthus vitis-idaea* Koenig ex Roxb., Fl. Ind 3: 665. 1832. *Melanthesa microphylla* Kurtz ex Teijsm. & Binnend. in Nat. Tijdschr. Ned. Ind. 27: 49. 1864, syn. nov. *Breynia microphylla* (Kurtz ex Teijsm. & Binnend.) Muell.Arg. in DC., Prodr. 15(2) : 442. 1866. *B. rhamnoides* (Retz.) Muell.Arg. l.c. 440. 1866; Hook.f., Fl. Brit. India 5: 329. 1887; Gamble, Fl. Pres. Madras 2: 1304. 1925. **(Fig. 126).**

Tamil: Manipulanti, Sithuruvam, Seppulaa

Trees or shrubs to 3 m high; branches horizontal. Leaves elliptic-ovate, 1.5-3 x 0.8-1.2 cm, greenish black on drying, rounded or cuneate at base, subacute at apex, entire at margin, penninerved; lateral nerves 5-6 pairs, obscure; petioles to

3 mm, slender; stipules narrowly lanceolate, to 2 mm. Male flowers: In proximal axils. Pedicels to 3 mm, slender. Calyx turbinate, to 1.5 x 1.7 mm, obscurely 6 lobed; lobes bent inwards, greenish. Staminal column to 2.3 mm; anthers adnate to column. Female flowers: Pedicels ca 1 mm. Calyx campanulate, ca 4.8 x 4 mm, connate at base, above 6 lobed; lobes ca 2 mm, equal, acute, apiculate. Ovary truncate, ca 6 x 3.8 mm, exserted; styles 3, to 0.5 mm. Capsules subglobose or truncate-globose, ca 5 x 3.2 mm, minutely beaked, greenish red when mature.

Common, on the borders of moist deciduous forests between 300-600 m.

Fl.: & Fr.: June onwards.

Specimens examined: XCH 10325, 10472, 10506, 10530, 10602, 10782, 11003, 11943, 13525, 14802, 15444, 15638, 16093, 16374, 16658, 17371, 18107, 18989, 19444. MH 3233, 4526, 47473, 86093.

Distribution: India, Sri Lanka to Taiwan, Ryukyu islands and Philippines and South to Malay Peninsula.

11. BRIDELIA Willd.

Old world, tropical Africa, Asia and Australia, mainly Asian, ca 49 species; 11 species in India.

Note: Here, we have used Willdenow's original spelling for the genus. Sprengel's correction at match the name of the botanist honoured (S.E. Bridel) is contrary to the current rules of nomenclature (Webster, 1994).

Key to the species

1a. Shrubs; flowering spike with small, distichous leaves .... 2. **B. stipularis**
1b. Trees; flowering spike without distichous leaves .... 1. **B. retusa**

1. **Bridelia retusa** (L.) A.Juss., Euphorb. Gen. 109. t. 7. f. 22. 1824; Dressler in Blumea 41(2): 289. 1996; N.P.Balakr. & T.Chakrab., Fam. Euphorb. India 319. 2007; T.Chakrab. & M.Gangop. in N.P. Balakr. *et al.* Fl. India 23: 380. 2012. *Clutia retusa* L., Sp. Pl. 104 2. 1753. *C. squamosa* Lam., Encycl. Meth. Bot. 2: 54. 1786. *C. spinosa* Roxb., Pl. Cor. 2(3): t. 172. 1802 (as *Cluytia*). *Bridelia spinosa* (Roxb.) Willd., Sp. Pl., 979. 1806. *B. crenulata* Roxb., Fl. Ind., 3: 734. 1832. *B. hamiltoniana* var. *glabra* Muell.Arg. in Linnaea 34: 77. 1865. *B. retusa* var. *roxburghiana* Muell.Arg. in DC., Prodr. 15(2): 493. 1866, nom. superfl.; Hook.f., Fl. Brit. India 5: 268. 1887; Gamble, Fl. Pres. Madras 2: 1280. 1925. *B. retusa* var *squamosa* (Lam.) Muell.Arg., l.c. 493.1866. *B. retusa* var. *glauca* Hook.f. l.c. 286.1887. *B. roxburghiana* (Muell.Arg.) Gehrm. in Engler, Bot. Jahrb. Syst. 41. Beibl. 95: 30. 1908. *B. squamosa* (Lam.) Gehrm., l.c. 30. 1908. *B. cinerascens* Gehrm., l.c. 30. 1908. *Bridelia retusa* var. *glabra* Gehrm., l.c. 30. 1908. *B. squamosa* var. *meeboldii* Gehrm., l.c. 30. 1908. **(Fig. 127).**

Tamil: Mulvaengai, Adamaruthu, Marivaengai

Monoecious or dioecious trees to 3 m; branchlets terete, stiff, woody, puberulous to glabrous; bark brownish, sparsely silver-grained. Leaves elliptic-oblong or obovate, 4-8.5 x 2.2-4.5 cm, glabrous above, glaucous beneath, pale greenish above on drying, brownish beneath; truncate or subcordate at base, obtuse to retuse at apex, subentire or sinuate-crenulate at margin; lateral nerves 13-20 pairs, parallel, often forked; cross-nervules parallel, compact; petioles to 6 mm, thick, pulvinate, pubescent; stipules lanceolate, to 0.8 mm, puberulous, deciduous. Inflorescence axillary, glomerulate, pedicellate. Male flowers: Calyx 5 lobed; lobes ovate-lanceolate, ca 3 x 1.3 mm, shortly connate, acute. Corolla 5 lobed; lobes obovate, ca 2 x 1.6 mm, cuneate at base, fimbriate; disc pulvinate. Stamens 5; filaments ca 0.6 mm; anthers oblong, to 2 mm. Female flowers: Pedicels to 0.15 mm. Calyx 5 lobed; lobes lanceolate, to 2 x 1.5 mm, greenish. Petals 5; lobes to 2 x 0.75 mm, connate, fimbriate; disc of 5 membranous, fimbriate corona, to 2 x 1 mm, enveloping the base of ovary, unequal. Ovary elliptic-globose, ca 3.5 x 1.5 mm; styles 2, to 1 mm long, forked. Drupes globose, ca 0.8 x 0.8 mm.

Common, on bare slopes of hills in dry deciduous forests.

Fl.: & Fr.: April - June.

Specimens examined: XCH 10723, 10762, 11730, 13817, 19751. MH 502, 3603.

Distribution: India, Sri Lanka, Nepal, Myanmar, Indo-China, Malay Peninsula and Sumatra.

2. **Bridelia stipularis** (L.) Blume, Bijdr. 597. 1826; Hook.f., Fl. Brit. India 5: 270. 1887; T.Chakrab. & M.Gangop. in N.P. Balakr. *et al.* Fl. India 23: 382. 2012. *Clutia stipularis* L., Mant. Pl. 127. 1767 (as *'Cluytia'*). *C. scandens* Roxb., Pl. Cor. 2(3): 39 t. 173. 1802 (as *'Cluytia'*). *Bridelia scandens* (Roxb.) Willd., Sp. Pl. 4: 979. 1806; Gamble. Fl. Pres. Madras 2: 1281. 1925. **(Fig. 128).**

Dioecious, scandent shrubs to 3 m, tomentose to puberulous; stems often thorny; bark dark brown to grayish, lenticellate; branchlets terete, flexuous, tomentose, when young densely fulvous tomentose. Leaves elliptic-oblong, 5.5-10 x 1.7-5 cm, glabrous above except nerves, densely fulvous tomentose beneath, brownish on drying, rounded or subcordate at base, obtuse or slightly retuse at apex, subentire or crenulate at margin; lateral nerves ca 10 pairs, faint above, prominent and raised beneath; petioles to 8 mm, fulvous pubescent; stipules linear-lanceolate, to 4 mm, fulvous pubescent. Inflorescence axillary, solitary or glomerulate, borne on slender spikes to 15 cm, with small distichous leaves, pubescent. Male flowers: Calyx 5 lobed; lobes lanceolate, acuminate. Corolla like calyx; disc pulvinate. Stamens 5. Female flowers: Sessile. Calyx 5 lobed; lobes triangular, ca 3.5 x 2 mm, fulvous tomentose within, tomentose to pubescent without, acute at apex, incurved. Corolla 5 lobed; lobes spathulate, ca 2 x 1.6 mm, fimbriate. Ovary ellipsoid, 1.8 x 1.5 mm; styles 2, to 1.5 long. Drupes oblong-ellipsoid to ovoid, ca 8.5 x 8 mm, seated on the persistent sepals, subsessile.

Common, in moist-deciduous forests.

Fl.: & Fr.: October - February.

Specimens examined: XCH 10883, 11925, 14385, 14403, 14438, 18000, 20451 MH 1235, 14569, 5005, 5039, 5436, 88790.

Distribution: India, Sri Lanka, Malay Islands, Philippines and Tropical Africa

12. CLAOXYLON A.Juss.

Paleotropical, spread-over Africa, Asia and Australia; extending towards Pacific islands up to society island, 70-80 species; 9 species in India.

Key to the species

1a. Disc lobes of female flowers elongated, strap shaped and prominent ... 1. **C. anomalum**
1b. Disc lobes of the female flowers short, indistinct 2
2a. Shrubs or undershrubs 3
2b. Herbs .... 3. **C. mercurialis**
3a. Hypogynous glands rectangular .... 2. **C. beddomei**
3b. Hypogynous glands linear-oblong .... 4. **C. wightii**

1. Claoxylon anomalum Hook.f., Fl. Brit. India 5: 412. 1887; Gamble, Fl. Pres. Madras 2: 1327. 1925; Chandrab. in A.N. Henry *et al.,* Fl. Tamil Nadu 2: 223. 1987; Balakr. & T.Chakrab., Fam. Euphorb. India 130. 2007; S.R.M. Susila Rani & N.P. Balakr. in N.P. Balakr. *et al.,* Fl. India 23: 107. 2012. **(Fig. 129).**

Dioecious shrubs 1-2 m; branchlets terete, at ends flattened, obscurely zig-zag, fistular, granular, white, glabrous to sparsely pubescent. Leaves alternate, elliptic or elliptic-lanceolate, or oblong-obovate, 6.5 - 12(-18) x 2.3(-3) - 4(-5.5) cm, membranous to chartaceous, punctate, scabrous, finely, sparsely hirsute beneath, subglabrous above, greenish on drying, cuneate to attenuate at base, acuminate-mucronate at apex, glandular dentate at margin, penninerved; lateral nerves 6-9 pairs, arcuate; midrib raised beneath, reticulation prominent; petioles 2-3(-7) cm, usually horizontal, flattened, sparsely hirsutulous; stipules usually caducous. Inflorescence axillary, racemose. Male flowers: Minute, in short racemes to 1.5 cm. Pedicels to 1.2 mm, puberulous. Calyx 3 lobed; lobes acute, ca 2 x 2 mm, pubescent without; anthers sessile, in several series around the margin of a flat, naked receptacle, erect, oblong, to 0.75 mm, longitudinally dehiscent. Female flowers: In short racemes to 1.5 cm. Calyx ca 0.95 x 0.75 mm, slightly split into three lobes, tomentose without. Ovary triangular, ca 0.6 x 0.3 mm, closely 3 lobed at apex, no distinction between styles and ovary, 3 locular, tomentose; ovules 3, pendulous; styles free, spreading, each to 1.5 mm. Fruit capsular, depressed globose, ca 9.5 mm across, 3 lobed, tomentose, with distinct 3 ringed styles; styles stalk to 3.5 mm, tomentose.

Very rare, in semi-evergreen forests at ca 600 m.

Fl.: & Fr.: April - June.

Specimens examined: XCH 19141, 19201, 19208 (Courtallam).

Distribution: Southern Western Ghats.

2. **Claoxylon beddomei** Hook.f., Fl. Brit. India 5: 413.1887; N.P. Balakr. & T.Chakrab., Fam. Euphorb. India 130. 2007; S.R.M. Susila Rani & N.P. Balakr. in N.P. Balakr. *et al.,* Fl. India 23: 109. 2012. *Micrococca beddomei* (Hook.f.) Prain in Ann. Bot. (Usteri) 25: 630. 1911; Gamble, Fl. Pres. Madras 2: 1328. 1925; Chandrab. in A.N.Henry *et al.* Fl. Tamil Nadu 2: 235. 1987. **(Fig. 130).**

Undershrubs to 1 m. Leaves elliptic-lanceolate, 9-15.5(-17) x 4-5.5 cm, scabrous, glabrous to sparsely hispid, green above on drying, paler beneath, acuminate to attenuate at base, acuminate at apex, coarsely glandular-dentate at margin, penninerved; lateral nerves 6-7 pairs, distantly arranged, arcuate, prominent beneath; cross nervules often dot like; petioles 2-4 cm, often two small glands at the point of insertion, winged. Inflorescence racemes, axillary. Flowers clustered, clusters often catkin like with many imbricate bracts. Bracts ca 1 x 0.4 mm, hispid. Male flowers: Pedicels to 6 mm, filiform. Calyx 3 lobed; lobes ovate or elliptic-ovate, 2.5(-2.6) x 1.5 mm, delicate, yellowish white, valvate. Stamens 14 or more, on a central receptacle, intermixed with erect, stipitate glands; glands to 0.6 mm, villous at apex; anthers ca 0.5 mm. Female flowers: Pedicels to 6(-10) mm. Calyx lobes ca 1.8 x 0.7 mm, often ciliate at apex; disc 1.5 x 1.9 mm. Ovary ca 1.5 mm across, pubescent; styles 3, each ca 1.2 mm. Capsules ca 4.5 mm across, minutely puberulous; seeds ca 4.3 x 3.3 mm.

Occasional, in semi evergreen to evergreen forests between 600-900 m.

Fl.: & Fr.: April – July.

Specimens examined: XCH 10580, 11916, 12874, 15083, 17175, 17927, 17951, 18924, 18925, 19053, 19065, 19589, 19641, 19665. MH 47468, 4468, 16326, 93222.

Distribution: Peninsular India.

2. **Claoxylon mercurialis** (L.) Thwaites, Enum. Pl. Zeyl. 271. 1861; Hook.f., Fl. Brit. India 5: 412. 1887; N.P. Balakr. & T.Chakrab., Fam. Euphorb. India 131. 2007; S.R.M. Susila Rani & N.P. Balakr. in N.P. Balakr. *et al.,* Fl. India 23: 114. 2012. *Tragia mercurialis* L., Sp. Pl. 980. 1753. *Micrococca mercurialis* (L.) Benth. in Hook., Niger Fl. 503. 1849; Gamble, Fl. Pres. Madras 2: 1328. 1925; Rani in Matthew, Fl. Tamil. Carnatic 3: 1460. 1983; Chandrab. in A.N. Henry *et al.,* Fl. Tamil Nadu 2: 235. 1987; Philcox in Dassan. & Clayton, Rev. Handb. Fl. Ceylon 11: 146. 1997. *Microstachys mercurialis* (L.) Dalzell & Gibson, Bombay Fl. 227. 1861. **(Fig. 131).**

Appressed pubescent herbs to 30 cm, erect. Leaves arising directly from stems or from branchlets, opposite, subpposite or alternate, ovate or rhombic-ovate, 3.2-5.2 x 1.5-2.5 cm, glabrous above, sparsely hirsute to subglabrous beneath acute or truncate at base, acute to acuminate at apex, serrate-crenate at margin; petioles 1-2.5 cm, pubescent; stipules linear-lanceolate, to 4 mm. Inflorescence racemose, androgynous, 2.5-4.5 cm, axillary, erect. Male flowers: Pedicels to 0.25 mm, puberulous. Calyx 3 lobed; lobes obovate, ca 0.8 x 0.5 mm, apiculate, greenish. Stamens 3-4, alternate with sepals; filaments free to 0.17 mm, anthers ca 0.25 mm; hypogynous glands obovate, to 0.4 x 0.25 mm, reflexed. Female flowers: Pedicels to 4 mm. Calyx lobes ca 1.8 x 0.4 mm, margin appressed hairy; disc ca 1.6 mm; ovary depressed-globose, ca 2.5 mm across, appressed pubescent; styles 3, plumose-laciniate. Capsules globose, ca 5 mm across, pubescent; seeds subglobose, ca 2 x 1.5 mm.

Common weed of plains and dry deciduous forests.

Fl.: & Fr.: November - January

Specimens examined: XCH 17830 (Papanasam). MH 13653.

Distribution: India, Andaman Islands, Sri Lanka, Saudi Arabia, Yemen, throughout tropical and subtropical Africa.

3. **Claoxylon wightii** Hook.f., Fl. Brit. India 5: 413. 1887; Susila & N.P. Balakr. in Rheedea 5: 133. 1995; Jothi *et al.* in Rheedea 12: 150. 2002; N.P. Balakr. & T.Chakrab., Fam. Euphorb. India 132. 2007; S.R.M. Susila Rani & N.P. Balakr. in N.P. Balakr. *et al.*, Fl. India 23: 117. 2012. *Micrococca wightii* (Hook.f.) Prain in Ann. Bot. (Usteri) 25: 630. 1911; Gamble, Fl. Pres. Madras 2: 1328. 1925.

Key to the varieties

1a. Plants laxly hirsute to glaberulous; leaves glabrous above, sparsely hirsute beneath .... 3. var. **wightii**

1b. Plants densely yellowish tomentose or hirsute; leaves densely hirsute on both surfaces 2

2a. Internodes 10-30 mm long .... 2. var. **hirsutum**

2b. Internodes 3-6 mm long .... 1. var. **angustatum**

1. var. **angustatum** Susila & N.P. Balakr. in J. Econ. Taxon. Bot. 16: 736. 1992; & in Rheedea 5: 134. f.6. 1995; Jothi *et al.* in Rheedea 12: 150. 2002; N.P. Balakr. & T.Chakrab., Fam. Euphorb. India 132. 2007; S.R.M. Susila Rani & N.P. Balakr. in N.P. Balakr. *et al.*, Fl. India 23: 118. 2012. *Micrococca wightii* (Hook.f.) Prain var. *angustata* (Susila & N.P. Balakr.) Radcl.-Sm. & Govaerts in Kew Bull. 52: 480. 1997. **(Fig. 132).**

Branches and branchlets densly, closely hirsute. Leaves linear-lanceolate, 1-5 x 0.3-1.5 cm, densely yellowish tomentose, acute to cuneate at base, subacuminate at apex, mucronate, glandular-dentate at margin; nerves obscure. Male flower: Pedicles to 0.25 mm. Calyx 3 lobed; lobes ca 1 x 0.7 mm, pubescent without. Stamens ca 10, exserted; filaments to 0.2 mm; anthers to 0.25 mm. Female flowers: Pedicles to 2 mm, densely hirsute. Calyx 3 lobed; each 1.2 x 0.5 mm, hirsute without; disc of 3 linear glands, to 0.85 mm, pubescent at apex. Ovary ca 2 x 1.6 mm; styles 3-partite, each to 0.7 mm, plumose-laciniate. Capsules ca 5 x 5.8 mm, hispid; seeds ovoid-globose, ca 4.5 x 2.2 mm; testa crustaceous.

In semi evergreen forests at ca 600 m.

Fl.: & Fr.: October – December.

Specimens examined: XCH 11593, 11597 (Kalakad).

Distribution: **Endemic** to Tirunelveli hills. **Endangered**.

2. var. **hirsutum** (Hook.f.) Susila & N.P. Balakr. in J. Econ. Taxon. Bot. 16: 735. 1992 & in Rheedea 5: 136. 1995; N.P. Balakr. & T.Chakrab., Fam. Euphorb. India 133. 2007; S.R.M. Susila Rani & N.P. Balakr. in N.P. Balakr. et al., Fl. India 23: 118. 2012. *Claoxylon hirsutum* Hook.f., Fl. Brit. India 5: 413. 1887; Jothi et al. in Rheedea 12: 148. 2002. *Micrococca wightii* (Hook.f.) Prain var. *hirsuta* (Hook.f.) Prain in Ann. Bot. (Usteri) 25: 630. 1911. **(Fig. 133).**

Shrubs to 2 m; branchlets prominently densely yellowish tomentose. Leaves lanceolate to ovate-lanceolate, 6-1 x 1.5-2.5 cm, acute to cuneate at base, acuminate at apex, mucronate, irregularly glandular-dentate at margin, golden yellow or yellowish green on drying, shining, penninerved to plinerved; lateral nerves 5-7 pairs, facing towards apex; petioles 1.5-6 cm, yellowish tomentose. Inflorescence racemose, axillary, 2.5-7 cm, yellowish tomentose, flexuous. Male flowers not seen. Female flowers: Pedicels to 4.5 cm, densely tomentose. Calyx 3 lobed; lobes 2 x 1 mm, greenish-white, hirsute without; disc ca 2 mm. Ovary ca 2 mm across, yellowish hirsute; styles 3, each to 1 mm. Capsules ca 7 mm across; seeds ca 3 x 1.5 mm, dotted with white pits, thinly crustaceous.

Rare, in moist deciduous forests.

Fl.: & Fr.: October – March.

Specimens examined: XCH 17762, 18557 (Mahendragiri). MH 88628.

Distribution: Southern W. Ghats

Note: First discovery of female plant.

3. var. **wightii** **(Fig. 134).**

Shrubs to 1 m, laxly white hirsute to glaberulous. Leaves lanceolate or oblong-lanceolate, membranous, sparsely hirsute beneath, less so to glabrous above, pale greenish on drying, subcordate or truncate, or rounded to acute at base, acute to acuminate at apex, glandular-crenate to serrate at margin, ca 6 x 2.5 cm; lateral nerves ca 6 pairs, at maturity very clear; petioles to 1.5 cm, two glandular at apex. Inflorescence to 8 cm, hispid. Male flowers: Pedicels to 3 mm. Calyx 3 lobed; lobes ovate, ca 2 x 1.5 mm, hirsute without. Stamens 9; filaments free; anthers erect, to 0.5 mm. Female flowers and fruits not seen.

In partially shaded moist-deciduous forests at ca 400 m.

Fl.: July onwards.

Distribution: **Endemic** to Tirunelveli hills. **Endangered**.

Specimens examined: XCH 16486, 16495 (Thirukurungudi).

13. CLEISTANTHUS Hook.f. ex Planch.

Old world tropics, from Africa to Asia, Australia and Melanesia, ca 100 species; 6 species in India, 1 cultivated.

Cleistanthus malabaricus (Muell.Arg.) Muell.Arg. in DC. Prodr. 15(2): 508. 1866; Hook.f., Fl. Brit. India 5: 276. 1887; Gamble, Fl. Pres. Madras 2: 1283. 1925; N.P. Balakr. & T.Chakrab., Fam. Euphorb. India 324. 2007; T.Chakrab. & M.Gangop. in N.P. Balakr. *et al.* Fl. India 23: 388. 2012. *Lebidiera malabarica* Muell.Arg. in Linnaea 32: 81. 1863. **(Fig. 135).**

Dioecious or monoecious, small trees to 3 m; branchlets terete, thick, spreading, minutely grained, rufous-villous when young, tomentose to puberulous at maturity. Leaves alternate, bifarious, elliptic-oblong or ovate-oblong, 5.3-8.8 x 1.8-3.2 cm, chartaceous, glaucous, rusty-villous to pubescent beneath, pubescent to subglabrous above, pale greenish or greenish brown on drying, acute to rounded at base, obtusely acuminate at apex, entire at margin; lateral nerves ca 12 pairs, prominent, raised beneath, faint above; petioles to 5 mm long, rusty villous; stipules linear-lanceolate, to 3 mm long, setaceous, villous. Inflorescence axillary, glomerulate. Male flowers: Sessile. Calyx 5 lobed; lobes ovate, ca 2 x 1.2 mm, pubescent without, valvate. Corolla 5 lobed; lobes minute, orbicular, ca 1 x 1 mm, slightly clawed-crenulate, alternate with sepals; disc pulvinate, crenulate. Stamens 5, on erect gonophore; filaments to 1.5 mm; pistillodes 3-lobed, to 1.2 mm, sparsely ciliate. Female flowers: Sessile, cupular. Calyx 5 lobed; lobes ovate, ca 1.5 x 1 mm, pubescent without, valvate, persistent. Corolla 5 lobed; lobes minute, orbicular, ca 1 x 1 mm, minutely clawed-crenulate, alternate with sepals; disc pulvinate, enclosing the ovary, obscurely crenulate. Ovary ellipsoid-globose, ca 3 x 1 mm, hirsute, 3 locular; ovules 6, 2 per locule, on axile placentae; styles 3, free, bifid, spreading. Fruits capsular, ellipsoid-globose, ca 9 x 7 mm, deeply 3 lobed, sessile, sparsely fulvous-hairy, yellow; seeds triangular, ca 5 x 3.5 mm, smooth, brown.

Occasional, on the edges of deciduous to moist deciduous forests, between 300-500 m .

Fl.: & Fr.: June - August.

Specimens examined: XCH 16504, 18387, 19564, MH 15158, 14610, 33897, 12424. MH. 15158, 14610, 33897, 12424.

Distribution: Western Peninsular India.

14. CROTON L.

Tropics and subtropics of old and New world, the majority of the species being in South America and West Indies. ca 800 species in the world; about 150 species in Asia occurring in S. China, Indo-China, South Asia and Malaysia; in India 16 species are found.

Key to the species

1a. Herbs 5
1b. Shrubs or trees 2

2a. Leaves white lepidote beneath .... 6. **C. zeylanicus**
2b. Leaves stellate pubescent beneath 3

3a. Leaves penninerved .... 5. **C. klotzschianus**
3b. Leaves 3 nerved at base 4

4a. Capsules up to 1.5 cm long .... 1. **C. aromaticus**
4b. Capsules more than 1.5 cm long .... 3. **C. caudatus**

5a. Fruits warty .... 2. **C. bonplandianus**
5b. Fruits stellate .... 4. **C. glandulosus** var. **hirtus**

1. Croton aromaticus L., Sp. P1. 1005. 1753; Hook.f., Fl. Brit. India 5: 338. 1887; T.Chakrab. & N.P. Balakr. in Bull. Bot. Surv. India 26: 200. 1984; N.P. Balakr. & T.Chakrab. Fam. Euphorb. India 212. 2007; T.Chakrab. & N.P. Balakr. in N.P. Balakr. *et al.,* Fl. India 23: 229. 2012. *C. laccifer* L., Sp. Pl. 1005. 1753; Gamble, Fl. Pres. Madras 2: 1315. 1925. **(Fig. 136).**

Shrubs to 3 m; branchlets scabrous, fulvous tomentose. Leaves broadly ovate, 3.5-6 x 2.5-3.5 cm, stellate-pubescent on both surfaces, blackish-brown above, pale yellowish green beneath on drying, subcordate or truncate at base, obtuse or occasionally acute at apex, serrate-dentate or sub-entire at margin, 3 nerved at base; lateral nerves 4-5 pairs, raised transverse nervules prominent; petioles 0.5-1 cm, densely stellate-pubescent. Racemes 2.5-8 cm, softly stellate hairy. Male flowers: Pedicels to 4.5 mm, densely stellate hairy. Calyx 5 lobed; lobes spathulate, ca 3.5 x 1.5 mm, ciliate. Filaments ca 4 mm; anthers to 0.5 mm. Female flowers: Pedicels to 1 mm, thick. Calyx 5 lobed; lobes elliptic, to 4.3 x 1.8 mm, pulvinate at base, stellate-tomentose without. Ovary ca 5 x 3 mm, densely pubescent; styles often tetrafid, ca 2 mm, spreading. Capsules ca 1.2 x 1 cm, densely stellate-tomentose; seeds ellipsoid, to 4.5 x 3 mm, smooth.

Occasional, in semi-evergreen forests mostly in rocky places between 500-900 m.

Fl.: & Fr.: March - May.

Specimens examined: XCH 10208, 10212, 10615, 11785, 12206, 12434, 13547, 15930, 19305, 19996. MH 460, 4506, 16376.

Distribution: India and Sri Lanka.

Note: Chakrabarty & Balakrishnan (Bull. Bot. Surv. India 26: 200. 1985) have followed Beddome and J. Hooker, in merging *C. laccifer* L. and *C. aromaticus* L. under the latter binomial. Philcox (Dassan. & Clayton, Rev. Handb. Fl. Ceylon 11: 99. 1997) and A.N. Henry *et al.*, (Fl. Tamil Nadu 2: 225. 1987) have retained them as distinct species.

2. **Croton bonplandianus** Baill. in Adansonia 4: 339. 1864; Rani in Matthew, Fl. Tamil. Carnatic 3: 1420. 1983; Chandrab. in A.N. Henry *et al.*, Fl. Tamil Nadu 2: 224. 1987; Philcox in Dassan. & Clayton, Rev. Handb. Fl. Ceylon 11: 89. 1997; Matthew, Fl. Palni Hills 2: 1118. 1999; N.P. Balakr. & T.Chakrab., Fam. Euphorb. India 212. 2007; T.Chakrab & N.P. Balakr. in N.P. Balakr. *et al.*, Fl. India 23: 231. 2012. *C. sparsiflorus* Morong in Ann. New York Acad. Sci. 7: 221. 1893; Gamble, Fl. Pres. Madras 2: 1316. 1925. **(Fig. 137).**

Woody herbs to 30 cm; tender-parts with white warty out growths. Leaves narrowly ovate-lanceolate, 3-5.5 x 1.5-3 cm, sparsely stellate hairy beneath, cuneate at base, acute at apex, serrate at margin, 3 veined at base; petioles to 1 cm, stellate-hairy. Racemes to 10 cm, often white-warted. Male flowers: Pedicels to 1.5 mm. Sepals ovate, ca 1 x 0.3 mm. Petals oblong, ca 2 x 1 mm, villous within, white. Stamens 12; filaments to 2 mm, free; anthers to 0.5 mm. Female flowers: Pedicels to 0.8 mm, stellate-hairy. Calyx 5 lobed; lobes lanceolate, acute, ca 1.5 x 0.5 mm. Ovary ellipsoid, ca 1.5 x 1 mm, warted; styles sparsely stellate hairy; seeds ellipsoid, ca 3.2 x 1.7 mm, minutely foveolate.

Introduced. Common in dry and exposed areas at low altitudes.

Fl.: & Fr.: Throughout the year.

Specimens examined: XCH 11185 (Papanasam), 16285 (Thirukurungudi).

Distribution: India and Sri Lanka, Native of S. America.

3. **Croton caudatus** Geiseler, Croton Monogr. 73. 1807; Hook.f., Fl. Brit. India 5: 388. 1887; Gamble, Fl. Pres. Madras 2: 1315. 1925; N.P. Balakr. & T.Chakrab., Fam. Euphorb. India 213. 2007; T.Chakrab. & N.P. Balakr. in N.P. Balakr. *et al.*, Fl. India 23: 232. 2012. *C. caudatus* var. *obovoideus* N.P. Balakr. & T.Chakrab. in Bull. Bot. Surv. India 25: 190. 1983 (1985); Chandrab. in A.N. Henry *et al.*, Fl. Tamil Nadu 2: 224. 1987. **(Fig. 138).**

Straggling shrubs to 2 m; branches stout, stellate-tomentose, scabrous when young. Leaves broadly ovate-cordate or ovate-oblong, 4.7-8.2 x 2.5-4.3 cm, scaberulous above, scabrid or stellate-tomentose beneath, reddish brown or blackish on drying; cordate or truncate-orbicular at base, acuminate at apex, denticulate at margin, 3 nerved at base; lateral neves 4-6 pairs, faintly arcuate; transverse nervules prominent; petioles 1-3 cm, densely stellate-hairy, scabrous. Racemes to 14(-23) cm, scabrous. Male flowers: Pedicels 3-5 mm. Calyx 5 lobed; lobes narrowly elliptic-oblong or spathulate, 2.5 x 0.7 mm, woolly without; filaments to 6 mm, far exserted; anthers ca 1.6 x 0.8 mm. Female flowers: Pedicels to 3 mm. Calyx 5 lobed; lobes triangular or ovate, subacute, ca 4 x 2 mm, stellate-tomentose without, persistent. Petals 0 or very minute. Ovary subglobose, ca 3 x 2 mm, densely stellate-tomentose; styles 0.5-1 cm, twisted. Capsules ellipsoid-globose, ca 2 x 1.5 cm, woody, with 6 slender ridges, densely rusty, scurfy-stellate hairy; seeds elliptic-ovoid, ca 0.75 x 0.45 cm, brown.

Rare, in semi-evergreen forests (near stream sides) at ca 700 m.

Fl.: May – July; *Fr.:* August onwards.

Specimens examined: XCH 13032, 16868 (Papanasam).

Distribution: India, Sri Lanka, S. China, Thailand and West Malaysia.

4. **Croton glandulosus** L., Syst. ed. 10, 2: 1275. 1795. var. **hirtus** (L'Her.) Muell.Arg. in DC. Prodr. 15(2): 684. 1866. T.Chakrab. & N.P. Balakr. in. Bull. Bot. Surv. India 34: 48. 1992 (publ. 1997); N.P. Balakr. & T.Chakrab., Fam. Euphorb. India 213. 2007. T.Chakrab. & N.P. Balakr. in N.P. Balakr. *et al.*, Fl. India 23: 234. 2012. *C. glandulosus* L. subsp. *hirtus* (L'Her.) Croizat in Bull. Torrey Bot. Club 75: 401. 1948. **(Fig. 139).**

Herbs 30-60 cm, stellate-hispid hairy. Leaves broadly triangular-ovate, 3.8-6.5 x 2.5-6 cm, densely hirsute; obtuse to subtruncate at base, acute at apex, doubly glandular crenate-serrate at margin, 3-5 nerved at base; lateral nerves 3-5 pairs; petioles 1-2.5 cm, densely strigose hairy. Racemes to 2.5 cm, densely stellate-hirsute. Bracts linear, to 2 mm, fringed with 2-6, slender, stalked, capitate glands, ca 2 mm. Male flowers: Pedicels to 0.5 mm, hirsute. Calyx 5 lobed; lobes lanceolate, ca 1.75 x 0.6 mm, hirsute without. Corolla 5 lobed; lobes linear-lanceolate, ca 2 x 0.4 mm, hirsute without. Filaments to 1.5 mm; anthers to 0.5 mm. Female flowers: Pedicels to 1 mm. Sepals spathulate or oblong, ca 2-5 x 1-2 mm, hirsute without. Ovary globose, ca 2 x 2 mm, densely hirsute; styles to 1.5 mm, erect, apex reflexed. Capsules globose, ca 5 x 5 mm, with persistent, enlarged, bracteate glands to 8 mm, stellate-hispid; seeds 3, oblong, ca 3.6 x 2.2 mm, often minutely foveolate-reticulate, blackish-brown.

Occasional, on fringes of deciduous forests mostly near streams or rivers at ca 200 m.

Fl.: & Fr.: Throughout the year.

Specimens examined: XCH 20018 (Papanasam). MH 92638.

Distribution: India, Sri Lanka, Malaysia, Indonesia, West Africa, Tanzania, Jamaica, Trinidad, Guiana and Brazil.

Note: This plant has been recently introduced into Peninsular India (obviously from Sri Lanka).

5. **Croton klotzschianus** (Wight) Thwaites, Enum. Pl. Zeyl. 276. 1861; Hook.f., Fl. Brit. India 5: 392.1887; Gamble, Fl. Pres. Madras 2: 1315. 1925; Chandrab. in A.N. Henry *et al.*, Fl. Tamil Nadu 2: 225. 1987; N.P. Balakr. & T.Chakrab., Fam. Euphorb. India 214. 2007; T.Chakrab. & N.P. Balakr. in N.P. Balakr. *et al.,* Fl. India 23: 236. 2012. *Tiglium klotzschianum* Wight, Ic. t. 1914. 1852. **(Fig. 140).**

Shrubs to 2 m; young shoots sparsely stellate hairy. Leaves ovate-oblong or oblong-elliptic, 5-11 x 2-3(-5) cm, often glossy, glabrous above, sparsely stellate-hairy beneath, pale yellow to green on drying, obtuse at base, acute or shortly, bluntly-acuminate at apex, serrate at margin, 3 or weakly 5-nerved at base; lateral nerves 5-6 pairs, arcuate; petioles 1.5-3 cm, glabrous or stellate-hairy. Racemes to 11 cm. Male flowers: Pedicels to 4 mm. Calyx 5 lobed; lobes elliptic-ovate, ca 2.5 x 1 mm, hairy at apex. Corolla 5 lobed; lobes oblong or subspathulate, ca 2 x 1 mm; disc 5 glandular, globose. Filaments to 2 mm; anthers ca 1 mm. Female flowers. Pedicels ca 7.5 mm. Calyx 5 lobed; lobes linear, 1.8(-3.5) x 0.5(-1) mm, hairy at apex. Petals 0. Ovary ca 2 mm, white stellate-tomentose; styles to 2.5 mm. Capsules ca 7 x 7 mm, stellate-pubescent; seeds hemispheric, ca 4.6 x 4 mm, smooth, marbled.

Common, in deciduous forests between 100-300 m.

Fl.: & Fr.: Throughout the year.

Specimens examined: XCH 11422, 12297, 12979, 13070, 13243, 13271, 13310, 13365, 13769, 13880, 14169, 14422, 16258, 16293, 16336, 16393, 16516, 16558, 16939, 16967, 17595, 17684, 19716. MH 90619, 3877, 9572.

Distribution: India and Sri Lanka.

6. **Croton zeylanicus** Muell.Arg. in Linnaea 34: 107. 1865; N.P. Balakr. in Bull. Bot. Surv. India 3: 39. 1963; Chandrab. in A.N. Henry *et al.*, Fl. Tamil Nadu 2: 225. 1987; Philcox in Dassan. & Clayton, Rev. Handb. Fl. Ceylon 11: 91. 1997; N.P. Balakr. & T.Chakrab., Fam. Euphorb. India 216. 2007; T.Chakrab. & N.P. Balakr. in N.P. Balakr. *et al.,* Fl. India 23: 243. 2012. *C. reticulatus* Heyne ex Muell.Arg. in DC., Prodr. 15(2): 580. 1866; Hook.f., Fl. Brit. India 5: 386. 1887; Gamble, Fl. Pres. Madras 2: 1314. 1925. **(Fig. 141).**

Shrubs to 2 m, stellate hairy; branchlets densely coppery silvery lepidote. Leaves elliptic-oblong or ovate-oblong, 7.5-11.5(-14) x 3-3.5(-4.5) cm, sparsely stellate-pubescent or glabrous, densely white lepidote beneath, coppery lepidote on main nerves, pale greenish above on drying, obtuse to acute at base, often unequal, caudate-acuminate at apex, to 1 cm, entire at margin; lateral nerves 5-7 pairs, prominently raised beneath, arcuate; petioles 0.5-2(-4) cm, rusty coppery to glabrescent. Racemes to 3 cm. Male flowers: Pedicels to 4 mm, stellate-pubescent.

Calyx 5 lobed; lobes ovate, ca 1.5 x 0.75 mm, stellate-pubescent without. Corolla 5 lobed; lobes spathulate-obovate, ca 1.5 x 0.7 mm; disc 5 glandular, ca 0.5 x 0.6 mm. Filaments to 1.5 mm; anthers to ca 0.7 mm. Female flowers. Pedicels to ca 2 mm, thick; calyx 5 lobed; lobes elliptic-oblong or ovate, ca 3 x 1.5 mm, sparsely stellate-pubescent without, densely stellate hairy within, occasionally glabrous. Petals 0. Ovary ca 2.4 x 2.4 mm, coppery lepidote; styles 3, each deeply tetrafid, to 4.4 mm, twice or thrice bifid above. Capsules globose, ca 1 x 1 cm; seeds oblong, ca 8 x 6 mm, often marbled.

Occasional, in semi-evergreen forests between 600-1000 m.

Fl.: & Fr.: April - June

Specimens examined: XCH 10080, 17524, 18921, 19204, 19585, 19858. MH 383, 16979, 93349, 99937.

Distribution: Peninsular India and Sri Lanka.

15. DALECHAMPIA Plum. ex L.

Tropical America, Africa and Asia, majority in America, few in Africa and in Asia, ca 110 species; 6 species in India.

Dalechampia indica Wight, Ic. t. 1882. 1852; Hook.f., Fl. Brit. India 5: 467. 1888; Gamble, Fl. Pres. Madras 2: 1335. 1925; Rani in Matthew, Fl. Tamil. Carnatic 3: 1421. 1983; Chandrab. in A.N. Henry *et al.*, Fl. Tamil Nadu 2: 225. 1987; Philcox in Dassan. & Clayton, Rev. Handb. Fl. Ceylon 11: 176. 1997; Matthew, Fl. Palni Hills 2: 1119. 1999. N.P. Balakr. & T.Chakrab., Fam. Euphorb. India 175. 2007; T.Chakrab. & N.P. Balakr. in N.P. Balakr. *et al.*, Fl. India 23: 189. 2012. *D. bidentata* Thwaites, Enum. Pl. Zeyl. 270. 1861, excl. syn., non Blume 1825. *D. ternata* var. *zeylanica* Muell.Arg. in DC., Prodr. 15, 2: 1241. 1866. **(Fig. 142).**

Monoecious, pubescent vines, slender. Leaves 3-foliolate; leaflets unequal; central one obovate, 6-7 x 2-3 cm, penninerved; lateral nerves 6-7 pairs; lateral ones smaller, obliquely ovate-lanceolate or obovate, 4-6.2 x 2-2.7 cm, pubescent beneath, puberulous above, 3-5 nerved at base, above lateral nerves 3-4 pairs, cuneate to rounded at base, shortly acuminate at apex, serrate at margin, with short petiolules; petiolules to 3 mm; stipellae 2, lanceolate, to 3 mm; petioles 2-4 cm, thick, pubescent; stipules lanceolate, to 5 mm. Inflorescence axillary or terminal, long peduncled racemes or heads, fascicled, androgynous, subtended by leafy, involucral bracts. Bracts pale yellow, often large, 3-fid, 3-4.8 mm, acute to acuminate at apex, serrulate at margin, pubescent without, strongly nerved; peduncles to 2-5 cm, pubescent. Flowers in short, cupular involucre; male central; female lateral and basal. Male flowers: Pedicels to 1.4 mm, puberulous. Calyx 5 lobed; lobes spathulate or lanceolate, to 3.4 x 1.6 mm, apiculate at apex, reflexed, folded, pubescent without, valvate. Stamens many, on a convex receptacle, exserted; filaments to 0.6 mm, free or basally connate, filiform, ciliate; anthers oblong, to 0.9 mm, parallel, longitudinally dehiscent. Female flowers: Pedicels to 2 mm, pubescent. Calyx 9 lobed; lobes linear-oblong, to 3 mm, rigid, pinnatifid at margin, glandular hairy, pubescent, persistent. Ovary globose, ca 2.4 x 2.4 mm, 3 locular, densely glandular hairy; ovules 3, pendulous; style to 1.3 cm, fleshy, with stout beak; stigma subcupular. Capsules depressed globose, to 1 cm across, 3 lobed, pubescent.

Very rare, in open low altitude forests.

Fl.: & Fr.: January - March.

Specimens examined: XCH 15101 (Mancholai). MH 552, 14572, 9900, 40145.

Distribution: Penisular India and Sri Lanka.

16. DIMORPHOCALYX Thwaites

Tropical South India, Andamam & Nicobar islands, Sri Lanka, Indo-China, Hainan, West Malaysia, New Guinea and Lesser Sunda islands.

Ca 15 species; 4 species in India.

Key to the species

1a. Stamens <10 .... 1. **D. beddomei**
1b. Stamens ≥10 .... 2. **D. glabellus**

1. **Dimorphocalyx beddomei** (Benth.) Airy Shaw in Kew Bull. 23: 124. 1969; Chandrab. in A.N. Henry *et al.*, Fl. Tamil Nadu 2: 225. 1987; Jothi *et al.* in J. Econ. Taxon. Bot. 25(3): 721. ff. 1.2.2001; N.P. Balakr. & T.Chakrab., Fam. Euphorb. India 204. 2007; T.Chakrab. & N.P. Balakr. in N.P. Balakr. *et al.,* Fl. India 23: 222. 2012. *Tritaxis beddomei* Benth. in J. Linn. Soc. Bot. 17: 221. 1878; Hook.f., Fl. Brit. India 5: 384. 1887. *Trigonostemon beddomei* (Benth.) N.P. Balakr. in Bull. Bot. Surv. India 10: 245. 1968 (1969).

Glabrous shrubs to 2 m. Leaves ovate-oblong to elliptic-lanceolate, 3.5-11.5 x 1-3.5 cm, acute to cuneate at base, acuminate at apex, shallowly sinuate to subentire at margin; lateral nerves 6-18 pairs, looping, obscure to faint; petioles 0.5-1 cm. Male flowers: Peduncles to 1 cm. Pedicels ca 0.5 mm. Calyx lobes suborbicular, 1.8-2.4 x 1.5-1.8 mm, pubescent. Petals linear-oblong, 3.8-4.8 x 0.7-1.7 mm, often emarginate at apex, glabrous to puberulous without; disc glands oblong, ca 0.5 mm. Stamens 5 + 2 or 5 + 3 or 5 + 4; outer filaments 2.4-3.5 mm; inner column to 3.3-5 mm; anthers ca 0.75 mm. Female flowers: Pedicels to 1.5 mm, puberulous. Calyx lobes oblong to obovate, ca 5 mm, obtuse to emarginate at apex, puberulous without. Petals twisted, oblong to spathulate, 3.8-4.6 x 1.2-1.6 mm; disc wavy along margin. Ovary 1.6-2.5 x 1.8-2 mm. Capsules ca 1 x 1 cm; seeds trigonous, ca 7.5 x 5 mm, smooth, mottled.

Rare, in evergeen forests, occasionally found in moist-deciduous forests between 400-900 m.

Fl.: & Fr.: April – June.

Specimens examined: XCH 16541, 16542, 16628, 18852, 18853 (Papanasam), 18986, 18951 (Mahendragiri). MH 8425, 47484.

Distribution: Peninsular India.

2. Dimorphocalyx glabellus Thwaites, Enum. Pl. Zeyl. 278. 1861; Hook.f., Fl. Brit. India 5: 403. 1887; Gamble, Fl. Pres.Madras 2: 1337. 1925; Chandrab. in A.N. Henry *et al.,* Fl. Tamil Nadu 2: 225. 1987; Philcox in Dassan. & Clayton, Rev. Handb. Fl. Ceylon 11: 107. 1997; N.P. Balakr. & T.Chakrab., Fam. Euphorb. India 205. 2007; T.Chakrab. & N.P. Balakr. in N.P. Balakr. *et al.,* Fl. India 23: 222. 2012.

Key to the varieties

1a. Fruiting calyx lanceolate or elliptic-oblong .... 2. var. **lawianus**
1b. Fruiting calyx obovate .... 1. var. **glabellus**

1. var. **glabellus** **(Fig. 143).**

Trees to 3 m high. Leaves ovate to elliptic-lanceolate, 6.5-11 x 3.2-4 cm, sometimes oblong-lanceolate, occasionally suborbicular, reddish to brown beneath on drying, black with bluish tinge above, obtuse, or acute to cuneate at base, acute or obtuse or obtusely acuminate at apex; acumen to 1 cm; crenate-serrate to entire at margin; petioles to 1 cm. Male flowers: Pedicels to 5 mm, filiform. Calyx cupular, to 2.5 mm, shortly 4 or 5 lobed; lobes triangular-orbicular, each ca 1.5 x 1.5 mm, margin ciliate; petals oblong, ca 8 x 2.5 mm, reflexed. Stamens 10-16; outer filaments to 2.5 mm; inner column 2.5-3 mm, unequal; anthers to 0.1 mm; disc glands ovate, to 0.7 mm; puberulous. Female flowers: Pedicels to 1.2 cm, puberulous; Calyx spathulate to obovate, ca 1.5 x 0.85 cm, obtuse to obscurely emarginate at apex. Petals caducous; disc slightly cupular-annular. Ovary ca 2.5 x 2.5 mm, densely puberulous; styles to 2 mm.

Rare, in moist deciduous to semi-evergreen forests, often along river banks between 400-800 m.

Fl.: & Fr.: May - July.

Specimens examined: XCH 11252, 14333, 15891, 18852, 18853, 19833. MH 89543.

Distribution: Peninsular India and Sri Lanka.

2. var. **lawianus** (Hook.f.) T.Chakrab. & N.P. Balakr. in Proc. Ind. Acad. (Pl. Sci) 100: 296. f. 4. 1990. *Dimorphocalyx lawianus* Hook.f., Fl. Brit. India 5: 404. 1887; Gamble, Fl. Pres. Madras 2: 1337. 1925; Chandrab. in A.N.

Henry *et al.*, Fl. Tamil Nadu 2: 226. 1987; N.P. Balakr. & T.Chakrab., Fam. Euphorb. India 205. 2007; T.Chakrab. & N.P. Balakr. in N.P. Balakr. *et al.*, Fl. India 23: 223. 2012. *Trigonostemon lawianus* Muell.Arg. in Linnaea 34: 212. 1865. **(Fig. 144).**

Trees to 2 m high. Leaves elliptic-lanceolate or lanceolate, 10.5-16.5 x 3.5-4.3 cm, blackish-green above on drying, pale brown beneath, acute to cuneate at base, acute to acuminate at apex; acumen to 1.5 cm; crenate-undulate to subentire at margin; lateral nerves ca 15 pairs, arcuate, faint to prominent; petioles 0.5 - 1.5 cm. Male flowers: Pedicels to 3.6 mm. Calyx cupular, to 2 mm, shortly 5 lobed; lobes to 2 x 1.2 mm, margin ciliate. Petals 5, oblong, ca 1 x 0.4 cm, apices prominently reflexed, white, glabrous; disc glands 5, ciliate. Stamens 11, 5 + 6, outer free; filaments to 3.75 mm; inner monadelphous, 6.25 - 8.75 mm; anthers to 1.25 mm. Female flowers: Sepals elliptic-lanceolate, ca 1.3 x 0.3 cm. Petals broadly oblong, ca 1 x 0.4 cm, acute, obtuse or emarginate at apex, white; disc cupular-annular, to 1.2 mm, obscurely 5 lobed. Ovary ca 4.4 x 4.4 mm, villous; styles to 6 mm long, minutely papillate. Capsules ca 8.5 x 8 mm, 3 lobed, 6 valved, pubescent.

Occasional, in semi-evergreen forests.

Fl.: & Fr.: March onwards.

Specimens examined: XCH 10875, 11252, 15923, 16847, 16628, 18635, 18719. MH 19897, 90504.

Distribution: Peninsular India.

17. DRYPETES Vahl

A large and variable genus of ca 200 species, poorly represented in the New World, with ca 10 species; ca 60 species in Africa and Madagascar; the remainder in Asia and Australia (Webster, 1994). 20 species in India.

Key to the species

1a. Ovary and fruit 1 locular..2. **D. sepiaria**
1b. Ovary and fruit 2 locular..1. **D. oblongifolia**

1. **Drypetes oblongifolia** (Bedd.) Airy Shaw in Kew Bull. 23: 56. 1969; T.Chakrab. *et al.*, in J. Econ. Taxon. Bot. 21: 270. 1997; N.P. Balakr. & T.Chakrab., Fam. Euphorb. India 332. 2007; T.Chakrab. & N.P. Balakr. in N.P. Balakr. *et al.* Fl. India 23: 400. 2012. *Laneasagum oblongifolium* Bedd. in Madras J. Sci. n.s. 6: 71. 1861. *Cyclostemon macrophyllus* Blume var. *sessiliflorus* Bedd., Fl. Sylv. S. India 2: 278. 1872. *C. macrophyllus* var. *peninsularis* Hook.f., Fl. Brit. India 5: 341. 1887. *C. macrophyllus* auct. non Blume, 1825; Gamble, Fl. Pres. Madras 2: 1302.1925. *Drypetes longifolia* Pax & K.Hoffm. in Engl., Pflanzenr. IV. 147. XV: 245. 1922, p.p. (non

Cyclostemon longifolius Blume, 1825); Chandrab. in A.N. Henry *et al.*, Fl. Tamil Nadu 2: 226. 1987; Philcox in Dassan. & Clayton, Rev. Handb. Fl. Ceylon 11: 266. 1997. **(Fig. 145).**

Trees to 8 m; branches stout, pale yellowish white, often sparsely granular. Leaves oblong, occasionally elliptic-oblong, 10-14 x 3.5-5.5 cm, sparsely ciliate to glabrous, brownish black on drying, acute to obtuse at base, oblique, obtusely acuminate at apex to 5 mm long, coarsely, sharply to bluntly serrate at margin; lateral nerves 8-10 pairs, reticulation prominent, raised; petioles to 6 mm long, thick. Male flowers: Flattened, subsessile. Calyx 5 lobed (3 + 2); lobes broadly orbicular, unequal, 5.5-7 x 5.5-7 mm, concave, coriaceous. Stamens 45-50; filaments filiform, ca 5.5 mm; anthers ca 1.5 mm; disc ca 1 cm across, obscurely radiately ribbed with inflexed marginal lobes, pubescent, coriaceous. Female flowers and fruits not seen.

Rare, along river banks above 1000 m.

Fl.: January - March. *Fr.:* April onwards.

Specimens examined: XCH 14949 (Kothayar). MH 68879, 60750, 70386, 60679.

Distribution: South India and Borneo.

2. **Drypetes sepiaria** (Wight & Arn.) Pax & K.Hoffm. in Engl., Pflanzenr. IV. 147x V. 271. 1922; Rani in Matthew, Fl. Tamil. Carnatic 3: 1423. t. 101g. 1983; Chandrasek. in A.N. Henry *et al.*, Fl. Tamil Nadu 2: 226. 1987; Philcox in Dassan. & Clayton, Rev. Handb. Fl. Ceylon 11: 261. 1997; Matthew, Fl. Palni Hills 2: 1120. 1999; N.P. Balakr. & T.Chakrab., Fam. Euphorb. India 333. 2007; T.Chakrab. & N.P. Balakr. in N.P. Balakr. *et al.* Fl. India 23: 401. 2012. *Hemicyclia sepiaria* Wight & Arn. in Edinburgh New Philos. J. 14: 297. 1833; Wight, Ic. t. 1872, 1852; Hook.f., Fl. Brit. India 5: 337. 1887; Gamble, Fl. Pres. Madras 2: 1299. 1925.

Tamil: Vellilambu, Veerai, Kaayalakkamaram

Shrubs to 4 m; branchlets puberulous to glabrous. Leaves oblong, occasionally elliptic-ovate, 3-6 x 1.7-3.2 cm, tawny pubescent to glabrous beneath, shining above, pale yellow to brown on drying, obtuse at base, unequal, obtuse or retuse at apex, entire to crenulate at margin; lateral nerves 6-7 pairs, very faint; petioles to 7 mm, tawny pubescent to glabrous. Male and Female flowers not seen. Drupes subglobose, ca 8 x 1 cm, brownish on drying; epicarp crinkled; one seeded; pedicels to 2.5 mm, tawny pubescent.

Occasional, in deciduous forests around 150 m.

Fl.: & Fr.: June onwards.

Specimens examined: XCH 16493 (Thirukurungudi). MH 12153, 15836, 42141.

Distribution: Peninsular India and Sri Lanka.

18. EPIPRINUS Griff.

A rather variable genus of ca 6 species to tropical Asia, from Assam to Malaya (Webster, 1994); 1 in India.

Epiprinus mallotiformis (Muell.Arg.) Croizat in J. Arnold Arbor. 23: 53. 1942; Rani in Matthew, Fl. Tamil. Carnatic 3: 1424. 1983; Chandrab. in A.N. Henry *et al.*, Fl. Tamil Nadu 2: 227. 1987; N.P. Balakr. & T.Chakrab., Fam. Euphorb. India 170. 2007; N.P. Balakr. in N.P. Balakr. *et al.*, Fl. India 23: 184. 2012. *Symphyllia mallotiformis* Muell.Arg. in Linnaea 34: 156. 1865; Gamble, Fl. Pres. Madras 2: 1323. 1925. *Cephalocroton indicus* Bedd., Fl. Sylv. t. 261. 1872. *Adenochlaena indica* (Bedd.) Bedd. ex Hook.f., Fl. Brit. India 5: 418. 1887. **(Fig. 146).**

Monoecious trees to 4 m; branchlets terete, stellately-pubescent. Leaves alternate, elliptic or oblong or broadly elliptic-ovate, 8.5-19 x 3-6.2 cm, shining above, sparsely stellate-pubescent beneath, blackish-green or pale brown on drying, rounded to acute at base, acute or acuminate at apex, entire at margin; lateral nerves 7-9 pairs, arching, strong beneath; petioles 0.6-4 cm, stellate-pubescent; stipules 2, ovate, to 1 mm. Inflorescence axillary or terminal spiciform racemes to 5 cm long. Flowers sessile. Male flowers: Many, fascicled. Calyx 4 lobed, lobes ovate, elliptic or oblong, 2-2.5 x 1-1.5 mm, acute or obtuse at apex, unequal, puberulous without, 1-3 nerved, valvate. Stamens 5, alternate with sepals, exserted, inflexed in bud; filaments free, to 3 mm; anthers oblong, to 1 mm, dorsifixed, longitudinally dehiscent; pistillode to 0.8 mm. Female flowers: Solitary or few together, at the base of inflorescence. Calyx cupular, 6 lobed; lobes lanceolate, inequal, 2.5-4 x 0.3-1 mm, acute at apex, connate at base, distantly yellow-glandular at margin, densely stellate-puberulous without, sparsely within. Ovary globose, ca 3 x 3 mm, 3-locular, stellate-pubescent; ovules 3; styles 3, stout, basally connate, 3-partite, each to 5.8 mm, twice forked at apex. Capsules depressed globose, ca 2.6 x 3 cm, 3 lobed; lobes minutely beaked at apex, slightly scabrous, minutely stellate-puberulous, woody.

Common, in evergreen forests between 600-1000 m.

Fl.: January - March. *Fr.:* Throughout the year.

Specimens examined: XCH 10268, 10327, 10492, 10697, 10702, 11057, 11632, 11866, 12168, 12210, 12954, 13556, 14992, 15086, 15342, 17345, 19026, 19603, 19970, 20002. MH 13459.

Distribution: Peninsular India.

19. EUPHORBIA L.

A vast genus of over 1000 species, subdivided in to many subgenera and sections; distributed throughout the world (Webster, 1994); 82 in India.

Key to the species

1a. Plants succulent 2
1b. Plants not succulent 8

2a. Leaves caducous 3
2b. Leaves persistent.. 9. **E. nivulia**

3a. Branches armed, 2-10 winged 4
3b. Branches unarmed, terete.. 16. **E. tirucalli**

4a. Involucre cupular 5
4b. Involucre turbinate/saucer-shaped 6

5a. Male flowers in fascicles of 5 .... 14. **E. susan-holmesiae**
5b. Male flowers in 5 groups of 6-8 each .... 1. **E. antiquorum**

6a. Branches spirally twisted.. 17. **E. tortilis**
6b. Branches straight 7

7a. Sub-branches flat, 2 winged, wings not prominently lobed............ 18. **E. vajravelui**
7b. Sub-branches terete, 4-10 winged, wings prominently lobed.......... 13. **E. santapaui**

8a. Leaves alternate below, opposite to whorled above 9
8b. Leaves opposite 12

9a. Branchlets dichotomously umbellate.. 10
9b. Branchlets not dichotomously umbellate.. 11

10a. Stem leaves, pesudoumbel leaves and ray leaves almost similar in size and shape .... 5. **E. dracunculoides**
10b. Stem leaves, pesudoumbel leaves and ray leaves always differ in size and shape .... 12. **E. rothiana**

11a. Upper leaves sinuate, scarlet red .... 4. **E. cyathophora**
11b. Upper leaves entire, green .... 6. **E. heterophylla**

12a. Erect or ascending herbs; leaves 0.5-1.5 cm across.. 13
12b. Prostrate herbs; leaves 0.3-0.5 cm across 15

13a. Floral leaves concealing the involucre 14
13b. Floral leaves never concealing the involucre.. 7. **E. hirta**

14a. Limbs of glands deeply pectinately laciniate-fimbriate........................ 3. **E. cristata**
14b. Limbs of glands suborbicular .... 8. **E. indica**

15a. Limbs of glands petaloid, conspicuous.. 11. **E. rosea**
15b. Limbs of glands obscure.. 16

16a. Stamens extrorse.. 2. **E. corrigioloides**
16b. Stamens introrse.. 17

17a. Cocci prominently keeled .... 10. **E. prostrata**
17b. Cocci obscurely keeled.. 15. **E. thymifolia**

1. **Euphorbia antiquorum** L., Sp. Pl. 450. 1753; Hook.f, Fl. Brit. India 5: 255. 1887; Gamble, Fl. Pres. Madras 2:1277. 1925; Rani in Matthew, Fl. Tamil. Carnatic 3: 1428. t. 104 a. 1983; Chandrab. in A.N. Henry *et al.*, Fl. Tamil Nadu 2: 227. 1987; Philcox in Dassan. & Clayton, Rev. Handb. Fl. Ceylon 11: 192. 1997; Matthew, Fl. Palni Hills 2: 1122. 1999; N.P. Balakr. & T.Chakrab., Fam. Euphorb. India 274. 2007; Binojk. & N.P. Balakr. in N.P. Balakr. *et al.* Fl. India 23: 319. 2012. **(Fig. 147).**

Glabrous shrubs 4-15 m; branches 3-6 winged, subsucculent; spines to 3.5 mm long, paired. Leaves obovate, to 5 mm, early deciduous. Cyathia axillary, 3, rarely more in peduncled cymes; central one sessile. Peduncles ca 5 mm. Bracts ca 3.5 mm, lobed at apex, irregularly dentate. Pedicels to 6 mm. Involucres cupular, to 8 mm; lobes 6, each to 3 mm, laciniate at apex; glands 6, fleshy, semi-circular, ca 2 x 0.5 mm, pale yellow, pitted above. Male florets: In 5 groups of 6-8 each, few sterile, mixed with many laciniate bracteoles; stalks to 3 mm. Anthers globose, to 0.5 mm across, longitudinally dehiscent. Female florets: Erect. Ovary subglobose, to 3 x 3.5 mm, 3 lobed, subsessile; styles 3, to 2 mm, bifid, connate to above middle, often recurved. Capsules subglobose, ca 1 x 0.8 cm, deeply 3-gonous; cocci rounded, bluntly angled, glabrous, reddish when mature; seeds globose-ovoid, ca 3 x 2.7 mm, smooth, white.

Common in foot hills, on rocky and stony places, also grown as hedge plant in plains. Few individuals to 15 m tall, observed in dense evergreen forest of Kalakad hills at 1100 m.

Fl.: & Fr.: December - March.

Specimens examined: XCH 15037, 16497, 18274, 18292, 19568. MH 15300.

Distribution: India, Pakistan, Bangladesh, Myanmar, Sri Lanka, Java and Malaya.

2. **Euphorbia corrigioloides** Boiss., Cent. Euphorb. 11. 1860 & in DC., Prodr. 15(2): 32. 1862; Hook.f., Fl. Brit. India 5: 251. 1887; Gamble, Fl. Pres. Madras 2: 1275. 1925; Rani in Matthew, Fl. Tamil. Carnatic 3: 1046, 1429. 1983; Chandrab. in A.N. Henry *et al.*, Fl. Tamil Nadu 2: 227. 1987; Matthew, Fl. Palni Hills 2: 1123. 1999; N.P. Balakr. & T.Chakrab., Fam. Euphorb. India 252. 2007; Binojk. & N.P. Balakr. in N.P. Balakr. *et al.* Fl. India 23: 282. 2012. **(Fig. 148).**

Herbs, prostrate, glabrous or pubescent. Leaves oblong, pubescent, obliquely cordate at base, subacute at apex, margin crenate-serrate, ca 6 x 4 mm, obscurely 3-nerved at base, areolate-reticulate; petioles minute; stipules sagittate, acuminate, ca 2 mm, ciliate. Cyathia axillary. Involucre campanulate, ca 2 mm; lobes 4, linear-lanceolate, ca 5 mm, lacerate; glands 4, semilunate, stipitate, to 2.3 mm. Male florets 3. Bracteoles 0. Filaments ca 1.5 mm; anthers globose, longitudinally dehiscent. Female florets: Gonophore to 0.7 mm. Ovary ovoid-globose, ca 1.7 x 1.5 mm, keeled, pubescent; styles 3, ca 0.5 mm, forked at apex; stigmas papillose.

Capsules ca 2.5 x 2 mm, sparsely pubescent; seeds oblong, 3 angled, ca 1.25 mm, crisped-puberulous.

Common, near dam sites.

Fl.: & Fr.: Throughout the year.

Specimens examined: XCH 16985 (Papanasam). MH 5657.

Distribution: Peninsular India.

3. **Euphorbia cristata** B.Heyne ex Roth, Nov. Pl. Sp. 226. 1821; Boiss. in DC., Prodr. 15(2): 19. 1862; Hook.f., F1. Brit. India 5: 247. 1887; Gamble, Fl. Pres. Madras 2: 1274.1925; Rani in Matthew, Fl. Tamil. Carnadic 3: 1429. t. 104 C. 1983; Philcox in Dassan. & Clayton, Rev. Handb. Fl. Ceylon 9: 197. 1997; Matthew, Fl. Palni Hills 2: 1123. 1999; Binojk. & N.P. Balakr. in N.P. Balakr. *et al.* Fl. India 23: 282. 2012. *E. fimbriata* auct. non. B. Heyne ex Roth, 1821, non Scop. 1788: Boiss. in DC., Prodr. 15(2): 19. 1862. Gamble, Fl. Pres. Madras 2: 1274. 1925. *E. laciniata* Panigrahi in Kew Bull. 30: 531. 1975; Chandrab. in A.N. Henry *et al.*, Fl. Tamil Nadu 2: 228. 1987; N.P. Balakr. & T.Chakrab., Fam. Euphorb. India 253. 2007. **(Fig. 149).**

Dwarf herbs to 20 cm, much branched, with thick rootstock; stems clothed with long, white, flexuous hairs. Leaves obliquely oblong-ovate, 0.8-1.4 x 0.5-0.8 cm, margin hirsute on both surfaces, unequally cordate or truncate at base, obtuse at apex, minutely serrulate, ca 4 veined at base; petioles to 2 mm, hispid, stipulate; stipules short, subulate, to 0.2 mm. Cyathia within distichously imbricate leaves. Involucres subsessile, ca 2 mm, densely long hirsute; glands 4, semicircular; with large, spreading petaloid limbs, deeply pectinately laciniate, subglabrous to hirsute. Male florets: 6, stalk ca 0.8 mm. Filaments ca 0.2 mm; anthers oblong-globose, transversely dehiscent. Female florets: Laterally pendulous. Ovary ca 1 mm, softly pubescent; styles 3, ca 1.3 mm, deeply bifid; stigmas capitate. Capsules to 3 mm across, hirsute; cocci globose; seeds oblong-ellipsoid, to 1.5 mm, with transverse, parallel ridges, reddish brown.

Occasional, on foot paths and in rocky crevices of low altitude grasslands at ca 300 m.

Fl.: & Fr.: June - August

Specimens examined: XCH 11735, 19755 (Kalakad), MH 13149, 12406, 13337, 15251 40146. MH 13149, 12406, 13337, 559, 15251, 40146.

Distribution: India, Sri Lanka and Myanmar.

4. **Euphorbia cyathophora** J.Murr., Commentat. Soc. Regiae Sci. Gott. 7: 81. 1786; Rani in Matthew, F1. Tamil. Carnatic 3: 1430. 1983; Chandrab. in A.N. Henry *et al.*, Fl. Tamil Nadu 2: 228. 1987; N.P. Balakr. & T.Chakrab., Fam. Euphorb. India 278. 2007; Binojk. & N.P. Balakr. in N.P. Balakr. *et al.* Fl. India 23: 329. 2012.

Laticiferous, glabrous shrubs to 2 m; branchlets terete. Leaves ovate-lanceolate, ca 16 x 10 cm, cuneate at base, acuminate at apex, entire at margin, with 1 or 2 blunt, lateral teeth; petioles to 5 cm. Caythia in terminal, branched cymes, subtended by crimson bracts. Bracts lanceolate or oblanceolate, ca 6 x 1.5 cm. Peduncles stout, to 1 cm. Involucres globose to ovoid, ca 0.8 x 0.5 cm; lobes 0; glands 5, conical, ca 4 x 2 cm. Male florets: Many. Filaments to 2 mm; anthers globose, ca 0.5 mm, surrounded by sterile flowers. Female florets: 1 or 2 per cyathia. Ovary subglobose, 3-angled, ca 4 x 3 mm; styles 3, to 3 mm; stigma bifid.

Fl.: & Fr.: Throughout the year.

Specimens examined: XCH 11735.

Distribution: Central America, now a pantropical weed.

5. **Euphorbia dracunculoides** Lam., Encycl. 2: 428. 1788; Hook.f., Fl. Brit. India 5: 262. 1887; Gamble, Fl. Pres. Madras 2: 1277. 1925; Chandrab. in A.N. Henry *et al.*, Fl. Tamil Nadu 2: 228. 1987; N.P. Balakr. & T.Chakrab., Fam. Euphorb. India 266. 2007; Binojk. & N.P. Balakr. in N.P. Balakr. *et al.* Fl. India 23: 305. 2012. **(Fig. 150)**.

Herbs to 90 cm, glaucous, tender parts pubescent, striate. Leaves alternate, linear-lanceolate, 2-5.5 x 0.25-0.5 cm, acute at base, subacute at apex, entire on margin, tender leaves pubescent, 1 nerved, subsessile; floral leaves ca 5 x 4.5 mm. Cyathia terminal, usually solitary or few aggregated. Involucres turbinate, ca 1.4 mm; lobes ovate, ciliolate, minute; glands 4, semi-lunate, ca 1.3 x 0.5 mm, minutely 2 horned; limb 0. Male florets: 5. Stalk to 0.6 mm, ciliolate. Filaments to 0.1 mm; anthers globose, ca 0.5 x 0.5 mm, transversely dehiscent. Female florets: Laterally pendulous, gonophore very short. Ovary ellipsoid, ca 1.6 x 0.7 mm, puberulous; styles 3, bifid at apex, ca 0.4 mm; stigma obtuse. Capsules ellipsoid, ca 5 x 3.6 mm, puberulous, hardly depressed; seeds oblong; ca 4.5 x 2 mm, with white, leprous, tuberculate testa, carunculate, brown.

Very rare, in open forests, ca 300 m.

Fl.: & Fr.: April – June.

Specimen examined: XCH 15941 (Donavoor).

Distribution: India, Arabia and Tropical Africa.

Note: A new record for Tirunelveli Hills.

6. **Euphorbia heterophylla** L., Sp. Pl. 453. 1753; Rani in Matthew, Fl. Tamil. Carnatic 3: 1432. t. 104 f. 1983; Philcox in Dassan. & Clayton, Rev. Handb. Fl. Ceylon 11: 200. 1997; Matthew, Fl. Palni Hills 2: 1123. 1999; N.P. Balakr. & Chakrab, Fam. Euphorb. India 278. 2007; Binojk. & N.P. Balakr. in N.P. Balakr. *et al.* Fl. India 23: 330. 2012. *E. geniculata* Orteg., Nov. Rar. Pl. Matr. 18. 1797; Fischer in Gamble, Fl. Pres. Madras 3: 1884.1928; Chandrab. in A.N. Henry *et al.,* Fl. Tamil Nadu 2: 228. 1987. *E. prunifolia* Jacq., Hort.

Schoenbr. 3: 15. t. 277.1798; Hook.f., Fl. Brit. India 5: 266. 1887, in adnot. *Poinsettia heterophylla* (L.) Klotzsch & Garcke in Klotzsch, Monatsber Akad. Wiss. Berlin 1859: 253. 1859. **(Fig. 151)**.

Erect herbs, to 50 cm; branches terete, hirsute to subglabrous. Leaves opposite or alternate below, rarely whorled above, rhomboid or oblanceolate, 4-8 x 1-4 cm, glabrous to pubescent on both surfaces, or sericeous beneath, cuneate at base, acute at apex, entire to serrate or sinuate at margin; petioles to 1.5 cm. Cyathia ca 8 in terminal clusters. Involucres campanulate, ca 3.5 mm; lobes 5, lanceolate, laciniate at apex; gland stipitate, more or less funnel shaped with circular opening. Male florets: Ca 12, mixed with white bracteoles. Stalk to 1 mm. Anthers subglobose, ca 0.5 mm, longitudinally dehiscent. Female florets: Laterally pendulous; gonophore to 2 mm. Ovary subglobose, ca 2 x 3 mm; styles 3, to 1 mm, erect, bifid at apex. Capsules ca 5 x 6 mm, glabrous; seeds angled, ca 2.5 x 2 mm, minutley tuberculate, black.

Weed of cultivated fields or in wastelands, often gregarious.

Fl.: & Fr.: Throughout the year.

Specimens examined: XCH 13933 (Courtallam). MH 5037.

Distribution: Throughout the tropics and subtropics. Originally introduced from the New world.

7. **Euphorbia hirta** L., Sp. Pl. 454. 1753; Gamble, Fl. Pres. Madras 2: 1275.1925; Rani in Matthew, Fl. Tamil. Carnatic 3: 1434. t. 104 b. 1983; Chandrab. in A.N. Henry *et al.*, Fl. Tamil Nadu 2: 228. 1987; Philcox in Dassan. & Clayton, Rev. Handb. Fl. Ceylon 11. 198. 1997; Matthew, Fl. Palni Hills 2: 1123. 1999; N.P. Balakr. & T.Chakrab., Fam. Euphorb. India 256. 2007; Binojk. & N.P. Balakr. in N.P. Balakr. *et al.* Fl. India 23: 287. 2012. *E. pilulifera* L., Sp. Pl. 454. 1753; Hook.f., Fl. Brit. India 5: 250. 1887. *Chamaesyce pilulifera* (L.) Small, Fl. SEUS. 714. 1903. *C. hirta* (L.) Millspaugh, Paul. Field Mus. Nat. Hist. Bot. Ser. 2: 303. 1909. **(Fig. 152)**.

Hispid herbs to 50 cm; stems sparingly branched above base, unbranched at top. Leaves elliptic-oblong or oblong-lanceolate, 1.5-3.7(-4) x 0.5-1.5 cm, scattered puberulous above, hispid hairy, glaucous beneath, obliquely truncate at base, obtuse or acute at apex, serrate at margin, 3 nerved at base; lateral nerves 1-2 pairs, raised beneath; petioles to 3 mm long, hispid. Cyathia in axillary or terminal, globose clusters, pedunculate. Primary peduncles to 1 cm; secondary peduncles to 3 mm, puberulous. Pedicels to 0.25 mm. Involucres campanulate, ca 0.6 mm, pubescent; lobes 5, acutely fringed, to 0.25 mm; glands 5, stipitate, minute, reddish. Male florets: 5, ebracteolate. Stalk to 0.3 mm; anthers ca 2 mm across, longitudinally dehiscent. Female florets: Laterally pendulous, pubescent. Gonophore to 0.7 mm. Ovary broadly globose, ca 0.4 x 0.7 mm, keeled; styles 3, to 0.25 mm, erect, bifid from base; stigma capitate. Capsules trigonous, ca 1.8 mm across, pubescent.

Common, in plains to 600 m.

Fl.: & Fr.: Throughout the year.

Specimens examined: XCH 11118, 13147, 15708.

Distribution: Tropics and subtropics worldwide.

8. Euphorbia indica Lam., Encycl. 2: 423. 1786; Chandrab. in A.N. Henry *et al.*, Fl. Tamil Nadu 2: 228.1987; Philcox in Dassan. & Clayton, Rev. Handb. Fl. Ceylon 11: 196.1997; Matthew, Fl. Palni Hills 2: 1124.1999; N.P. Balakr. & T.Chakrab., Fam. Euphorb. India 257. 2007; Binojk. & N.P. Balakr. in N.P. Balakr. *et al.* Fl. India 23: 290. 2012. *E. hypericifolia* auct. non L. 1753; Hook.f., Fl. Brit. India 5: 249. 1887; Gamble, Fl. Pres. Madras 2: 1275. 1925; Rani in Matthew, Fl. Tamil. Carnatic 3: 1434. t. 104 i. 1983. *E. parviflora* L., Syst. Nat. ed. 10, 2: 1047. 1759. **(Fig. 153).**

Erect herbs to 30 cm, sparsely crisped-pubescent; stem terete; branches striate. Leaves broadly or narrowly elliptic-oblong, 1.7-2.4 x 0.65-0.75 cm, glabrous to minutley, sparsely crisped pubescent beneath, scabrous on drying, oblique, unequally rounded or cordate at base, obtuse or minutely acute at apex, serrulate at margin, 3-4 nerved at base, forked; lateral nerves 2-3 pairs, distinct, areolate-reticulate; petioles to 2 mm long, filiform, pubescent. Cyathia in axillary or terminal, capitate cymes, many, shortly pedunculate, or subsessile, often subtended by two small floral leaves, sparsely pubescent. Peduncles ca 1 cm, filiform. Pedicels to 1.5 mm. Involucres camparulate, ca 2.2 x 1.5 mm, sparsely puberulous to glabrous without; lobes 4, linear, ciliolate; glands 4, minute, with distinct petaloid limbs; limbs suborbicular, 3-4 times larger than glands. Male florets: Many, few sterile, fertile floret to 0.53 mm; anthers globose, ca 0.25 mm, longitudinally dehiscent. Female florets: Laterally pendulous, subsessile. Ovary subglobose, ca 1.5 x 1.3 mm, 3-locular, sparsely crisped-pubescent; styles 3, ca 0.5 mm, erect, deeply bifid. Capsules subglobose, ca 1.25 x 0.95 mm, keeled, edges ribbed, slightly pubescent to subglabrous; seeds 4 angled, ca 0.26 x 0.16 mm, smooth, minutely to broadly transversely furrowed.

Occasional, in foot hills.

Fl.: & Fr.: Throughout the year.

Specimens examined: XCH 13850 (Courtallam), 18501 (Sivagiri), MH 12138, 8543.

Distribution: India, Sri Lanka, China, Africa, Arabia, Iran and Afghanistan.

9. Euphorbia nivulia Buch.-Ham. in Trans. Linn. Soc. London 14: 286.1825; Hook.f., Fl. Brit. India 5: 255.1887; Gamble, Fl. Pres. Madras 2: 1277.1925; Rani in Matthew, Fl. Tamil. Carnatic 2: 1435 : 1983; Chandrab. in A.N. Henry *et al.*, Fl. Tamil Nadu 2: 229. 1987; N.P. Balakr. & T.Chakrab., Fam. Euphorb. India 276. 2007; Binojk. & N.P. Balakr. in N.P. Balakr. *et al.* Fl. India 23: 324. 2012. **(Fig. 154).**

Glabrous, laticiferous trees to 6 m; branches whorled, subcylindrical, with pairs of sharp stipular spines; spines to 7 mm, inserted on flat, corky bases, tuberculate; tubercles spirally arranged, distant, conical, truncate. Leaves obovate-oblong, 10-25 x 3-5.7 cm, subsessile, glaucous beneath, cuneate at base, retuse at apex, entire at margin; midrib stout beneath; lateral nerves absent. Cyathia usually paired or in lax cymes, subterminal. Involucres cupular, ca 0.8 x 1.3 mm, thick; lobes 0; glands 5, transversely oblong, ca 3 x 7 mm, thick. Male florets: In 5 groups of 8 each with sterile florets. Stalk to 3 mm long; anthers to 0.7 mm across, longitudinally dehiscent. Female florets: Ovary subglobose, ca 4 x 6 mm; styles 3, to 3 mm, connate at base, bifid at apex, erect; stigmas capitate. Capsules subglobose, ca 5 x 7 mm; seeds 4 angular, to 4.5 mm, smooth.

Occasional, on dry rocky hills upto 900 m, often cultivated as hedge plant in hills.

Fl.: & Fr.: November – January.

Specimens examined: XCH 11125, 14217. MH 5598.

Distribution: India and Madagascar.

10. Euphorbia prostrata Aiton, Hort. Kew. ed. 1. 2: 139. 1789; Hook.f., Fl. Brit. India 5: 266. 1887; Gamble, Fl. Pres. Madras 2: 1276. 1925; Rani in Matthew, Fl. Tamil. Carnatic 3: 1436. t. 105 1.1983; Matthew, Fl. Palni Hills 2: 1124. 1999; N.P. Balakr. & T.Chakrab., Fam. Euphorb. India 259. 2007; Binojk. & N.P. Balakr. in N.P. Balakr. *et al.* Fl. India 23: 295. 2012. *E. chamaesyce* auct. non. L., Sp. Pl. 455. 1753; Chandrab. in A.N. Henry *et al.*, Fl. Tamil Nadu 2: 227.1987. **(Fig. 155).**

Herbs to 30 cm; branchlets spreading, hispid pubescent. Leaves obliquely elliptic, ca 7.3 x 3.5 mm, subcordate to truncate at base, unequal, obtuse at apex, crenate to serrate towards apex, bluntly pointed, 3 nerved at base, nerves prominent, areolate-reticulate; floral leaves 6, elliptic, ca 2.5 x 1 mm, sparsely pubescent beneath; stipules linear-lanceolate, ca 1.25 mm. Cyathia in axillary, solitary or fasciculate cymes, often two central, 4 lateral, pinkish, subsessile, bracteate. Involucres tubular, ca 1 mm; lobes 3, ovate-acute, ca 1.2 x 0.8 mm, lacerate at apex; glands 4, stipitate. Male florets: 2. Bracteoles 0. Filaments to 0.4 mm; anthers globose, longitudinally dehiscent. Female florets: Ovary ellipsoid, ca 0.8 x 0.6 mm, minutely, sparsely pubescent; styles 3, to 0.5 mm, forked at apex. Capsules ovoidly globose, ca 1.2 x 1 mm, deeply lobed, prominently keeled; keels with stiff, spreading hairs; seeds tetragonous, ca 0.7 x 0.4 mm, transversely furrowed, brown.

Occasional, in primary deciduous forests, often near Dam site.

Fl.: & Fr.: September - December.

Specimens examined: XCH 13853, 13855 (Courtallam), 20054 (Papanasam).

Distribution: Tropics and subtropics, Introduced from West Indies.

11. Euphorbia rosea Retz., Obs. Bot. 4: 26. 1786; Hook.f., Fl. Brit. India 5: 251. 1887; Gamble, Fl. Pres. Madras 2: 1275. 1925; Chandrab. in A.N. Henry *et al.*, Fl. Tamil Nadu 2: 229. 1987; Philcox in Dassan. & Clayton, Rev. Handb. Fl. Ceylon 9: 204. 1997; N.P. Balakr. & T.Chakrab., Fam. Euphorb. India 260. 2007; Binojk. & N.P. Balakr. in N.P. Balakr. *et al.* Fl. India 23: 296. 2012. *Chamaesyce rosea* (Retz.) Webster in J. Arnold Arbor. 48: 423. 1967, in adnot. *C. auriculata* (Boiss.) V.S. Raju & P.N. Rao in Phytologia 37: 453. 1977. **(Fig. 156).**

Prostrate herbs to 30 cm, much branched; branches straggling, leafy, glabrous or puberulous. Leaves obovate or linear-spathulate, rarely oblong-elliptic, 6(-10) x 5(-4) mm, glabrous above, sparsely puberulous beneath, cuneate to truncate at base, unequal, subacute at apex, entire or occasionally minutely serrulate at margin, 1 nerved, areolate-reticulate; petioles very short, to 2 mm; stipules minute; triangular fimbriate, ca 0.75 mm, intrapetiolar. Cyathia ca 10, aggregated in axillary fascicles or subterminal. Pedicels to 1.3 mm, puberulous. Involucres campanulate or subcampanulate, to 1.3 mm, glabrous; lobes 0; glands 4, semicircular, with large, leafy, coloured limbs; limbs 4, unequal; the two lateral obliquely oblong, ca 2.5-2.7 x 1 mm, conspicuous, petaloid, pinkish; central two obovate, ca 0.5 x 0.6 mm. Male florets: Ca 10, bracteolate. Bracteoles ciliate. Stalks unequal, 0.7-1.4 mm; filaments to 0.4 mm; anther cells globose, transversely dehiscent. Female florets: Gonophore to 1.5 mm. Ovary subglobose, ca 2 x 2 mm, rough, punctate-pubescent; styles 3, to 1.3 mm, forked; stigma obtuse. Capsules ellipsoid-globose, ca 3.4 x 4 mm, prominently keeled, punctate-pubescent, with persistent styles; seeds ovoidly trigonous, ca 2.65 x 2 mm, with transverse ribs, rugose; testa white.

Rare, on the floor in low altitude grasslands at ca 300 m.

Fl.: & Fr.: October - December.

Specimens examined: XCH 20035 (Kalakad), MH 3341, 13613.

Distribution: Peninsular India, Sri Lanka and Afghanistan.

12. Euphorbia rothiana Spreng., Syst. Veg. 3: 796.1826; Hook.f., Fl. Brit. India 5: 263.1887; Gamble, Fl. Pres. Madras 2: 1278.1925; Rani in Matthew, Fl. Tamil. Carnatic 3: 1438. t. 1050. 1983; Chandrab. in A.N. Henry *et al.*, Fl. Tamil Nadu 2: 229. 1987; Philcox in Dassan. & Clayton, Rev. Handb. Fl. Ceylon 11: 205.1997; Matthew, Fl. Palni Hills 2: 1125. 1999; N.P. Balakr. & T.Chakrab., Fam. Euphorb. India 270. 2007; Binojk. & N.P. Balakr. in N.P. Balakr. *et al.* Fl. India 23: 314. 2012. *E. laeta* Heyne ex Roth, Nov. Pl. Sp. 230. 1821, non Ait. 1789. **(Fig. 157).**

Glabrous, subshrubs to 1 m, glaucous; upper branchlets dichotomously umbellate. Leaves alternate below, whorled above, narrowly oblong to oblanceolate, 4.5-10 x 0.5-1.2 cm; subchartaceous, pale greenish on drying; attenuate at base, acute at apex, apiculate, entire at margin; nerves obscure; midrib impressed beneath. Flowering branches whorled, with conspicuous leaves.

Cyathia solitary, in the axils of leafy bracts. Bracts broadly triangular-ovate, to 1.5 x 1 cm, cordate at base. Involucres turbinate to shortly campanulate, ca 2.5 x 2 mm, glabrous; glands semilunar, to 0.7 x 1.8 mm, unequally 2 horned, mouth villous. Male florets: 12, with fimbriate bracteoles. Stalks to 0.4-1.5 mm, ciliate, mixed with sterile stamens. Female florets: Suberect. Ovary ellipsoid, 2.1 x 2.4 mm, smooth; styles 3, bifid above from middle; stigmas obtuse. Capsules ellipsoid, ca 4.5 x 5.5 mm; lobes rounded, smooth; seeds oblong to oblong ovoid, ca 3 x 2 mm, smooth, bluish.

Occasional, on montane grasslands upto 1200 m.

Fl.: & Fr.: November - March.

Specimens examined: XCH 15395 (Kalakad), MH 13218, 14514.

Distribution: India, Sri Lanka.

13. Euphorbia santapaui A.N. Henry in Bull. Bot. Surv. India 6: 329. 1964 (1965); Chandrab. in A.N. Henry *et al.*, Fl. Tamil Nadu 2: 229.1987; N.P. Balakr. & T.Chakrab., Fam. Euphorb. India 276. 2007; Binojk. & N.P. Balakr. in N.P. Balakr. *et al.* Fl. India 23: 325. 2012. **(Fig. 158).**

Glabrous shrubs or small trees to 5 m; branches 4-9(-10) winged with thin lobulate wings bearing spines, subsucculent; spines to 4 mm, paired. Leaves caducous. Cyathia solitary, above the spines. Involucre turbinate, ca 3.5 x 2.3 mm; lobes 6, ca 1.8 x 2 mm, irregularly toothed at apical margin; glands 6, transversely oblong, ca 2.5 x 1 mm, fleshy, pitted above. Male florets: 5-8. Bracteoles laciniate, ca 2 mm. Stalks to 1 mm; filaments to 0.45 mm, articulate over the stalks, pinkish in colour; anthers subglobose, longitudinally dehiscent. Female florets: Gonophore very short, to 1.4 mm. Ovary subglobose, ca 4 x 5 mm, pinkish; styles 3, ca 3 mm, connate upto middle, free above; stigma papillose. Capsules subglobose, ca 5 x 5 mm, pinkish; seeds oblong, ca 3 x 2.6 mm, smooth.

In rocky places in evergreen montane forests at ca 1500 m.

Fl.: & Fr.: November - February.

Specimens examined: XCH 17997 (Poongulum). MH 66059, 93258, 17353, 68814.

Distribution: Agasthiyamalai hills, Peninsular India.

Note: **Endemic** to Agasthiyamalai hills, Western Ghats. **Endangered**.

14. Euphorbia susan-holmesiae Binojk. & Gopalan in Cactus & Succulent J. 65(6): 277. 278. f.1. 1993; N.P. Balakr. & T.Chakrab., Fam. Euphorb. India 277. 2007; Binojk. & N.P. Balakr. in N.P. Balakr. *et al.* Fl. India 23: 326. 2012. **(Fig. 159).**

Shrubs to 3 m; branches crowded at apex, narrowly 4 winged or rarely 3-5 winged, often loosely twisted; spines 2 to 3.5 mm, divaricate. Leaves very small, orbicular, early caducous. Cyathia solitary or in triads, above the spines, pedunculate. Involucres cupular, ca 3.5 x 4 mm; lobes 5, ca 1.75 x 1.5 mm, lacerate at margin; glands 5, transversely oblong, ca 2 x 1.2 mm, yellow. Male florets: In fascicles of 5. Stalks 1.5-2.2 mm; filaments 0.3-0.5 mm; anthers subglobose, transversely dehiscent. Female florets: Gonophore to 3 mm. Ovary globose, to 1.8 x 1.8 mm, glabrous; styles 3, ca 1 mm, each bifid from middle above; stigmas plumose. Capsules globose, to 3 x 3.5 mm; cocci acutely keeled; seeds ovoid, ca 2.5 x 2 mm, smooth, caruncułate.

Occasional, in deciduous forests and usually in semi-evergreen forests in shallow soil over bedrock, 200 - 1300 m.

Fl.: & Fr.: July - September.

Specimens examined: XCH 17921, 19050 (Papanasam).

Distribution: Peninsular India and Sri Lanka.

15. **Euphorbia thymifolia** L., Sp. Pl. 454. 1753; Hook.f., Fl. Brit. India 5: 252. 1887; Gamble, Fl. Pres. Madras 2: 276 1925; Rani in Matthew, Fl. Tamil. Carnatic 3: 1439. t. 105 q. 1983; Chandrab. in A.N. Henry *et al.,* Fl. Tamil Nadu 2: 230. 1987; Philcox in Dassan. & Clayton, Rev. Handb. Fl. Ceylon 9: 201. 1997; Matthew, Fl. Palni Hills 2: 1126. 1999; N.P. Balakr. & T.Chakrab., Fam. Euphorb. India 261. 2007; Binojk. & N.P. Balakr. in N.P. Balakr. *et al.* Fl. India 23: 299. 2012. *Chamaesyce thymifolia* (L.) Millspaugh, Publ. Field Mus. Nat. Hist., Bot. Ser. 2: 412. 1916. **(Fig. 160).**

Herbs to 15 cm; stems slender, prostrate, spreading, densely hispid-hairy. Leaves oblong or elliptic-oblong, 4-8 x 3-5 mm; sparsely pubescent; obliquely truncate at base, obtusely pointed at apex, serrulate at margin, 3-4 nerved at base, obscurely areolate-reticulate, subsessile or with minute petiole; stipules intrapetiolar, subulate, to 1 mm, ciliate at margin. Cyathia in axillary fascicles or in upper axils. Involucres campanulate, to 2.5 mm, puberulous; lobes 4, ovate-acute, ca 1.8 mm, ciliate; glands 4, stipitate. Male florets: 1. Stalk 1-2.4 mm; filaments to 0.6 mm; anthers globose, to 0.25 mm across, didymous, transversely dehiscent. Female florets: Ovary subglobose, ca 0.6 x 0.5 mm, pubescent; styles 3, deeply bifid. Capsules subglobose, ca 1.5 x 1.5 mm; keels obscure, hirsute, often subglabrous; seeds ca 0.8 x 0.3 mm, 4 angled, minutely furrowed, pale brown.

Common, in foot hills up to 700 m.

Fl.: & Fr.: Throughout the year.

Specimens examined: XCH 11128 (Courtallam). MH 13104, 2958, 8438, 39457.

Distribution: India, Sri Lanka, Tropical Asia and Australia.

16. **Euphorbia tirucalli** L., Sp. Pl. 452. 1753; Hook.f., Fl. Brit. India 5: 254. 1887; Gamble, Fl. Pres. Madras 2: 1276.1925; Rani in Matthew, Fl. Tamil.

Carnatic 3: 1440. t. 105 r. 1983; Chandrab. in A.N. Henry *et al.,* Fl. Tamil Nadu 2: 230. 1987; Philcox in Dassan. & Clayton, Rev. Handb. Fl. Ceylon 11: 195. 1997; Matthew, Fl. Palni Hills 2: 1126. 1999; N.P. Balakr. & T.Chakrab., Fam. Euphorb. India 277. 2007; Binojk. & N.P. Balakr. in N.P. Balakr. *et al.* Fl. India 23: 335. 2012. **(Fig. 161).**

Decumbent, ascending to suberect shrubs to 3 m; branchlets succulent, terete, articulate, often leafless. Leaves linear-spathulate, 8-15 x 1.2-1.5 mm, glabrous, fleshy, early deciduous, narrowly cuneate at base, obtuse at apex, entire on margin; stipules minute. Cyathia terminal, few, subsessile. Involucres cupular, to 3 mm, thick; glands 5, peltate, suborbicular, ca 1 mm across. Female florets: Ovary subglobose, ca 1.8 x 1.4 mm; styles 3, to 3 mm, spreading, bifid above middle; arms curved. Capsules subglobose, ca 4 x 6 mm, subglabrous to pubescent; seeds globose, ca 3.5 x 3.5 mm, smooth, carunculate.

Occasional, from foot hills to dry deciduous forests, often grown as hedge plant.

Fl.: & Fr.: April - May.

Specimen examined: XCH 20070 (Mancholai).

Distribution: Native of Tropical Africa; otherwise widespread in old world tropics, often becoming naturalized.

17. **Euphorbia tortilis** Rottl. ex Ainslie, Mat. Ind. 2: 424.1826 : Hook.f., Fl. Brit. India 5: 256. 1887; Gamble, Fl. Pres. Madras 2: 1277. 1925; Rani in Matthew, Fl. Tamil. Carnatic 3: 1440. 1983; Chandrab. in A.N. Henry *et al.*, Fl. Tamil Nadu 2: 230. 1987; N.P. Balakr. & T.Chakrab., Fam. Euphorb. India 277. 2007; Binojk. & N.P. Balakr. in N.P. Balakr. *et al.* Fl. India 23: 326. 2012. **(Fig. 162).**

Glabrous shrubs to 3 m; branches 3 angled, compressed, spirally twisted; spines to 5 mm long. Leaves deciduous. Cyathia axillary, paired or in groups of 3 or 4. Involucres cupular, ca 7 x 5 mm; lobes 5, sheathing, fimbriate, to 1.5 x 0.6 mm; glands 5, semicircular, ca 2 x 1.4 mm, thick. Male florets: In 5 groups of 5 each, bracteolate. Bracteoles sheathing, 3-5 lobed, deeply fimbriate. Stalk to 1.5 mm; filaments to 1 mm; anthers globose, ca 0.5 x 0.5 mm, longitudinally dehiscent. Female florets: Straight or slightly pendulous. Gonophore to 4 mm, curved, cylindrical. Ovary globose, ca 1.5 x 1.7 mm, prominently keeled; styles 3, to 1.5 mm, connate at base, slightly recurved at apex, wavy. Capsules subglobose, ca 5 x 5 mm, smooth; seeds trigonous, ca 3.6 x 2.4 mm, smooth; stalks to ca 5 mm, filiform, often twisted.

Common from foot hills to dry deciduous forests.

Fl.: & Fr.: September - December.

Specimens examined: XCH 13845, 17773 (Courtallam), MH 15225.

Distribution: Peninsular India and Sri Lanka.

18. Euphorbia vajravelui Binojk. & Balakr. in Cactus & Succ. 63(3): 229. f.1. 1991; N.P. Balakr. & T.Chakrab., Fam. Euphorb. India 277. 2007; Binojk. & N.P. Balakr. in N.P. Balakr. *et al.* Fl. India 23: 327. 2012. **(Fig. 163).**

Small trees to 4 m; main branches quadrangular or pentangular; sub-branches flat, 2 winged, rarely 3 winged; spines 2 to 4 mm, divaricate. Leaves obovate to suborbicular, ca 3 x 2.5 mm, fleshy, sessile, caducous. Cyathia axillary, solitary or in triads of simple dichasium; the central sessile, bearing male flowers only; lateral ones pedunculate, to 3 mm, bisexual. Involucre saucer shaped, to 5 mm; lobes 5, orbicular, ca 2 x 1.6 mm, lacerate; glands 4 or 5, ca 3.8 x 1.7 mm, strongly-pitted, pale yellow. Male florets: In fascicles of 5 or 6, each surrounded by bracteoles. Bracteoles deeply laciniate, ca 2.5 mm, setaceous. Stalks to 2.7 mm; filaments to ca 0.7 mm; anthers subglobose, 0.75 mm across, transversely dehiscent. Female florets: Gonophore to 1.8 mm. Ovary globose, ca 3 x 3 mm; styles 3, to 1.5 mm, bifid at apex; stigmas papillose. Capsules globose, ca 6 x 6 mm, obtusely keeled; lobes rounded; seeds ovoid, ca 3 x 2.5 mm, smooth.

Occasional, in open rocky places in dry deciduous forests.

Fl.: & Fr.: December - March.

Specimen examined: XCH 18353 (Sivagiri).

Distribution: Peninsular India.

20. EXCOECARIA L.

A paleotropical genus of ca 40 species, mostly in tropical Asia but a few taxa in Africa, Madagascar and tropical Australia (Webster, 1994); 6 in India.

Key to the species

1a. Branchlets mostly flattened or tetragonous; male inflorescences 3.5-8 cm long; fruits 1-3 cm in diameter .... 1. **E. crenulata**

1b. Branchlets mostly terete; male inflorescences 5-22 cm long; fruits 4-6 cm in diameter .... 2. **E. oppositifolia**

1. **Excoecaria crenulata** Wight, Ic. Pl. Ind. or. 5(2): 20.t. 1865. 1852; Hook.f., Fl. Brit. India 5: 473. 1888; Gamble, Fl. Pres. Madras 2: 1345. 1925; Chandrab. in A.N. Henry *et. al.,* Fl. Tamil Nadu 2: 230.1987; Philcox in Dassan. & Clayton, Rev. Handb. Fl. Ceylon 11:187.1997; T.Chakrab. & N.P. Balakr. in N.P. Balakr. *et al.* Fl. India 23: 344. 2012. *E. crenulata* var. *crenulata* (Wight) T.Chakrab. & M.Gangop. in J. Econ. Taxon. Bot. 18: 208. 1994; N.P. Balakr. & T.Chakrab., Fam. Euphorb. India 291. 2007. **(Fig. 164).**

Tamil: Tillai, Paal maram

Shrubs upto 3 m. Leaves elliptic-lanceolate, 5.5-10 x 2-2.8 cm, cuneate at base, acute to obtusely acuminate at apex, faintly crenate at margin; lateral nerves 9-12 pairs, obscure above, faint to prominent beneath, arcuate; petioles to 1 cm;

stipules sometimes falcately lanceolate, to 0.4 mm. Bracts rhomboid-ovate, ca 0.65 x 0.5 mm, toothed at margin, with large, two glands at sides. Bracteoles subulate, ca 0.6 x 0.1 mm, toothed at margin. Male flowers: Calyx oblong or ovate, ca 1 x 0.9 mm, margin inwardly curved, irregularly toothed. Stamens 0.75-1 mm. Female flowers: Pedicels to 1.8 mm. Calyx ovate, acuminate, ca 1.5 x 0.8 mm, ciliate at margin. Ovary ca 2 x 1.5 mm. Capsules ca 1.5 cm; seeds smooth, oblong-ovoid, ca 7 mm.

Common, in moist-deciduous to evergreen forests, between 400-700 m.

Fl.: May - July. *Fr.:* February - March.

Specimens examined: XCH 10261, 15580, 15998, 17938, 16026, 18074, 18168, 18230, 18343, 18449, 19785. MH 32950, 99160, 70621.

Distribution: Peninsular India and Sri Lanka.

2. **Excoecaria oppositifolia** Griff. in Calcutta J. Nat. Hist. 4: 386. 1844; Hook.f., Fl. Brit. India 5: 474. 1888; N.P. Balakr. & T.Chakrab., Fam. Euphorb. India 290. 2007; T.Chakrab. & N.P. Balakr. in N.P. Balakr. *et al.* Fl. India 23: 345. 2012. *E. robusta* Hook.f., l.c. 474. *E. aboriana* R.N. De in Indian For. 69: 395. 1943; *E. crenulata* var. *oppositifolia*, T.Chakrab. & M.Gangop. in J. Econ. Taxon. Bot. 18: 208. 1994; N.P. Balakr. & T.Chakrab., Fam. Euphorb. India 291. 2007. **(Fig. 165).**

Small trees to 3 m. Leaves oblong to oblong-lanceolate, 11.5-15 (-18.5) x 3-5 cm, acute to obtuse at base, sometimes unequal, acute or shortly sub-acuminate at apex, entire or sinuate-subserrate at margin; lateral nerves 13-20 pairs. Spikes to 10 cm. Bracts rhomboid-ovate, ca 3.2 x 2.8 mm, margin dentate, ciliate, with two sinuate glands; bracteoles ovate, dentate at margin. Male flowers: Pedicels to 0.6 mm. Calyx ovate, ca 1.8-2.3 x 0.8-1.5 mm, irregularly toothed, somewhat cucullate at apex. Stamens ca 2 mm. Female flowers and fruits not seen.

Occasional, in evergreen forests.

Fl.: & Fr.: May - July.

Specimen examined: XCH 12159, 19123, 19673, 16001.

Distribution: Peninsular India.

21. FALCONERIA Royle

Monotypic genus distributed in India, China, Vietnam and Sri Lanka.

Falconeria insigne Royle, Illus. Bot. Himal. Mts 354. t. 84a or 98, f.2. 1839; N.P. Balakr. & T.Chakrab., Fam. Euphorb. India 292. 2007; T.Chakrab. & N.P. Balakr. in N.P. Balakr. *et al.* Fl. India 23: 347. 2012. *F. wallichiana* Royle, Ill. Bot. Himal.: 354. t.84 aq or 98 f.3. 1839. *F. malabarica* Wight, Ic. Pl. Ind. or. 5(2): 20, t.1866. 1852. *Sapium insigne* (Royle) Kurz, For. Fl. Brit. Burma 2: 412. 1877;

Hook.f., Fl. Brit. India 5: 471.1888; Gamble, Fl. Pres. Madras 2: 1346. 1925; Rani in Matthew, Fl. Tamil. Carnatic 3: 1472. 1983; Chandrab. in A.N. Henry *et al.*, Fl. Tamil Nadu 2: 239. 1987. *S. insigne* var. *malabarica* (Wight) Hook.f., l.c. 472. **(Fig. 166).**

Tamil: Sakkarakkalli

Laticiferous, glabrous trees to 6 m, deciduous; branchlets spreading at apex; bark grayish-white, smooth, shrunk when dry. Leaves alternate, crowded at ends of branchlets, broadly elliptic or oblong-lanceolate, 9-17 x 4-5.5 cm; glabrous, penninerved, green or brown when dry, acute or obtuse at base, acuminate at apex, margin crenate-serrulate; lateral nerves 12-15 pairs, arcuate, prominent beneath; cross nervules prominent beneath; petioles 2.3-4 cm long, flattened when dry. Inflorescence terminal, spicate, to 12 cm long, erect, robust, rigid. Male flowers: Many, minute, in circular clusters, sessile, green, minutely bracteate. Calyx globose, split into two lobes, each obovate, ca 1 x 0.75 mm, connate at base, concave, glabrous, persistent. Stamens 2, to 1.5 mm; filaments very minute; anther-cells oblong, longitudinally dehiscent. Female flowers: Many, solitary, sessile. Calyx 3 lobed; lobes ovate-acuminate, ca 1.5 x 1.2 mm. Ovary ovoid-globose, ca 2 mm across, 3 locular; ovules 3, one per locule; styles 3, spreading to recurved, very short, persistent. Capsules ovoid-globose, ca 9 x 7 mm.

Occasional, in moist deciduous to evergreen forests in rocky places between 900-1200 m.

Fl.: & Fr.: January - March.

Specimens examined: XCH 15442, 17943, 18125.

Distribution: India, Sri Lanka, Myanmar and Indochina; also possibly in the Malay Peninsula.

Note: New record for Tirunelveli Hills.

22. FLUEGGEA Willd.

A genus of 14 species, wide spread in the tropics and temperate eastern Asia, but with a distinct relict distribution. The genus was merged with *Securinega* by Mueller (1866), who was followed by many subsequent workers. However Baillon (1858), Bentham (1880), and J.D. Hooker (1887) maintained *Flueggea* as distinct; and indeed it is not closely related to *Securinega*, as indicated by both pollen and seed characters (Webster, 1994); 2 species in India.

Key to the species

1a. Branchlets armed .... 1. **F. leucopyrus**

1b. Branchlets unarmed .... 2. **F. virosa**

1. **Flueggea leucopyrus** Willd., Sp. Pl. 4: 757. 1806; Hook.f., Fl. Brit. India 5: 328. 1887; Gamble, Fl. Pres Madras 2: 1296.1925; Philcox in Dassan. & Clayton, Rev. Handb. Fl. Ceylon 11: 235. 1997; Matthew, Fl. Palni Hills 2: 1127. 1999; N.P. Balakr. & T.Chakrab., Fam. Euphorb. India 342. 2007; T.Chakrab., M.Gangop. & N.P. Balakr. in N.P. Balakr. *et al.* Fl. India 23: 412. 2012. *Securinega leucopyrus* (Willd.) Muell.Arg. in DC., Prodr. 15(2): 451. 1866; Rani in Matthew, Fl. Tamil. Carnatic 3: 1475. t. 109r. 1983; Chandrab. in A.N. Henry *et al.*, Fl. Tamil Nadu 2: 240. 1987. **(Fig. 167).**

Tamil: Vellai poolan, Mudpulanti.

Shrubs to 2 m. Leaves obovate, 1-2 x 0.5-1 cm, acute to cuneate at base, retuse to emarginate or occasionally subacute at apex, entire on margin; lateral nerves 3-5 pairs, obscure; petioles to 5 mm. Male flowers: Pedicels to 3.2 mm. Sepals orbicular or sublanceolate, 0.5-1 x 0.25-0.75 mm; disc ca 0.25 x 0.3 mm. Filaments unequal, 1.5-1.75 mm; anthers to 0.5 mm. Female flowers: Pedicels to 1 mm. Sepals orbicular, 0.8-1 x 0.7-0.9 mm. Ovary ca 0.5 mm across; styles to 0.6 mm long. Capsules ca 2.5 x 2.5 mm; seeds ca 1.5 x 0.5 mm.

Very common along the foothills to deciduous forests.

Fl.: September - December. *Fr.:* Throughout the year.

Specimens examined: XCH 11141, 11305, 11496, 13230, 13678, 13871, 16278, 17588, 17810. MH 15136, 8384, 9565, 15180, 42132.

Distribution: Peninsular India and Sri Lanka.

2. **Flueggea virosa** (Roxb. ex. Willd.) Voigt, Hort. Suburb. Calc. 152. 1845; Baill., Etud. Gen. Euphorb. 593. 1858; Gamble, Fl. Pres. Madras 2: 1296. 1925; Matthew, Fl. Palni Hills 2: 1128. 1999; N.P. Balakr. & T.Chakrab., Fam. Euphorb. India 342. 2007; T.Chakrab., M.Gangop. & N.P. Balakr. in N.P. Balakr. *et al.* Fl. India 23: 413. 2012. *Phyllanthus virosus* Roxb. ex Willd., Sp. Pl. 4: 578. 1805; Rani in Matthew, Fl. Tamil. Carnatic 3: 1476. 1983; Chandrab. in A.N. Henry *et al.,* Fl. Tamil Nadu 2: 240. 1987. *Flueggea microcarpa* Blume, Bijdr. 580. 1826; Hook.f., F1. Brit. India 5: 328. 1887 p.p. *Securinega virosa* (Willd.) Baillon in Adansonia 6: 334. 1866 excl. descr.

Tamil: Pula, Irubulai, Varadbula

Malayalam: Perinkluv

Shrubs to 3 m high. Leaves obovate , 1.8-4.2 x 1.2-2.4 cm, cuneate at base, obtuse or shallowly retuse at apex, entire to subentire at margin; lateral nerves 4-6 pairs, obscure; petioles 5-7 mm. Male flowers (in bud): Pedicels to 5.2 mm. Sepals, disc, stamens and pistillode like *F. leucopyrus*. Female flowers: Pedicels to 5 mm. Sepals 1-1.5 x 0.5-0.75 mm; disc annular. Ovary ca 1.5 x 1.5 mm; styles similar to that of *F. leucopyrus.* Capsules globose, ca 2 x 2 mm; seeds ca 2 x 1 mm.

Common from foot hills to deciduous forests.

Fl.: October - December. *Fr.:* May onwards.

Specimens examined: XCH 12048, 12643, 12707, 13264, 13867, 13893, 15728, 15779, 15849, 16183, 16387, 16434, 16962, 18390, 18543. MH. 3393, 10980, 2883, 3626, 27920.

Distribution: India, Sri Lanka, Tropical Africa, throughout S. Asia to Japan, and Malaysia, Australia and Polynesia.

23. GIVOTIA Griff.

A genus of 4 species, 1 from Africa, 2 from Madagascar, and 1 from India to Sri Lanka.

Givotia rottleriformis Griff. ex Wight, Ic. t. 1889. 1852; Hook.f., Fl. Brit. India 5: 395. 1887; Gamble, Fl. Pres. Madras 2: 1342. 1925; Rani in Mathew, Fl. Tamil. Carnatic 3: 1443. t. 107g. 1983; Chandrab. in A.N. Henry *et al.,* Fl. Tamil Nadu 2: 230. 1987; Philcox in Dassan. & Clayton, Rev. Handb. Fl. Ceylon 11: 103. 1997; N.P. Balakr. & T.Chakrab., Fam. Euphorb. India 232. 2007; T.Chakrab. & N.P. Balakr. in N.P. Balakr. et al. Fl. India 23 264. 2012. *Croton moluccanus* L., Sp. Pl. 1005. 1753, p.p. *Givotia moluccana* (L.) Sreem. in Taxon 24: 696. 1975. **(Fig. 168).**

Dioecious, deciduous trees to 7 m, openly branched; branches thick, prominently leaf scarred, stout, softly white stellate-tomentose; wood white. Leaves alternate, orbicular-deltoid to broadly ovate-deltoid, 6.7-8 x 7-9.5 cm (or more), densely fulvous stellate-tomentose when young, at maturity minutely sparsely stellate hairy above, densely fulvous stellate-tomentose beneath, cordate at base, acute at apex, sinuate-dentate at margin, 5-7 nerved at base; lateral nerves above 3-5 pairs, forked; petioles striate, 4-6 cm, often with one pair of glands at point of insertion or beneath, stellate-hairy. Inflorescence axillary or terminal racemes to 19 cm, widely, openly branched, often pendulous, densely fulvous stellate-tomentose, pedunculate; peduncles to 5 mm. Flowers aggregated at the ends of branches. Male flowers: Subsessile. Calyx 5 lobed; lobes oblong, ca 5 x 2.5 mm, cucullate, stellate-tomentose within and without, imbricate, basally connate. Petals 5 + 5; outer truncate, to 6 x 3 mm, narrowed at base, base villous, revolute, cohering with inner petals; inner petals spathulate to obovate, ca 3.5 x 2.5, villous; disc 5 glandular, truncate, ca 0.75 x 0.75 mm, affixed with inner petals. Stamens 14, on a dense villous receptacle; filaments basally connate, erect, 3-4 mm; anthers oblong, to 1 mm, basifixed, longitudinally dehiscent. Female flowers: Ovary globose, 2-3 locular; ovules 2 or 3, pendulous; styles 2, deeply bifid. Fruits drupaceous, ellipsoid, ca 2.2 x 1.7 cm, pale greenish, hoary, smooth; seeds ellipsoid, ca 1.6 x 1 cm, fleshy.

Common in dry deciduous forests and low altitude grass lands.

Fl.: June - August. *Fr.:* September - November.

Specimens examined: XCH 11199, 11348, 12345, 13293, 13676, 13827, 14031, 14231, 16224, 16355, 16552, 17091, 17590, 19799. MH 2811, 3686, 27945, 38842, 3274.

Distribution: India, Western Himalaya to Sri Lanka, Myanmar, Indochina, Malay peninsula.

24. GLOCHIDION J.R.Forst. & G.Forst.

The genus consists of about 300 species distributed in tropical Asia to Northern Australia and Polynesia. A few species also occur in Madagascar and tropical America. Of the 32 species occurring in the Indian subcontinent, 14 species are of restricted distribution whereas 2 species *G. bourdillonii* and *G. lanceisepalum* exhibit disjunct distribution. Of the widely distributed species, *G. zeylanicum* is the most widespread, occurring in Sri Lanka, Peninsular India, East India to North-east India and South-east Asia, Andaman & Nicobar Islands, through Malaysia to Northern Australia (Chakrabarty & Gangopadhyay, 1995). 22 species in India.

Key to species

1a Capsules not lobed .... 5. **G. zeylanicum**
1b. Capsules 4-8 lobed 2

2a. Branchlets zig-zag at ends .... 3. **G. ellipticum**
2b. Branchlets straight at ends 3

3a. Female flowers sessile .... 4. **G. hohenackeri**
3b. Female flowers pedicellate 4

4a Anthers subulate, free .... 2. **G. candolleanum**
4b. Anthers oblong, united .... 1. **G. bourdillonii**

1. **Glochidion bourdillonii** Gamble in Bull. Misc. Inf. Kew 1925: 330. 1925 & Fl. Pres. Madras 2: 1308. 1925; Chandrab. in A.N. Henry *et al.*, Fl. Tamil Nadu 2: 231. 1987; N.P. Balakr. & T.Chakrab., Fam. Euphorb. India 347. 2007; T.Chakrab., M.Gangop. & N.P. Balakr. in N.P. Balakr. *et al.* Fl. India 23: 417. 2012. **(Fig. 169).**

Glabrous trees to 3 m; branchlets grayish, terete, angular when young. Leaves obovate or oblong or ovate-oblong, 4-7.5 x 3-3.5 cm, blackish green or brown on drying, paler beneath, attenuate at base, often decurrent into petioles, shortly acute, rarely obtuse or shallowly retuse, entire on margin; lateral nerves ca 6 pairs, widely separated towards apex, looping, reticulation prominent; petioles to 5 mm; stipules triangular, persistent, to 1 mm. Flowers supra-axillary, up to 5 in a cluster. Male flowers: Pedicels to 7 mm, filiform. Calyx 6 lobed; lobes 3 + 3, unequal; inner lobes cucullate in bud, oblong-obtuse, ca 3 x 1.2 mm, reflexed. Anthers 3, ca 1 mm; connectives to 0.2 mm. Female flowers: Pedicels ca 2.4 mm. Calyx subequally 6 lobed; lobes oblong-obtuse, ca 2.6 x 1 mm, tubular, occasionally spreading, connate at base. Ovary ca 1 x 2 mm, hard, similar width to style, distinguished from it by small grooves; styles columnar, ca 2.6 x 2 mm, shallowly 5 lobed at apex, half exserted. Capsules not seen.

In evergreen forests and meadows between 900-1200 m.

Fl.: & Fr.: March onwards.

Specimens examined: XCH 17334 (Kothayar), 18739 (Mahendragiri).

Distribution: Peninsular India. **Endangered**

Note: The following are the distinctive features of this species.

(i) Apex of the leaves obtuse or shortly acute.

(ii) In female flowers, calyx prominently 6 lobed.

(iii) The ovary is grooved.

2. **Glochidion candolleanum** (Wight & Arn.) T.Chakrab. & M.Gangop. in J. Econ. Taxon. Bot. 19: 191. f. 10 E-I. 1995; N.P. Balakr. & T.Chakrab., Fam. Euphorb. India 347. 2007; T.Chakrab., M.Gangop. & N.P. Balakr. in N.P. Balakr. *et al.* Fl. India 23: 418. 2012. *Lobocarpus candolleanus* Wight & Arn., Prodr. 1: 7. 1834. *Glochidion neilgherrense* Wight, Ic. Pl. Ind. or. 5(2): 29. t. 1907. 1852; Hook.f., Fl. Brit, India 5: 316. 1887; Gamble, Fl. Pres. Madras 2: 1307. 1925. *G. arboreum* Wight, l.c. 29. t.1907. 1852; Hook.f., l.c. 316; Gamble, Fl. Pres. Madras 2: 1307. 1925. *Phyllanthus neilgherrense* (Wight) Muell.Arg. in Flora 48: 385. 1865. *P. arboreus* (Wight) Muell.Arg., 11. cc. 380. 1865. *Glochidion pycnocarpum* (Muell.Arg.) Bedd., For. Man. Bot. 194.1873. *G. pauciflorum* Gamble, Fl. Pres. Madras 2: 1307. 1925. *G. sisparense* Gamble, l.c. 1307. **(Fig. 170).**

Trees to 4 m; branchlets pubescent, grayish-brown on drying. Leaves broadly ovate-oblong, or ovate-lanceolate, 6.5-10 x 3-4.7 cm, coppery brown to blackish brown on drying, paler beneath, unequally attenuate at base, acute or obtusely acuminate at apex, ca 5 mm; lateral nerves ca 5 pairs, prominently looping, obscure to faint above, faint to prominent beneath; petioles to 5 mm. Flowers slightly supra-axillary, mostly male flowers very few at proximal axils; female at distal axils, fascicled. Male flowers: Pedicels to 1.3 cm, filiform. Calyx 6 lobed; 3 + 3; inner 3 cucullate, margin shallowly crenate, unequal, subspathulate, 3.5-5 x 2-2.7 mm. Anthers 3; column to 1 mm; connectives subulate, to 0.5 mm, free. Female flowers: Pedicels to 1.5 mm, puberulous. Calyx connate at base, unequally 6 lobed; lobes ca 1 x 0.8 mm, puberulous without. Ovary not distinguished; style hard, clavate, ca 2.3 x 1.3 mm, finely 4 lobed.

Occasional, in evergreen forests at ca 1000 m.

Fl.: January - March.

Specimens examined: XCH 14866 (Kothayar). 10961, 11149, 11694, 14322, 15195, 17380, 17547,18564. MH 17343.

Distribution: Peninsular India.

Note: This species is characteristically different for the subulate connectives.

3. **Glochidion ellipticum** Wight., Ic. Pl. Ind. Or. 5(2): t. 1906. 1852; Hook.f., Fl. Brit. India 5: 321. 1887; Gamble, Fl. Pres. Madras 2: 1308. 1925; Rani in Matthew, Fl. Tamil. Carnatic 3: 1445. t. 108h. 1983; Chandrab. in A.N. Henry *et al.*, Fl. Tamil Nadu 2: 231. 1987; T.Chakrab. & M.Gangop. in J. Econ. Taxon. Bot. 19: 199. 1995; Matthew, Fl. Palni Hills 2: 1129. 1999; N.P. Balakr. & T.Chakrab., Fam. Euphorb. India 348. 2007; T.Chakrab., M.Gangop. & N.P. Balakr. in N.P. Balakr. *et al.* Fl. India 23: 420. 2012. *G. malabaricum* (Muell.Arg.) Bedd., For. Man. Bot. 194. 1873; Hook.f., Fl. Brit. India 5: 319. 1887; Gamble, Fl. Pres. Madras 2: 1308. 1925. G. *ralphii* Hook.f., Fl. Brit. India 5: 314. 1887. *G. ellipticum* var. *ralphii* (Hook.f.) Gamble, Fl. Pres. Madras 2: 1308. 1925. *G. balakrishnanii* G.J. Jothi *et al.* in J. Econ. Taxon. Bot. 26: 114, f.l. 2002. **(Fig. 171).**

Glabrous trees to 5 m, sometimes puberulous; branchlets slender, hard, striate. Leaves elliptic or elliptic-oblong or lanceolate-oblong or subfalcate, 5.5-8 x 1.3-2 cm, chartaceous or coriaceous, shining, equal, subequal or unequally acute to attenuate, rarely obtuse at base, acuminate at apex; acumen to 1 cm; mucronate, entire at margin; lateral nerves 6-8 pairs, faint above, prominently raised beneath; petioles to 3 mm. Flowers in axillary fascicles. Male flowers: Pedicels to 1 cm, filiform. Sepals 3 + 3; lobes elliptic-spathulate, ca 3 x 1.4 cm, greenish yellow. Anthers 3, oblong, to 0.75 mm; connectives to 0.25 mm. Female flowers: Sessile or shortly pedicellate, to 1 mm. Calyx 6 lobed; lobes elliptic-spathulate or oblong, ca 0.75 x 0.35 mm, puberulous without, equal or unequal. Ovary subglobose, ca 0.6 x 0.75 mm, puberulous; style columnar, to 1 x 0.5 mm, minutely 4 lobed at apex, each bifurcated, exserted. Capsules globose or subglobose, ca 6.5 mm across, shallowly lobed or unlobed, glabrous to puberulous.

Common, in moist deciduous to semi evergreen forests, between 500-900 m.

Fl.: & Fr.: January - December.

Specimens examined: XCH 10540, 10593, 10672, 10733, 10780, 11938, 12158, 12187, 16078, 17173, 18113, 18118, 18153, 18204, 18807, 18947, 19046, 19511, 19856. MH 3130, 5018, 98281, 4519, 29205, 17399, 90639.

Distribution: Peninsular India, Orissa, W. & E. Himalayas, N. E. India, Bangladesh, Myanmar, Thailand, Indo-china, S. W. China, Hainan and Taiwan.

4. **Glochidion hohenackeri** (Muell.Arg.) Bedd., For. Man. Bot. 193. 1873; Hook.f., Fl. Brit. India 5: 314. 1887; T.Chakrab. & M.Gangop. in J. Econ. Taxon. Bot. 19: 208. 1995; N.P. Balakr. & T.Chakrab., Fam. Euphorb. India 349. 2007; T.Chakrab., M.Gangop. & N.P. Balakr. in N.P. Balakr. *et al.* Fl. India 23: 422. 2012. *Phyllanthus hohenackeri* Muell.-Arg. in Flora 48: 373. 1865 & in DC., Prodr. 15(2): 288. 1866. *P. fagifolius* Muell.-Arg. 11. cc

373. 1865 & 288. 1866, p.p., *tantum quoad specimena* Metz 1465 (incl. var. *concolor*). *Glochidion fagifolium* (Muell.Arg.) Bedd. l.c. 193. 1873; Hook.f., l.c. 312. 1887, p.p.; Gamble, Fl. Pres. Madras 2: 1308.1925; Chandrab. in A.N. Henry *et al.,* Fl. Tamil Nadu 2: 231 1987.

Key to the varieties

1a. Branchlets, petioles and flowers glabrous .... 1. var. **hohenackeri**

1b. Branchlets, petioles and flowers tawny tomentellous to sparsely puberulous .... 2. var. **johnstonei**

1. var. **hohenackeri** **(Fig. 172).**

Glabrous trees to 3 m; branches grayish, rough; branchlets smooth, spreading, coppery-brown on drying. Leaves oblong or ovate-lanceolate, acute to cuneate at base, unequal, acute or shortly obtusely acuminate at apex to 5 mm, often mucronate, entire at margin, coppery brown above on drying, paler beneath; lateral nerves ca 5 pairs. Flowers in axillary clusters. Male flowers: Many, pedicellate. Calyx 6 lobed; lobes subequal. Anthers 3, columnar. Female flowers: Sessile. Calyx 5 lobed; lobes subequal; style subglobose, finely 3-6 lobed, connate at base with ovary. Capsules subglobose.

Occasional, in semi evergreen forests at ca 600 m

Fl.: & Fr.: January - March.

Specimens examined: XCH 14840, 14646 (Mahendragiri), 10197, 10247, 14932 (Kothayar).

Distribution: Peninsular India.

2. var. **johnstonei** (Hook.f.) T.Chakrab. & M.Gangop. in J. Econ. Taxon. Bot. 19: 210. 1995; N.P. Balakr. & T.Chakrab., Fam. Euphorb. India 349. 2007; T.Chakrab., M.Gangop. & N.P. Balakr. in N.P. Balakr. *et al.* Fl. India 23: 423. 2012. *Glochidion johnstonei* Hook.f., Fl. Brit. India 5: 314. 1887; Gamble, Fl. Pres. Madras 2: 1307. 1925. **(Fig. 173).**

Trees to 4 m; branchlets tawny tomentellous to sparsely puberulous. Leaves similar to that of var. *hohenackeri*, except the fulvous puberulous major nerves beneath and petioles. Male flowers: Pedicellate. Calyx unequal, subspathulate, puberulous without. Anthers similar to those of var. *hohenackeri*. Female flowers: Sessile. Calyx 6 lobed; lobes lanceolate, subequal, puberulous without. Ovary and style similar to that of var. *hohenackeri*.

Fl.: January - March.

Specimen examined: XCH 14727 (Kothayar).

Distribution: Peninsular India.

5. **Glochidion zeylanicum** (Gaertner) A.Juss., Tent. Euphorb. 107, t. 3. 1824; Hook.f., Fl. Brit. India 5: 310. 1887; Gamble, Fl. Pres. Madras 2: 1306.1925; Rani in Matthew, Fl. Tamil. Carnatic 3: 446. 1983; Chandrab. in A.N. Henry *et al.,* Fl. Tamil Nadu 2: 232. 1987; T.Chakrab. & M.Gangop. in J. Econ. Taxon. Bot. 19: 226. 1995; Philcox in Dassan. & Clayton, Rev. Handb. Fl. Ceylon 11: 243. 1997; Matthew, Fl. Palni Hills 2: 1130. 1999; N.P. Balakr. & T.Chakrab., Fam. Euphorb. India 353. 2007; T.Chakrab., M.Gangop. & N.P. Balakr. in N.P. Balakr. *et al.* Fl. India 23: 429. 2012. *Bridelia zeylanica* Gaertner, Fruct. 2: 128. t. 109. 1791. **(Fig. 174).**

Tamil: Kumbalam.

Glabrous trees to 3 m; branchlets terete, towards apex flattened. Leaves broadly ovate-oblong or elliptic-oblong, 7-16.5 x 3.8-7.8 cm, coriaceous, reddish-brown or green on drying, rounded-subcordate at base, often with unequal sides, acute, occasionally retuse or subacute at apex, entire at margin; lateral nerves 6-8 pairs, raised beneath, arcuate, transverse nervules prominent beneath; petioles to 5 mm, stout; stipules falcately lanceolate or subulate, to 5 mm. Flowers supra-axillary, pedicelled, sometimes in shortly peduncled, umbellate clusters. Male flowers: Pedicels to 1.2 cm, filiform. Calyx 6 lobed; lobes 3 + 3, oblong, ca 4 x 2 mm, reflexed at apex, whitish-yellow in colour. Anthers 4, connate, to 1 mm; connectives to 0.6 mm. Female flowers: Pedicels to 5 mm, increasing upto 8 mm in fruit. Calyx tubular, to 3 mm, with 6 unequal lobes; lobes oblong, ca 0.5 x 0.3 mm. Ovary subglobose, ca 2.5 x 2.3 mm, hard; styles conical, to 1.5 mm, hard, 6 toothed, bent on one side at apex. Capsules subglobose, ca 5 mm across beaked, stony.

Common, in dry deciduous to moist deciduous forests along streams.

Fl.: & Fr.: January - December.

Specimens examined: XCH 10961, 11149, 11694, 14322, 15195, 17380, 17547, 18564. MH 16232, 3739, 29124.

Distribution: Peninsular India to E. India, Sri Lanka, N. E. India to S. E. Asia, Andaman and Nicobar Islands, throughout Malaysia to N. Australia.

25. JATROPHA L.

A diverse genus of ca 175 species of tropics and subtropics in the old world and New world; native species absent from Australia and the Pacific islands (Webster, 1994); 11 in India.

Key to the species

1a. Leaves shallowly 5 lobed, glandular; petals connate to the middle or so ...... 1. **J. curcas**
1b. Leaves deeply 3-5 lobed, glandular; petals free or nearly so .... 2. **J. gossypifolia**

1. **Jatropha curcas** L., Sp. Pl. 1006. 1753; Hook.f., Fl. Brit. India 5: 383. 1887; Gamble, Fl. Pres. Madras 2: 1340. 1925; Rani in Matthew, Fl. Tamil. Carnatic 3 : 1449. t. 108j. 1983; Chandrab. in A.N. Henry *et al.*, Fl Tamil Nadu 2: 232. 1987; Philcox in Dassan. & Clayton, Rev. Handb. Fl. Ceylon 11: 85. 1997; N.P. Balakr. & T.Chakrab., Fam. Euphorb. India 223. 2007; T.Chakrab. & N.P. Balakr. in N.P. Balakr. *et al.*, Fl. India 23: 250. 2012. **(Fig. 175).**

Tamil: Kattukkottai

Shrubs to 3 m. Leaves shallowly 5 lobed, ca 2.7-7.5 x 2-7 cm; lobes cordiform, glabrous beneath, puberulous above, cordate at base, acute at apex, 5 nerved. Cymes to 6 cm. Male flowers: Sepals elliptic, to 4.5 x 2.4 mm, pubescent. Petals oblong, to 6.3 x 3 mm, puberulous without at base; disc globose. Stamens monadelphous. Female flowers: Pedicels ca 6 mm. Sepals lanceolate, ca 5 x 3 mm, pubescent without. Petals 5, elliptic, ca 0.65 x 0.18 mm, pubescent; disc 6 lobed. Ovary subglobose, ca 4 x 3.6 mm; styles to 1.6 mm, deeply bifid. Capsules to 2.8 x 2 cm.

Widely cultivated as a hedge plant, naturalized in the old world.

Fl.: & Fr.: April onwards.

Specimen examined: XCH 15738 (Papanasam).

Distribution: Native to tropical America, naturalized throughout the tropics.

2. **Jatropha gossypifolia** L., Sp. Pl, 1006. 1753; Hook.f., Fl. Brit. India 5: 383. 1887; Gamble, Fl. Pres. Madras 2: 1340. 1925; Rani in Matthew, Fl. Tamil. Carnatic 3: 1450. t. 100k. 1983; Chandrab. in A.N. Henry *et al.*, Fl. Tamil Nadu 2: 232. 1987; Philcox in Dassan. & Clayton, Rev. Handb. Fl. Ceylon 11: 83. 1997; N.P. Balakr. & T.Chakrab., Fam. Euphorb. India 224. 2007; T.Chakrab. & N.P. Balakr. in N.P. Balakr. *et al.*, Fl. India 23: 252. 2012. **(Fig. 176).**

Tamil: Aatalai

Shrubs to 2 m; branchlets glandular-hairy. Leaves deeply 3-5 lobed; ca 6 x 11.5 x 5-13.5 cm; lobes obovate, subcordate at base, acute at apex, glandular hairy at margin; petioles 4-8 cm; stipules glandular. Inflorescence terminal cymose, ca 4 cm, densely pubescent. Male flowers: Sepals ovate, ca 4 x 2 mm glandular at margin. Petals free, ca 5.5-6 cm, purplish-red. Stamens 12. Female flowers: Sepals and petals similar to those of male flowers. Ovary pubescent; styles bifid. Capsules ca 1 cm across, glabrescent.

Common from the plains to dry deciduous forests.

Fl.: & Fr.: Throughout the year.

Specimen examined: XCH 16609 (Papanasam).

Distribution: Native of the New world, from the West Indies, Central and South America; Introduced into the old world tropics.

26. KOILODEPAS Hassk.

A genus of 10 species distributed in India and Southeastern Asia as far as Hainan and Borneo (Webster, 1994); 1 in India.

Koilodepas calycinum Bedd., Fl. Sylv. t. 320. 1873 & For. Man. 207. 1873 (as *Coelodepas*); Hook.f., Fl. Brit. India 5: 419. 1887; Gamble, Fl. Pres. Madras 2: 1324. 1925; Chandrab. in A.N. Henry *et al.*, Fl. Tamil Nadu 2: 224. 1987; N.P. Balakr. & T.Chakrab., Fam. Euphorb. India, 171. 2007; N.P. Balakr. in N.P. Balakr. *et al.*, Fl. India 23: 185. 2012. **(Fig. 177).**

Tamil: Kattu-pira

Monoecious, straggling trees to 3 m; branchlets terete, striate, angular, pubescent, fulvous tomentose when young; bark brownish red. Leaves alternate, elliptic or linear-oblong, 5.5-15 x 2-4.9 cm, chartaceous to sub-coriaceous, sparsely stellate tomentose or glabrous on both surfaces, pale greenish on drying, cuneate at base, acute or acuminate at apex with acumen to 1 cm long, crenate-serrate at margin; lateral nerves 7-13 pairs, raised beneath; cross-nervules prominent; petioles to 5 mm, pulvinate at base, scurfy-pubescent, thick; stipules deltoid-lanceolate, to 5 mm, fimbriate, pubescent. Inflorescence axillary, spicate, ca 3.5 cm, scurfy pubescent. Male flowers: Subsessile. Calyx campanulate, 4-5 lobed; lobes up to 0.35 mm, unequal, stellately pubescent without. Stamens 5, very short, exserted; filaments cuneately dilated from the base upwards, to 0.6 mm; anthers oblong, to 0.6 mm, longitudinally dehiscent. Female flowers: Subsessile. Calyx cupular, to 2.5 mm, 9 lobed; lobes up to 0.7 mm, unequal, stellately pubescent without. Ovary globose, ca 2 x 2 mm, 3 lobed, 3 locular, stellately pubescent, reddish; ovules 3, 1 per locule, pendulous; styles 3, spreading, to 2 mm, prominently, deeply lacerate, 2 - multifid. Capsules globose, ca 1.25 x 1.25 cm, 3 lobed, punctate, sparsely stellate-pubescent, rose-reddish, with persistent, accrescent calyx; stalks to 5 mm, stellate-pubescent.

Occasional, in shady places in moist deciduous forests often on the edges of the forests, between 400-600 m.

Fl.: & Fr.: March - May.

Specimens examined: XCH 18850 (Papanasam), 19947 (Sivagiri). MH 99162, 1867

Distribution: Southern Western Ghats

27. MACARANGA Thouars

A very large and diverse paleotropical genus of ca 300 species; about 50 species in Africa and Madagascar, over 200 in tropical Asia from India to New Guinea, and a few in Fiji and other pacific islands (Webster, 1994); 9 in India.

Key to the species

1a. Panicle branches zig-zag.. 1. **M. indica**

1b. Panicle branches straight ....2. **M. peltata**

1. **Macaranga indica** Wight, Ic. Pl. Ind. Or. 5, 2: tab. 1883. 1852 & 6 : tab. 1949. f. 2. 1853; Hook.f., Fl. Brit. India 5: 446. 1887; Gamble, Fl. Pres. Madras 2: 1326. 1925; Rani in Matthew, Fl. Tamil. Carnatic 3: 1453. 1983; Chandrab. in A.N. Henry *et al.,* Fl. Tamil Nadu 2: 233. 1987; Philcox in Dassan. & Clayton, Rev. Handb. Fl. Ceylon 11: 169. 1997; Matthew, Fl. Palni Hills 2: 1133. 1999; N.P. Balakr. & T.Chakrab., Fam. Euphorb. India 141. 2007; S.R.M. Susila Rani & N.P. Balakr. in N.P. Balakr. *et al.,* Fl. India 23: 132. 2012. **(Fig. 178).**

Tamil: Vattakkanni

Trees to 10 m. Leaves orbicular-ovate, 10-17 x 9-16 cm, glabrous above, tomentose, with minute granular glands beneath, rounded at base, acute or cuspidate-acuminate at apex, entire or toothed at margin, 6-8 palmately nerved at base, with two large elliptic glands up to 3 mm; petioles 5-10 cm; stipules lanceolate-acuminate, caducous. Male inflorescence upto 8 cm, much branched, pubescent. Male flowers: Sessile, many, fascicled, subtended by bracts; filaments to 0.3 mm; anthers to 0.2 mm. Female flowers: Pedicels 3-6 mm, subtended by bracts. Sepals ovate, obtuse at apex. Ovary subglobose, ca 2 x 2 mm; seeds spherical, pale brown.

Occasional, on borders of moist deciduous to semi-evergreen forests.

Fl.: December – March. *Fr.:* Throughout the year.

Specimens examined: XCH 18716, 15686.

Distribution: India, Sri Lanka.

2. **Macaranga peltata** (Roxb.) Muell.Arg. in DC., Prodr. 15, 2: 1010. 1866; Gamble, Fl. Pres. Madras 2: 1326. 1925; Chandrab. in A.N. Henry *et al.,* Fl. Tamil Nadu 2: 233. 1987; Philcox in Dassan. & Clayton, Rev. Handb. Fl. Ceylon 11: 170. 1997; N.P. Balakr. & T.Chakrab., Fam. Euphorb. India 142. 2007; S.R.M. Susila Rani & N.P. Balakr. in N.P. Balakr. *et al.,* Fl. India 23: 137. 2012. *Osyris peltata* Roxb., Fl. Ind. 3: 755. 1832. Wight, Ic. Pl. Ind. Or. 3, 2: t. 817. 1843. *Macaranga roxburghii* Wight, Ic. Pl. Ind. Or. 5, 2: 23. 1852 & 6 : t. 1949. f. 4.1853; Hook.f., Fl. Brit. India 5: 448. 1887. *M. tomentosa* Wight, Ic. Pl. Ind. Or. 5, 2: 23. 1852, & 6: t. 1949, f. 1. 1853. **(Fig. 179).**

Trees to 8 m. Leaves broadly ovate, 8-14 x 5.5-11 cm, base rounded, shortly acuminate at apex, entire at margin, densely orange brown, granular-glandular beneath, 5-7 palminerved at base, eglandular near insertion of petiole; petioles 3-10 cm; stipules lanceolate, caducous. Male inflorescence upto 10 cm, puberulous. Male flowers: Sessile fascicled, subtended by bracts. Female inflorescence to 3 cm, pubescent. Female flowers: Sepals lacking. Ovary globose, ca 2 mm, with persistent stigma on one side, black granular-glandular; seeds globose, rough.

Common in secondary moist deciduous forests.

Fl.: October – December. *Fr.:* Throughout the year.

Specimens examined: XCH 14719, 12529. MH 15184, 3624, 49637.

Distribution: India and Sri Lanka.

28. MALLOTUS Lour.

This large and complex paleotropical genus includes ca 150 species, of which there are only two in Africa and Madagascar, the rest in Asia and Australia to Fiji (Webster, 1994); 17 in India.

Key to the species

1a. Leaves red-glandular beneath .... 4. **M. philippensis**
1b. Leaves yellow-glandular beneath 2

2a. Leaves 2-4 nerved at base 3
2b. Leaves penninerved 6

3a. Leaves opposite .... 2. **M. distans**
3b. Leaves alternate and opposite 4

4a. Fruits echinate or tubercled .... 3. **M. ferrugineus**
4b. Fruits smooth 5

5a. Leaves opposite and alternate, subglabrous beneath .... 7. **M. rhamnifolius**
5b. Leaves alternate, tomentose beneath .... 5. **M. repandus**

6a. Leaves alternate or opposite in young shoots .... 1. **M. beddomei**
6b. Leaves strictly opposite .... 6. **M. resinosus**

1. Mallotus beddomei Hook.f., Fl. Brit. India 5: 438. 1887; Gamble, Fl. Pres. Madras 2: 1323. 1925; Chandrab. in A.N. Henry *et al.*, Fl. Tamil Nadu 2: 234. 1987; N.P. Balakr. & T.Chakrab., Fam. Euphorb. India 146. 2007. **(Fig. 180).**

Shrubs to 2 m; branchlets terete, at end angular, woolly to pubescent, compressed. Leaves broadly ovate or ovate-oblong or oblong, 7-19 x 3-7.4 mm, sparsely minutely yellow glandular beneath, glabrous except the nerves beneath, pale brownish green on drying, obtuse at base, caudate-acuminate at apex, often mucronate, sinuate, or entire at margin, penninerved; lateral nerves 6-8 pairs, distantly arranged, arcuate, raised beneath; cross nervules prominent; petioles of the pair unequal in length, 0.5-0.7 and 2-2.8 cm, pubescent, pulvinate at base and apex. Inflorescences racemose; male to 3 cm; female to 15.5 cm, distantly flowered; rachis pubescent; bracteate; bracts lanceolate, ca 2.7 x 2 mm, margin dentate, puberulous without. Male flowers: (in buds) Pedicels to 1.2 mm, pubescent. Calyx 3 lobed; lobes unequal, ovate-oblong, 1-2 x 1.3-1.6 mm, puberulous without.

Stamens ca 26; filaments free, unequal, 0.5-1 cm; anthers oblong, to 0.5 mm, cells separated by a narrow, apiculate connective. Female flowers: (in buds) Oblong. Pedicels to 5 mm, pubescent. Calyx 3-6 lobed; lobes connate or free, obtusely lanceolate, ca 8.5 x 1.8 mm, sparsely puberulous without. Ovary globose, ca 3.5 mm across echinate; each spine villous at apex; styles 3 fid, basally connate to 1.5 mm; each arm to 5.5 mm. Capsules globose, ca 1.5 cm across, echinate; each spine to 2.5 mm, villous at apex, granular, pubescent; seeds ovoid-globose, ca 6 x 5 mm, smooth.

Rare, in semi-evergreen forests between 600-800 m.

Fl.: & Fr.: January - March

Specimens examined: XCH 17178, 18133, 18134, 18137, 18808, 19999.

Distribution: Peninsular India.

Note: Rediscoved after Beddome (1872).

2. **Mallotus distans** Muell.Arg. in Linnaea 34: 194. 1865; Hook.f., Fl. Brit. India 5: 441. 1887; Gamble, Fl. Pres. Madras 2: 1322. 1925; Chandrab. in A.N. Henry *et al.*, Fl. Tamil Nadu 2: 234. 1987; Philcox in Dassan. & Clayton, Rev. Handb. Fl. Ceylon 11: 164. 1997; N.P. Balakr. & T.Chakrab., Fam. Euphorb. India 146. 2007. **(Fig. 181).**

Shrubs or trees to 3 m; branches minutely, densely stellate pubescent. Leaves opposite, broadly ovate or oblong, 6-23 x 3-10.5 cm, glabrous above, densely stellate pubescent beneath, intermixed with minute, glossy, yellow-granular glands, 3 veined at base, rounded or cordate at base, acuminate at apex, entire or sinuate at margin; lateral veins 3-5 pairs, raised beneath; cross nervules prominent beneath, 2-8 black or dark brown, elliptic glands present at or towards base of lamina above; petioles 2-5 cm, densely pubescent. Male inflorescence: Spicate, to 7 cm, densely pubescent. Male flowers: Sessile. Calyx 3 lobed; lobes ovate, ca 2 x 1.5 mm, puberulous. Filaments ca 1 mm; anthers oblong; connectives obscure. Female inflorescence ca 15 cm. Female flowers: Subsessile. Pedicels to 3 mm in fruit. Calyx 3 lobed; lobes ovate, ca 2.5 x 2 mm, acute, puberulous. Ovary ca 1 mm across, densely stellate-tomentose; styles ca 2 mm, coarsely plumose. Fruit ca 6 mm across, yellow stellate-pubescent; seeds globose, mottled, brown.

Rare, in semi-evergreen forests between 600-900 m.

Fl.: & Fr.: April - July

Specimens examined: XCH 12989, 13074, 13935, 14511, 16172, 16537, 18092, 19040. MH 14628, 19276, 463, 507, 15173, 93395.

Distribution: Peninsular India and Sri Lanka.

Note: Hooker (1887) & Gamble (1925) mentioned ‘The leaves eglandular beneath’ but our specimens have glands beneath.

3. **Mallotus ferrugineus** (Roxb.) Muell.Arg. in Linnaea 34: 188. 1865 & in DC., Prodr. 15(2): 982. 1866. *Rottlera ferruginea* Roxb., Fl. Ind. 3: 828. 1832. *Mallotus albus* sensu Muell.Arg. in Linnaea 34: 188. 1865; Hook.f., Fl. Brit. India 5: 429. 1887. *Mallotus albus* var. *occidentalis* Hook.f., l.c. 5: 429. *M. tetracoccus* Kurz in J. Asiat. Soc. Bengal 41(2): 245. 1873; Rani in Matthew, Fl. Tamil. Carnatic 3: 1457. 1983; Chandrab. in A.N. Henry *et al.,* Fl. Tamil Nadu 2: 234. 1987; Philcox in Dassan. & Clayton, Rev. Handb. Fl. Ceylon 9: 153. 1997; Matthew, Fl. Palni Hills 2: 1135. 1999. **(Fig. 182).**

Trees to 15 m; branchlets tomentose. Leaves alternate or opposite, broadly ovate to deltoid, 8-20 x 6.8-12 cm, fulvous tomentose beneath, truncate at base, acuminate at apex, entire at margin; lateral nerves ca 8 pairs; petioles ca 16 cm. Inflorescence ca 22 cm. Male flowers: Calyx 3 lobed; lobes elliptic-oblong, ca 2 x 1 mm, tomentose. Stamens ca 25; filaments ca 2 mm; anthers oblong. Female flowers: Calyx 4 lobed, ovate, ca 1.6 x 1.2 mm, tomentose without. Ovary ca 1.5 x 1 mm, tomentose, 3 locular; ovules 3; styles ca 1 mm, spreading, plumose. Capsules ca 7.5 mm across; seeds oblong, angular, ca 4 mm.

Common in evergreen forests, abundant at Kothayar.

Fl.: & Fr.: July – September.

Specimens examined: XCH 10509, 10579, 10674, 11058, 11817, 12837, 12967, 13634, 13960, 14804, 16137, 17340, 17341, 17570, 19857. MH 90642, 48135, 99482, 3137, 3138.

Distribution: India, Sri Lanka and Myanmar.

4. **Mallotus philippensis** (Lam.) Muell.Arg. (*Sphalm. philippines*, Corr. Merr.) in Linnaea 34: 196. 1865; Hook.f., Fl. Brit. India 5: 442. 1887; Gamble, Fl. Pres. Madras 2: 1322. 1925; Rani in Matthew, Fl. Tamil. Carnatic 3: 1456. t. 108k. 1983; Chandrab. in A.N. Henry *et al.,* Fl. Tamil Nadu 2: 234. 1987; Philcox in Dassan. & Clayton, Rev. Handb. Fl. Ceylon 11: 158. 1997; Matthew, Fl. Palni Hills 2: 1134. 1999; N.P. Balakr. & T.Chakrab., Fam. Euphorb. India 150. 2007. *Croton philippensis* Lam., Encycl. 2: 206. 1786.

Trees to 3 m, much branched; branchlets slender, rusty tomentose; bark greyish brown, grained. Leaves alternate to sub-opposite, ovate, ovate-lanceolate, obovate-lanceolate or ovate-oblong, 7.5-17.5 x 3.5-7.3 cm, red-glandular, stellate pubescent to subglabrous beneath, glabrous above, pale green or yellow or brownish green on drying, 3 nerved at base, acute or obtuse or rounded at base, acute to cuspidate-acuminate at apex, entire or sinuate toothed at margin; lateral nerves 3-5 pairs, distantly arranged; transverse nervules prominent beneath; petiole pairs unequal, 2-5 cm, slender, rusty pubescent; stipules lanceolate, to 2 mm. Inflorescence upto 16 cm, axillary or sub terminal; rachis rusty pubescent. Male flowers: Pedicels to 1 mm. Calyx 4-5 lobed; lobes lanceolate or ovate, 2.1-3.2 x 0.8-1.4 mm, subglabrous, glandular. Stamens ca 20; filaments free, unequal, 1-1.15 mm; anthers parallel, oblong, ca 0.6 mm. Female flowers: Solitary, subsessile. Calyx 5 lobed; lobes lanceolate or elliptic-lanceolate, ca 1.5 x 0.45 mm, puberulous. Ovary globose, ca 1.5 x 1.8 mm, pubescent, glandular; styles 3-partite, plumose, each to 1.2 mm. Capsules globose, ca 4.8 mm across, brick-crimson red; seeds elliptic-globose, ca 0.52 mm, dark brown.

Common in exposed rocky places in dry and moist deciduous forests, between 50-400 m.

Fl.: July - September. *Fr.:* Throughout the year.

Specimens examined: XCH 10664, 10675, 10798, 10888, 11097, 11227, 11375, 11401, 11940, 12019, 13767, 13799, 13804, 13955, 14035, 14112, 14511, 14743, 16147, 16640, 16641, 16812, 16908, 17413. MH 3310, 91735.

Distribution: India, Sri Lanka, Taiwan, South-east Asia, Malaysia, Australia and Pacific Islands.

5. **Mallotus repandus** (Rottl. ex Willd.) Muell.Arg. in Linnaea 34: 197. 1865; Hook.f., Fl. Brit. India 5: 442. 1887; Gamble, Fl. Pres. Madras 2: 1321. 1925; Chandrab. in A.N. Henry *et al.,* Fl. Tamil Nadu 2: 234. 1987; Matthew, Fl. Palni Hills 2: 1134. 1999; Philcox in Dassan. & Clayton, Rev. Handb. Fl. Ceylon 11: 165. 1997; N.P. Balakr. & T.Chakrab., Fam. Euphorb. India 150. 2007. *Croton repandus* Rottl. ex Willd., Ges. Naturf. Fr. Neue Schr. 11: 206. 1803. **(Fig. 183).**

Straggling shrubs to 3 m; branches yellowish-tomentose, stellate pubescent when young. Leaves alternate or opposite, broadly deltoid-ovate, 3.1-7.6 x 2.2-4.2 cm, finely stellate tomentose beneath, glabrous above, occasionally with minute, stellate scales, 3 nerved at base, truncate or subcordate at base, acuminate at apex, subentire to obscurely sinuate-dentate at margin; lateral nerves above 3-4 pairs, prominent beneath; petioles 1.5-4 cm, stellate-pubescent. Male inflorescence to 10 cm. Male flowers: Pedicels to 3 mm, pubescent. Calyx 5 lobed; lobes elliptic-ovate, ca 3.4 x 1.4 mm, pubescent, glandular; filaments ca 2 mm; anthers oblong; connective inconspicuous. Female inflorescence to 6 cm. Female flowers: Pedicels to 0.5 mm, densely pubescent. Calyx 6 lobed; lobes oblong-ovate, ca 1.5 x 0.3 mm, pubescent with stellate-hairs, glandular, appressed to ovary at base. Ovary subglobose, ca 2 x 2 mm, densely pubescent, glandular, 2 celled, 2 locular; ovules 2; styles 2, spreading. Capsules globose, ca 1 x 1 cm, densely pubescent; seeds subglobose, ca 6 x 5 mm, smooth.

Rare, in open deciduous to moist deciduous forests, between 200-400 m.

Fl.: & Fr.: March - June.

Specimens examined: XCH 15108 (Papanasam), 12294 (Manjolai), MH 15834.

Distribution: India, Sri Lanka, Myanmar, Thailand; Malaysia, China, Hong-kong, New Caledonia, New Guinea and North Australia.

6. **Mallotus resinosus** (Blanco) Merr., Sp. Blanco. 222. 1918; N.P. Balakr. & T.Chakrab. in Rheedea 1: 36-39. 1991 et Fam. Euphorb. India 150. 2007. *Adelia resinosa* Blanco, Fl. Filip. ed. 2. 562. 1845. *Claoxylon muricatus* Wight, Ic. Pl. Ind. or. 5, 2: 24. t. 1866. 1852. *Mallotus muricatus* (Wight) Muell.Arg. in Linnaea 34: 191. 1865; Hook.f., Fl. Brit. India 5: 436. 1887; Gamble. Fl. Pres. Madras 2: 1322. 1925; Chandrab. in A.N. Henry *et al.*, Fl. Tamil Nadu 2: 234. 1987. *M. stenanthus* Muell.Arg. in Linnaea 34: 191. 1865; Hook.f., l.c. 437; Gamble, Fl. Pres. Madras 2: 1322. 1925; Chandrab. in A.N. Henry *et al.,* Fl. Tamil Nadu 2: 234. 1987; Matthew, Fl. Palni Hills 2: 1135. 1999. *M. walkerae* Hook.f., l.c. 437. *M. andamanicus* Hook.f., l.c. 439. *M. intermedius* (Baill.) N.P. Balakr. in Bull. Bot. Surv. India 10: 245. 1969. *M. subramanyamii* J.L. Ellis in Bull. Bot. Surv. India 25: 199. 1985. **(Fig. 184).**

Shrubs or small trees to 3 m; branches angular, at ends flattened, pubescent when young, soon glabrous. Leaves opposite, ovate-lanceolate or rhombic-ovate, 5.5-14 x 2-4.5 cm, glabrous above, stellate-puberulous or glabrous, minutely yellow glandular beneath, acute to cuneate at base, acuminate or caudate-acuminate at apex to 2 cm, sinuate-dentate at margin; lateral nerves 6-10 pairs; two black granular glands usually present at the base of lamina; petioles 0.4-1.5 cm, often pubescent, yellow-glandular, pulvinate at both ends. Male inflorescence 2-3 cm, often finely puberulous. Male flowers: Pedicels to 1 mm. Calyx 4 lobed; lobes oblong-ovate, ca 4 x 1.2 mm, densely pubescent-glandular without. Filaments 1-2 mm; anthers ca 0.5 mm. Female inflorescence to 3 cm, puberulous-glandular. Female flowers: Pedicels to 0.75 mm, pubescent. Calyx 4 lobed; lobes elliptic-ovate, ca 2 x 1 mm, pubescent without. Ovary 1x 0.75 mm, closely, densely echinate; styles ca 1 mm, coarsely plumose. Capsules ca 5 mm across, yellow-glandular, with soft, deciduous spines.

Occasional, in moist deciduous to semi evergreen forests, between 300-800 m.

Fl.: & Fr.: March – June.

Specimens examined: XCH 10052, 10114, 11829, 12489, 13609, 13656, 14938, 14956, 16061.

Distribution: Peninsular India and Sri Lanka.

7. **Mallotus rhamnifolius** (Rottler ex Willd.) Muell.Arg. in Linnaea 34: 196. 1865; Hook.f., Fl. Brit. India 5: 440.1887; Gamble, Fl. Pres. Madras 2: 1322.1925; Chandrab. in A.N. Henry *et al.*, Fl. Tamil Nadu 2: 234. 1987; Philcox in Dassan. & Clayton, Rev. Handb. Fl. Ceylon 11: 161. 1997; N.P. Balakr. & T.Chakrab., Fam. Euphorb. India 151. 2007. *Croton rhamnifolius* Rottler ex Willd., Ges. Naturf. fr. Neue schr. 4: 190. 1803. *Mallotus micranthus* Muell.Arg. in Linnaea 34: 191. 1865. *Rottlera oppositifolia* Thwaites, Enum. Pl. Zeyl. 273. 1864, excl. Syn., non Blume. *Mallotus zeylanicus* Muell.Arg. in Linnaea 34: 195. 1865. *M. rhamnifolius* var. *ovatifolia* Hook.f., Fl. Brit. India 5: 440. 1887. **(Fig. 185).**

Shrubs to 12 m. Leaves opposite towards end of branches, alternate or subopposite below, lanceolate to ovate-lanceolate, 5.8-11.5 x 1.8-4.3 cm, glabrous above, minutely stellate-pubescent beneath, 3 veined at base; rounded at base, acute to long-acuminate at apex, entire, or subentire or minutely denticular at margin; lateral nerves ca 6 pairs. Male inflcresence to 8 cm. Male flowers: Pedicels to 1.5 mm. Calyx 4 lobed; lobes elliptic-lanceolate, ca 3 x 2 mm, pubescent without. Filaments 1-2.5 mm; anthers oblong; connectives broadly apiculate. Female infloresence to 8 cm. Female flowers: Pedicels to 1 mm, pubescent. Calyx 3-4 lobed; lobes lanceolate or oblong, ca 2.5 x 2 mm, pubescent without. Ovary ca 2 mm across, yellow-glandular; styles 3, papillate, spreading. Capsules ca 5 x 5 mm, minutely pubescent; seeds oblong, to 4 mm, smooth.

Common in dry and moist deciduous forests, mostly bordering the forest areas.

Fl.: & Fr.: July onwards.

Specimens examined: XCH 10678, 10714, 10773, 10901, 11179, 12271, 12372, 12784, 13101, 13175, 13028, 13030, 13678, 14178, 14410, 16543, 16544, 16887, 16977, 17699, 18848, 18849, 19226, 19557.

Distribution: Peninsular India and Sri Lanka.

29. MARGARITARIA L.f.

A genus of 14 species occurring in tropical America, Africa, Asia and Australia (Webster, 1994). 1 species in India.

Margaritaria indica (Dalz.) Airy Shaw in Kew Bull. 20: 387. 1966; Chandrab. in A.N. Henry *et al.*, Fl. Tamil Nadu 2: 235. 1987; N.P. Balakr. & T.Chakrab., Fam. Euphorb. India 358. 2007; T.Chakrab., M.Gangop. & N.P. Balakr. in N.P. Balakr. *et al.* Fl. India 23: 436. 2012. *Prosorus indicus* Dalz. in Hooker, J. Bot. Kew. Gard. Misc. 4: 346. 1852; Gamble, Fl. Pres. Madras 2: 1294. 1925. *Phyllanthus indicus* (Dalz.) Muell.Arg. in Linnaea 32: 52. 1863; Hook.f., Fl. Brit. India 5: 305. 1887. **(Fig. 186).**

Dioecious, glabrous, deciduous trees to 12 m; branchlets terete, prominently white-spotted. Leaves distichous, elliptic or elliptic-obovate, 5-9.5 x 2.5-4 cm, chartaceous to coriaceous, dull glaucous beneath, acute at base, shortly acuminate at apex, mucronate, entire at margin, penninerved; lateral nerves ca 9 pairs, arcuate, raised; petioles to 7 mm. Inflorescence axillary, pedunculate. Male and Female flowers not seen. Fruits capsular, depressed globose, ca 1 cm across, with persistent calyx; calyx 4 lobed; stalks to 7 mm; seeds 6, 2 per locule, subtrigonous, ca 7 x 3 mm, bluish.

Occasional, in deciduous forests at ca 400 m.

Fl.: & Fr.: Throughout the year.

Specimens examined: XCH 11163, 11272, 16662, 17771 (Courtallam). MH 38853.

Distribution: Peninsular India and Sri Lanka.

30. MEINECKIA Baill.

A genus of 20 species with a disjunct distribution in the New World and Old World, from Mexico to Columbia and Brazil, and from central Africa to Madagascar, Peninsular India, Sri Lanka and Assam (Webster, 1994); 4 species in India.

Meineckia parvifolia (Wight) Webster in Acta Bot. Neerl. 14: 342., ff. 5,11,19. 1965; Chandrab. in A.N. Henry *et al.*, 2: 235. 1987; N.P. Balakr. & T.Chakrab., Fam. Euphorb. India 360. 2007; T.Chakrab., M.Gangop. & N.P. Balakr. in N.P. Balakr. *et al.* Fl. India 23: 439. 2012. *Peltandra parvifolia* Wight, Ic. Pl. Ind. Or. 5: t. 1892. 1852. *Phyllanthus peltandrus* Muell.Arg. in Linnaea 32: 11.1863 & DC., Prodr. 15(2): 341. 1866. *Phyllanthus thwaitesianus* Muell.Arg. in Linnaea 32: 11. 1863; Hook.f., Fl. Brit. India 5: 287. 1887. *Phyllanthus suberosus* Muell.Arg., in Linnaea 32: 11. 1863; Hook.f. Fl. Brit. India 5: 289. 1887. *Neopeltandra suberosa* (Muell.Arg.) Gamble, Fl. Pres. Madras 2: 1286. 1925; *Neopeltandra parvifolia* (Wight) Alston in Trimen., Handb. Fl. Ceylon 6: 256. 1931. **(Fig. 187).**

Monoecious, glabrous herbs to 60 cm; branchlets subterete, striate, smooth, reddish on drying, paler when young, thick. Leaves simple, alternate, elliptic to elliptic ovate, 3-4 x 1.5-3.5 cm, hyaline, glaucous beneath, subglaucous above, pale greenish on drying, acute at both ends, apiculate at apex, entire at margin, penninerved; lateral nerves 5-6 pairs, faint above, obscure beneath; petioles filiform, to 3 mm; stipules lanceolate-acuminate, to 1.6 x 0.4 mm, finely denticulate at margin. Inflorescence axillary cymules, bisexual. Male flowers: Clustered. Pedicels filiform, to 5 mm. Calyx 5 lobed; lobes ovate-oblong to 1 x 0.65 mm, obscurely 1 nerved; disc orbicular, finely 5 lobed; stamens 5, ca 9 mm; filaments united to 0.5 mm at base, above free to 0.1 mm; anthers divaricate, to 0.2 x 0.25 mm. Female flowers: Solitary. Pedicels filiform, to 1.5 cm, basally clothed with bracts. Bracts ovate-acute, to 1 x 0.6 mm, denticulate. Calyx 5 lobed; lobes broadly ovate, ca 1.25 x 0.75, thin, obscurely 1 nerved; disc orbicular, finely lobed. Ovary globose, ca 1 mm across, 3 locular; ovules 6, 2 per locule, pendulous; styles 3, to 0.25 mm, each bipartite, erect, persistent; stigmas capitate. Fruits capsular, obscurely depressed globose, ca 2 mm across, 3 lobed, 6 valved; stalks filiform, similar to flowering pedicels; seeds reniform, ca 1 x 0.5 mm, obscurely transversely striate, minutely pitted, brown.

In the woodlands of foot hills.

Fl.: & Fr.: November - January.

Specimens examined: XCH 14154 (Papanasam). MH 3174.

Distribution: Peninsular and Sri Lanka. **Endangered.**

Note: After Barber's Collection, (15.06.1901) there is no other collection from Tirunelveli hills.

31. MICROSTACHYS Spreng.

A genus of ca 100 species, mainly neotropical but with outliers in temperate North America, Asia and Australia (Webster, 1994); 1 in India.

Microstachys chamaelea (L.) Muell.Arg. in Linaea 32: 95. 1863; N.P. Balakr. & T.Chakrab., Fam. Euphorb. India 393. 2007.; T.Chakrab. & N.P. Balakr. in N.P. Balakr. *et al.* Fl. India 23: 348. 2012. *Tragia chamaelea* L., Sp. Pl. 1391. 1753. *Sebastiania chamaelea* (L.) Muell.Arg. in DC., Prodr. 15(2): 1175. 1866; Hook.f., Fl. Brit. India 5: 475. 1888; Gamble, Fl. Pres. Madras 2: 1344. 1925; Rani in Matthew, Fl. Tamil. Carnatic 3: 1474. t.109 q.1983; Chandrab. in A.N. Henry *et al.,* Fl. Tamil Nadu 2: 240. 1987; Philcox in Dassan. & Clayton, Rev. Handb. Fl. Ceylon 11: 190. 1997; Matthew, Fl. Palni Hills 2: 1143. 1999. **(Fig. 188).**

Monoecious, glabrous, erect or ascending herbs to 50 cm; stems many, slender, terete, dichotomously branched. Leaves alternate, linear-lanceolate, 4-8 x 0.21-0.8 cm, glabrous above, minutely sericeous to glabrous beneath, acute at base, obtuse-apiculate at apex, finely serrate at margin, penninerved; lateral nerves ca 30 pairs, arcuate; petioles ca 3 mm; stipules lanceolate, to 0.6 mm, sericeous. Inflorescence axillary, spicate, to 2 cm, leaf opposed, bracteate. Bracts lanceolate, to 1 x 0.15 mm, minutely fimbriate, with a large, stipitate gland on each side, 1-2 flowered. Male flowers: Minute. Calyx 3 lobed; lobes ovate-oblong, to 7 x 0.4 mm, cilate; stamens 3, alternate with sepals, exserted; filaments free, connate at base to 0.38 mm; anther cells oblong, to 0.24 mm, basifixed, longitudinally dehiscent. Female flowers: 1-2 at the base of inflorescence or axillary, solitary. Calyx 3 lobed; lobes ovate-lanceolate, to 1.5 x 0.7 mm, acute at apex, ciliate. Ovary globose, ca 2 mm across, 3 lobed; each lobe echinate in rows, 3 locular; ovules 3, pendulous; styles 3, each to 0.5 mm, spreading. Fruit capsular, dull yellow, depressed globose, ca 5.5 x 6.25 mm, 3 lobed, each with 3 vertical rows of soft prickles.

Common in low altitude grasslands.

Fl.: & Fr.: Throughout the year.

Specimens examined: XCH 10830, 11035, 11539, 12694, 13349, 14023, 15172, 15246, 16835, 17859, 19323, 19739. MH 15249.

Distribution: India, Sri Lanka, Myanmar, Malacca, Singapore, West Africa; the Malaysian islands, New Guinea and Tropical and subtropical Australia.

32. PARACROTON Miq.

A tropical Asian genus of 4 species, found from Southern India and Sri Lanka to the Philippines and Borneo (Webster, 1994). 2 in India.

Paracroton pendulus (Hassk.) Miq. subsp. **zeylanicus** (Thwaites) N.P. Balakr. & T.Chakrab. in Kew Bull. 48: 719. f.2. 1993 *et.* Fam. Euphorb. India. 218. 2007. *Desmostemon zeylanicus* Thwaites, Enum. Pl. Zeyl. 278. 1861.

Ostodes zeylanicus (Thwaites) Muell.Arg. in Linnaea 34: 213. 1865; Hook.f., Fl. Brit. India 5: 400. 1887; Gamble, Fl. Pres. Madras 2: 1336. 1925. *Fahrenheitia zeylanica* (Thwaites) Airy shaw in Kew Bull. 20: 410.1967; Chandrab. in A.N. Henry *et al.*, Fl. Tamil Nadu 2: 230. 1981; Philcox in Dassan. & Clayton, Rev. Handb. Fl. Ceylon 11: 113. 1997; T.Chakrab. & N.P. Balakr. in N.P. Balakr. *et al.*, Fl. India 23: 245. 2012. **(Fig. 189).**

Dioecious trees to 10 m; young shoots densely minutely reddish-lepidote; branchlets robust, brownish, angled or striate, closely granular, puberulous to glabrous. Leaves alternate, elliptic-oblong or obovate, 13-36 x 4-12 cm, chartaceous to coriaceous, densely to sparsely lepidote, often glabrous except midrib above, sparsely lepidote to glabrous beneath, reddish-brown above on drying, pale green to brown beneath, acute or cuneate at base, cuspidate-acuminate (acumen to 2 cm) or subacuminate at apex, margin glandular-dentate, penninerved; lateral nerves 10-15 pairs, raised above; transverse nervules prominent; two stipitate glands present at base; petioles 4-9 cm, semiterete, pulvinate at apex, lepidote-pubsecent. Inflorescence axillary, racemose to 24(-28) cm, densely fulvous-tomentose, often black granular. Male flowers: Not seen. Female flowers: Solitary or few together. Pedicels to 4 mm, tomentose. Calyx 5 lobed; lobes ovate-oblong, ca 7.5 x 3.75 mm, dorsally hard-appendiculate, concave, often punctate, membranous, ciliate at margin, persistent. Petals 0. Ovary globose, ca 3 x 3 mm, densely hispid, 3 locular; ovules 3; styles 3, ca 2 mm, bifid, ribbed, spreading. Capsules globose, ca 1.4 x 1.4 cm, unlobed or scarcely 3 lobed, hard, densely appressed rusty tomentellous; stalks to 1 cm.

Rare, in semi-evergreen forests around 1000 m.

Fl.: & Fr.: July onwards.

Specimens examined: XCH 16733 (Papanasam). MH 90554.

Distribution: Peninsular India and Sri Lanka.

33. PHYLLANTHUS L.

A large and very diverse genus of 750-800 species, about 200 of which are in America, 100 in Africa, 70 in Madagascar and the remaining in Asia and Australasia (Webster, 1994); 53 species in India.

Note: The generic delimitation of *Phyllanthus* complex has been a matter of dispute. The genus was first established by Linnaeus (1753). Several allied genera like *Kirganelia* A. Juss., *Emblica* Gaertner, *Reidia* Wight, were added in to the *Phyllanthus* complex.

Key to the species

1a. Fruits berries or drupaceous, soft.......... 2

1b. Fruits capsular, hard.......... 3

2a. Fruits >1 cm across, green in colour .... 4. **P. emblica**
2b. Fruits <1 cm across, purple or bluish black in colour .... 16. **P. reticulatus**

3a. Mature capsules green 4
3b. Mature capsules brick red .... 20. **P. singampattianus**

4a. Seeds hirtelllous 5
4b. Seeds glabrous (rarely puberulous) 10

5a. Leaves falcate, prominently unequal at base 8
5b. Leaves not falcate, obscurely unequal at base 6

6a. Disc of male flowers 2-lobed .... 2. **P. baillonianus**
6b. Disc of male flowers 4-lobed 7

7a. Bracts of male flowers tufted .... 6. **P. gageanus**
7b. Bracts of male flowers not tufted .... 11. **P. macrocalyx**

8a. Glabrous undershrubs 9
8b. Tomentose undershrubs .... 14. **P. pinnatus**

9a. Leaves 1-2 cm long .... 7. **P. heyneanus**
9b. Leaves 3-4 cm long .... 5. **P. fimbriatus**

10a. Anthers longitudinally dehiscent 11
10b. Anthers transversely dehiscent 16

11a. Branches closely grained .... 15. **P. polyphyllus**
11b. Branches sparsely or not grained 12

12a. Capsules verrucose .... 22. **P. urinaria**
12b. Capsules smooth 13

13a. Stipules many .... 8. **P. lawii**
13b. Stipules 1 or 2 14

14a. Leaves membranous .... 9. **P. leschenaultii**
14b. Leaves chartaceous 15

15a. Lateral veins present on leaves .... 17. **P. rheedei**
15b. Lateral veins absent on leaves .... 12. **P. maderaspatensis**

16a. Anthers reniform 17
16b. Anthers subglobose or globose 19

17a. Seeds glabrous .... 1. **P. amarus**
17b. Seeds puberulous 18

18a. Styles papillate at apex .... 3. **P. debilis**
18b. Styles not as above .... 18. **P. rotundifolius**

19a. Branches arising from stout rootstock, reddish .... 13. **P. narayanswamii**
19b. Branches not arising from stout rootstock, not reddish 20

20a. Stamens five .... 21. **P. tenellus**
20b. Stamens three 21

21a. Leaves coriaceous, elliptic-ovate, margin glabrous .... 10. **P. macraei**
21b. Leaves chartaceous linear-oblong, margin ciliate .... 19. **P. simplex**

1. **Phyllanthus amarus** Schum. & Thonn., Kongel. Danske Vidensk.-selsk. Skr. 4: 195. 1829; Rani in Matthew, Fl. Tamil. Carnatic 3: 1463. t. 108 m. 1983; Chandrab. in A.N. Henry *et al.,* Fl. Tamil Nadu 2: 236. 1987; Philcox in Dassan. & Clayton, Rev. Handb. Fl. Ceylon 11: 226. 1997; Matthew, Fl. Palni Hills 2: 1137. 1999; N.P. Balakr. & T.Chakrab., Fam. Euphorb. India 386. 2007; T.Chakrab., M.Gangop. & N.P. Balakr. in N.P. Balakr. *et al.* Fl. India 23: 482. 2012. *P. niruri* auct. non L. (1753); Wight, Ic. t. 1894. 1852; Hook.f., Fl. Brit. India 5: 298. 1887; Gamble, Fl. Pres. Madras 2: 1290. 1925. **(Fig. 190).**

Glabrous herbs to 30 cm, often branched; stem terete. Leaves ca 1 x 0.45 cm, oblong, glaucous beneath, obtuse at base and apex, minutely apiculate, entire at margin; lateral nerves ca 4 pairs, faint. Male flowers: Pedicles to 1 mm, filiform. Sepals 5; obovate, ca 0.75 x 0.5 mm, apiculate, greenish-white; disc glands 5, globose. Stamens 3; filaments connate in a column to 0.48 mm; anthers sessile, transversely dehiscent. Female flowers: Solitary. Pedicels to 1.5 mm, filiform. Sepals 5; oblanceolate, ca 1x 0.5 mm, broadly white margined, acute at apex; disc annular, 5 lobed. Ovary ca 1 x 1 mm; styles 3, bifid at apex; arms recurved, short, spreading. Capsules ca 2 x 2 mm; seeds 6, trigonous, ca 0.85 x 0.35 mm.

Very common in plains and hills upto 900 m.

Fl.: & Fr.: Throughout the year.

Specimens examined: XCH 14025, 16583 (Papanasam).

Distribution: Apparently of American origin, but now pantropical.

2. **Phyllanthus baillonianus** Muell.Arg. in Linnaea 32: 47. 1863; Hook.f., F1. Brit. India 5: 300. 1887; Chandrab. in A.N. Henry *et al*., Fl. Tamil Nadu 2: 236. 1987; Philcox in Dassan. & Clayton, Rev. Handb. Fl. Ceylon 11: 229. 1997; N.P. Balakr. & T.Chakrab., Fam. Euphorb. India 371. 2007; T.Chakrab., M.Gangop. & N.P. Balakr. in N.P. Balakr. *et al.* Fl. India 23: 448. 2012. *Reidia latifolia* Wight, Ic. Pl. Ind. Or. 5: 28 t. 1904-2. 1852, non *Phyllanthus latifolius* Sw., 1800. *Epistylium latifolium* (Wight) Thwaites, Enum. Pl. Zeyl. 283.1861. *Reidia bailloniana* (Muell.Arg.) Gamble, Fl. Pres. Madras 2: 1293. 1925. *R. cordifolia* (Baill.) Alston in Trimen, Rev. Handb. Fl. Ceylon 6: 258. 1931. **(Fig. 191).**

Glabrous subshrubs to 60 cm; leaves on main stem reduced to scales; branchlets slender, terete, ca 30 cm long, deciduous, often leafless at ends. Leaves ovate-lanceolate, 2.2-4.5 x 1.3-1.7 cm, membranous to subcoriaceous, obtuse or subcordate at base, acute at apex, apiculate, entire at margin; lateral nerves ca 6 pairs, finely reticulate beneath; petioles filiform, to 2 mm; stipules deltoid-lanceolate, to 3.8 mm. Male flowers: Mostly solitary, in proximal axils. Pedicels to 7 mm, filiform. Sepals 4; orbicular or broadly deltoid-ovate, ca 2.8 x 2.2 mm, obscurely lacerate at margin, reddish; disc 2 or 4 glandular, ca 1.4 mm across, tuberculate. Stamens 2; filaments connate into short column to 0.5

mm; anthers with connectives transversely enlarged, ca 1.4 mm across; anthersacs widely separated. Female flowers: Pedicels to 1.8 cm, pulvinate at apex. Sepals 6; subequal, ovate, to 3 x 1.5-2 mm, entire at margin, enlarged in fruit to 4 mm long; disc large, flat, 6 lobed; lobes tuberculate. Ovary globose, ca 1.4 mm across; styles 3, bifurcate, ca 1.2 mm, spreading. Capsules ca 1.8 mm across; seeds trigonous, ca 5 mm long, appressed hairy.

Occasional, in evergreen forests between 600-900 m.

Fl.: & Fr.: February - April.

Specimens examined: XCH 14872, 16681, 19118. MH 398, 12451, 3638, 4489, 16309, 29200, 90137. MH 398, 12451, 3638, 4489, 16309, 29200, 90137.

Distribution: Peninsular India and Sri Lanka.

Note: Hooker stated that "the disc is very remarkable and this is certainly not in the Travancore plant figured by Wight (Ic. t. 1904.f) as *Reidia latifolia*". But we got the plant with remarkable disc characters. In Southern Floras, *P. baillonianus* is often confused with *P. macrocalyx*. These two are differentiated by the disc character of female plants viz., *P. baillonianus* has 6 lobed disc, whereas *P. macrocalyx* has saucer shaped, petaloid, sinuate disc, as mentioned by Hooker.

3. **Phyllanthus debilis** Klein ex Willd., Sp. Pl. 4: 582. 1804; Webster in J. Arnold Arb. 38: 307. t. 19. f. C & D. 1957, excl. syn. *P. debilis* Ham. ex Hook.f., Fl. Brit. India 5: 299. 1887; Philcox in Dassan. & Clayton, Rev. Handb. Fl. Ceylon 11: 224. 1997. non. *P. debilis* sensu Gamble, Fl. Pres. Madras 2: 1290. 1925; Rani in Matthew, Fl. Tamil. Carnatic 3: 1446. t. 1066. 1983; Chandrab. in A.N. Henry *et al.*, Fl. Tamil Nadu 2: 236. 1987; N.P. Balakr. & T.Chakrab., Fam. Euphorb. India 386. 2007; T.Chakrab., M.Gangop. & N.P. Balakr. in N.P. Balakr. *et al.* Fl. India 23: 484. 2012. *P. niruri* var. *javanicus* Muell. Arg. in Linnaea 32: 43 1863 *et* in DC., Prodr. 15(2): 407. 1866. *P. niruri* var. *debilis* (Klein ex Willd.) Muell.Arg. *ibid. P. mukerjeeanus* Mitra & Bennet in Bull. Bot. Soc. Bengal 19: 145. f.1. t.1. 1966. **(Fig. 192).**

Glabrous herbs, to 50 cm; stems prominently striate, sometimes much branched. Leaves linear-elliptic-lanceolate, 5-10 x 2-5 mm, glaucous beneath, dark green above when dry, acute at base, acute or subacute or obtuse at apex, entire at margin; petioles to 1 mm, easily detached; stipules lanceolate, acuminate, ca 1.1 x 0.4 mm. Male flowers: At proximal ends, bracteate. Bracts narrowly lanceolate-acuminate, ca 0.95 x 0.7 mm. Pedicels to 0.6 mm. Sepals 6, biseriate, obovate, or oblanceolate, ca 0.6 x 0.4 mm, broadly white margined, apex sometimes slightly cucullate, hyaline; disc 5 glandular, suborbicular, 0.15 x 0.1 mm. Filaments united in a column to 0.2 mm; anthers transversely dehiscent. Female flowers: At distal ends. Pedicels to 0.7 mm. Sepals 6, biseriate, oblong, ca 0.5 x 0.12 mm, enlarged in fruits; disc saucer shaped, shallowly 6 lobed. Ovary ca 0.3 mm across; styles 3, forked at apex; arms recurved, minutely papillate, spreading. Capsules ca 2 mm across; seeds trigonous, ca 1.25 x 0.5 mm, with 5-7 longitudinal ribs, transversely striate, sometimes sparsely puberulous.

Occasional near cultivated lands.

Fl.: & Fr.: Throughout the year.

Specimens examined: XCH 16193, 17666, 18123, 18528.

Distribution: India, Sri Lanka, introduced into Indonesia, the Pacific islands and the West Indies.

4. **Phyllanthus emblica** L., Sp. Pl. 982. 1753; Hook.f., Fl. Brit. India 5: 289. 1887; Rani in Matthew, Fl. Tamil. Carnatic 3: 1466. 1983; Chandrab. in A.N. Henry *et al.*, Fl. Tamil Nadu 2: 236. 1987; Philcox in Dassan. & Clayton, Rev. Handb. Fl. Ceylon 11: 219. 1997; Matthew, Fl. Palni Hills 2: 1138. 1999; N.P. Balakr. & T.Chakrab., Fam. Euphorb. India 369. 2007; T.Chakrab., M.Gangop. & N.P. Balakr. in N.P. Balakr. *et al.* Fl. India 23: 446. 2012. *Emblica officinalis* Gaertner, Fruct. 2: 122. 1791; Wight, Ic. t. 1896. 1852; Gamble, Fl. Pres. Madras 2: 1295. 1925. **(Fig. 193).**

Tamil: Nelli, Perunelli.

Malayalam: Nelli, Nellikamaram, Amalakam.

Glabrous, woody trees to 10 m, deciduous, with flaking bark; branchlets slender, glabrous to finely pubescent. Leaves sessile, linear-oblong, 1-18 x 0.3 cm, truncate to subcordate at base, apex apiculate, entire at margin; lateral nerves ca 7 pairs, raised beneath. Flowers in fascicles along branchlets, at ends leafy twigs, usually male and female flowers mixed. Male flowers: Pedicels to 2.5 mm, capillary. Sepals 6, oblong-obovate or oblanceolate, ca 1.5 x 0.7 mm, margin minutely puberulous. Filaments to 0.7 mm; anthers oblong; column to 1.3 mm. Female flowers: Sessile. Sepals 6, oblanceolate or spathulate, subequal, ca 2.2-3 x 1-1.5 mm, cucullate at apex, minutely puberulous at margin; disc cup-like, investing the ovary, unequally lacerate, ca 1.5 mm, puberulous without. Ovary subglobose, ca 1.5 mm; styles basally connate to, 1.8 mm, above 3-fid, each arm twice bifid, ca 2 mm; arms spreading, reflexed. Capsules depressed globose, ca 1.5 cm across, fleshy, juicy, greenish, sessile; seeds trigonous, ca 0.7 x 0.3 mm, smooth, brown.

Common in the low altitude grasslands.

Fl.: & Fr.: Throughout the year.

Specimens examined: XCH 12276, 17549, 18163, 19039.

Distribution: India and Sri Lanka to eastern Asia; widely cultivated.

5. **Phyllanthus fimbriatus** (Wight) Muell.Arg. in Linnaea 32: 47. 1863; Hook.f., Fl. Brit. India 5: 302. 1887; N.P. Balakr. & T.Chakrab., Fam. Euphorb. India 372. 2007; T.Chakrab., M.Gangop. & N.P. Balakr. in N.P. Balakr. *et al.* Fl. India 23: 451. 2012. *Reidia fimbriata* Wight, Ic. Pl. Ind. or. 5(2): 27, t. 1994. f.1. 1852; Gamble, F1. Pres. Madras 2: 1292. 1925. **(Fig. 194).**

Glabrous shrubs to 60 cm; branchlets terete, slender. Leaves elliptic, unequal, 3-4 x 1.3 cm, glaucous beneath, shining, prominently unequal at base (rounded on one side, cuneate on other side), acute at apex, apiculate, entire at margin; lateral nerves ca 6 pairs, prominently reticulate beneath; petioles to 1 mm; stipules subulate, to 4 mm. Male flowers: Pedicels to 6 mm. Sepals 4(-6), biseriate, deltoid-ovate, ca 3.5 x 1.5 mm, distal half dentate at margin; disc 4 glandular; glands ca 0.5 x 0.75 mm; filaments connate to 0.25 mm; anthers ca 0.5 mm across. Female flowers and fruits not seen.

Around 700 m.

Fl.: & Fr.: January - March.

Specimen examined: XCH 18271 (Courtallam).

Distribution: Sri Lanka, and Peninsular India. **Critically endangered**

6. **Phyllanthus gageanus** (Gamble) Mohanan in J. Econ. Taxon. Bot. 6: 480. 1985; Chandrab. in A.N. Henry *et al.*, Fl. Tamil Nadu 2: 237. 1987; N.P. Balakr. & T.Chakrab., Fam. Euphorb. India 372. 2007; T.Chakrab., M.Gangop. & N.P. Balakr. in N.P. Balakr. *et al.* Fl. India 23: 453. 2012. *Reidia gageana* Gamble in Kew Bull. 1925: 331. 1925. et Fl. Pres. Madras 2: 1292. 1925. **(Fig. 195).**

Glabrous undershrubs to 60 cm; branchlets slender, spreading. Leaves ovate-lanceolate, 2.8- 4 x 1-1.5 cm, glaucous beneath, shining, cuneate to acute at base, unequal, acute at apex, sometimes minutely apiculate, entire at margin; lateral nerves ca 5 pairs, obscure beneath; petioles to 1 mm long, filiform; stipules subulate, to 3 mm. Male flowers: In axillary glomerules, rarely solitary, arising from tuft of bracts. Pedicels to 5 mm, filiform. Sepals 4, deltoid-lanceolate, ca 3.5 x 1.8 mm, lacerate at margin, minutely pellucid-punctate; disc 4 glandular, broadly truncate, ca 1.2 x 1.5 mm. Stamens 2; filaments connate in a column to 0.5 mm; anthers ca 1.5 mm across. Female flowers: At the end of branchlets. Pedicels to 1.5 cm, pulvinate at apex; sepals 6, lanceolate, ca 4.5 x 2.2 mm, acute at apex, cucullate; disc saucer shaped, flattened at margin, wavy, tuberculate. Ovary globose, ca 1.3 mm across; styles 3, in upper half bifid to ca 1.8 mm; arms reflexed. Capsules globose, ca 4.5 mm across; seeds trigonous, ca 3 x 0.75 mm, appressed hairy.

In evergreen forests at ca 700 m.

Fl.: & Fr.: January - March.

Specimen examined: XCH 18081 (Courtallam).

Distribution: Peninsular India.

Note: This seems to be **endangered** in Tirunelveli hills. It has been collected after Gamble's (1925) collection from the type locality.

7. Phyllanthus heyneanus Muell.Arg. in Linnaea 32: 49. 1863 (non Muell. Arg. 1865); N.P. Balakr. & T.Chakrab., Fam. Euphorb. India 373. 2007; T.Chakrab., M.Gangop. & N.P. Balakr. in N.P. Balakr. *et al.* Fl. India 23: 455. 2012. *Reidia ovalifolia* Wight, Ic. Pl. Ind. or. 5(2): t. 1904. f.3. 1852 (non *P. ovalifolius* Forssk. 1775). *Phyllanthus nephradenius* Muell.Arg. in DC., Prodr. 15(2): 432. 1866. *P. longiflorus* B.Heyne in Wallich (Cat. no. 7905. 1847. *nom. nud.*) ex Hook.f., F1. Brit. India 5: 302. 1887. *Reidia longiflora* (B.Heyne ex Hook.f.) Gamble, Fl. Pres. Madras 2: 1293. 1925.

Glabrous undershrubs to 50 cm, rarely to 1 m; branchlets slender, spreading towards top, crowded at summit of stem, rough with many scars, 11-26 cm. Leaves sessile, obliquely ovate-oblong, 1-2 x 0.4-0.5 cm, somewhat falcate, glaucous beneath, pale brown or greenish brown on drying, unequally obtuse to acute at base, abruptly apiculate at apex, entire at margin; lateral nerves ca 6 pairs, faint; stipules minute, many, persistent. Male flowers: In axillary fascicles, at proximal axils. Pedicels to 3.5 mm, filiform. Sepals 4, broadly elliptic-orbicular, ca 2.1 x 1.4 mm, acute at apex, undulate at margin, white or yellow or pinkish white or yellowish pink in colour; disc 4 glandular; glands reniform, tuberculate. Stamens 4; filaments connate into a column to 1.2 mm; anthers extrorse. Female flowers: At distal axils, solitary. Pedicels to 1 cm, filiform, pulvinate at apex. Sepals 6, oblong-ovate, ca 3 x 1.5 mm, persistent; disc saucer shaped, thin, undulate at margin, often minutely tuberculate. Ovary globose, ca 1.5 mm across; styles 3, deeply bifid, erect; stigmas slightly capitate. Capsules depressed globose, ca 6 mm across, smooth, greenish; seeds trigonous, ca 3.5 x 2.5 mm, appressed hairy, occasionally glabrous.

Common in moist deciduous forests, abundant along the way sides from Courtallam to Thekkumalai estate, rare in other localities.

Fl.: & Fr.: March - July.

Specimens examined: XCH 11209, 13824, 13898, 16540, 17167, 17623, 18016. MH 3447, 13152, 27961, 2895.

Distribution: Peninsular India and Sri Lanka.

8. Phyllanthus lawii J.Graham, Cat. Pl. Bombay 181. 1839; Muell.Arg. in DC., Prodr. 15(2): 353. 1866; Hook.f., Fl. Brit. India 5: 290. 1887; Gamble, Fl. Pres. Madras 2: 1289. 1925; Chandrab. in A.N. Henry *et al.,* Fl. Tamil Nadu 2: 237. 1987; T.Chakrab., M.Gangop. & N.P. Balakr. in N.P. Balakr. *et al.* Fl. India 23: 465. 2012; *P. juniperinoides* Muell.Arg. in Linnaea 32: 18. 1863; N.P. Balakr. & T.Chakrab., Fam. Euphorb. India 378. 2007. *P. nemorum* Wallich ex Muell.Arg. in Linnaea 34: 70. 1865. **(Fig. 196).**

Glabrous, wiry shrubs ca 60 cm; branchlets terete, rough with stipule scars. Leaves elliptic or elliptic-oblong, ca 7 x 3 mm, obtuse at base, acute at apex, apiculate, entire at margin, penninerved; lateral nerves ca 5 pairs, obscure; stipules many, thorny, to 2 x 1 mm. Male flowers: Fascicled, bracteate. Bracts ca 1 x 0.6 mm, irregularly dentate at apex. Pedicels filiform, to 6 mm. Sepals 6, unequal, lanceolate-obovate, ca1 x 0.7 mm, curved at apex; disc 6 glandular; glands globose, ca 0.5 mm across; filaments basally connate to 0.6 mm, above free to 0.6 mm; anthers globose, longitudinally dehiscent. Female flowers and fruits not seen.

Rare, in rocky places, in evergreen forests, ca 1000 m.

Fl.: November onwards.

Specimen examined: XCH 12055 (Kothayar).

Distribution: Peninsular India.

9. Phyllanthus leschenaultii Muell.Arg. in Linnaea 32: 37. 1863 (excl. var. *tenellus*); Hook.f., Fl. Brit. India 5: 296. 1887; Chandrab. in A.N. Henry *et al.*, Fl. Tamil Nadu 2: 238. 1987; N.P. Balakr. & T.Chakrab., Fam Euphorb. India 307. 2007; T.Chakrab., M.Gangop. & N.P. Balakr. in N.P. Balakr. *et al.* Fl. India 23: 486. 2012. *P. missionis* Hook.f., l.c. 5: 297. 1887; Gamble, Fl. Pres. Madras 2: 1290. 1925. **(Fig. 197).**

Glabrous shrubs to 1 m; stems terete, grained, lenticellate reddish white. Leaves membranous, obovate or obovate-oblong, 1-1.7 x 0.5-1 cm, glaucous beneath, acute at base, obtuse or acute at apex, entire at margin; lateral nerves ca 6 pairs, arcuate, faint to subprominent beneath, faint above; petioles ca 2 mm long; stipules lanceolate-acuminate, ca 1 x 0.35 mm. Flowers: In axillary glomerules, bracteate. Bracts triangular-acuminate, to 0.55 mm. Male flowers: Pedicels to 4 mm, capillary. Sepals 6, elliptic-oblong, ca 1.5 x 1 mm; disc 6 glandular, surface of glands papillate. Filaments ¾ united, above free; anthers oblong, longitudinally dehiscent. Female flowers: Pedicels to 4.5 mm, pulvinate at apex. Sepals 6, subequal, obovate, 1.4-1.6 x 0.8 mm, apiculate; disc cushion-like, 6 lobed, margin crenulate. Ovary ca 1.5 x 1.5 mm; styles 3, deeply bifid; arms slightly recurved, spreading. Capsules ca 3 mm across, often white with granular testa; seeds trigonous, ca 1.5 x 0.4 mm, with vertical ribs and minutely transverse striae.

Occasional, on rocky slopes in dry deciduous to moist deciduous forests.

Fl.: & Fr.: Throughout the year.

Specimens examined: XCH 19853 (Courtallam). MH 12133, 3383, 3246.

Distribution: Peninsular India.

Note: Collected after Barber's (1915) collection.

10. Phyllanthus macraei Muell.Arg. in Linnaea 32: 29. 1863: Hook.f., F1. Brit. India 5: 296. 1887; N.P. Balakr. & T.Chakrab., Fam. Euphorb. India 378. 2007; T.Chakrab., M.Gangop. & N.P. Balakr. in N.P. Balakr. *et al.* Fl. India 23: 467. 2012. *Macraea rheedii* Wight, Ic. Pl. Ind. Or. 5(2): 27. t. 1901.1852, non *P. rheedii* Wight, 1852. **(Fig. 198).**

Glabrous herbs to 30 cm; stems flat, winged, rough with leaf scars. Leaves closely, imbricately arranged, elliptic-ovate, ca 2.9 x 1.3 cm, thick, coriaceous, brownish yellow when dry, rounded at base, acute at apex, entire at margin; petioles very minute; stipules subsagittate to lanceolate, ca 1 x 0.5 mm. Bracts broadly oblong, deeply laciniate. Male flowers: Pedicels to 6.5 mm, capillary; sepals 6; biseriate, subequal, obovate and ovate, 1.8 x 1-1.5 mm, obtuse, sometimes cucullate; disc 6 glandular; glands subspherical, surface papillate; stamens 3, free; filaments to 1 mm long; anthers globose, longitudinally dehiscent, transverse like. Female flowers: Pedicels to 1 cm, pulvinate at apex; sepals 6, ovate, ca 2 x 1 mm across, imbricate; disc cushion like, wavy; ovary ca 1.2 mm, tuberculate; styles 3, deeply bifid, spreading, often reflexed. Capsules ca 4 mm across; seeds trigonous, ca 2 x 0.6 mm, verruculose.

Occasional in low and high altitude grasslands.

Fl.: & Fr.: August to December in hills; June – August in foot hills.

Specimens examined: XCH 10215, 11834, 13650, 13952, 15920, 16042, 16161, 17270, 17554, 19419, 19455, 19569, 20154. This plant shows great variability with increase in altitude.

Distribution: Peninsular India.

11. Phyllanthus macrocalyx Muell.Arg. in Linnaea 32: 48. 1863; Hook.f., Fl. Brit. India 5: 301. 1887; Chandrab. in A.N. Henry *et al.*, Fl. Tamil Nadu 2: 237. 1987; N.P. Balakr. & T.Chakrab., Fam. Euphorb. India 373. 2007; T.Chakrab., M.Gangop. & N.P. Balakr. in N.P. Balakr. *et al.* Fl. India 23: 456. 2012. *Reidia latifolia* Wight, Ic. t. 1904. f.2. 1852 (non *phyllanthus latifolius* Sw. 1800). *Reidia macrocalyx* (Muell.Arg.) Gamble, Fl. Pres. Madras 2: 1292.1925. **(Fig. 199).**

Glabrous subshrubs to 90 cm, much branched; branches terete; slender, at ends leafless. Leaves elliptic, 2.2-5.5 x 1-2 cm, thinly coriaceous, glaucous beneath, obtuse at base, acute at apex, mucronate to 1 mm long, entire at margin, finely reticulate; lateral nerves ca 8 pairs, faint to prominent beneath; petioles to 2 mm long, filiform; stipules subulate, to 3.3 mm long. Male flowers: In proximal axils, mostly solitary or glomerulate. Pedicels to 1 cm, capillary. Sepals 4, deltoid-ovate, ca 3.5 x 2.3 mm, margin lacerate to denticulate, yellow; disc 4 glandular; glands broadly truncate, ca 1 x 1.5 mm, tuberculate. Stamens 2; filaments connate in a column to 0.8 mm; anther-sacs widely separated. Female flowers: Pedicels to 2.4 cm, pulvinate at apex. Sepals 6, elliptic-oblong, ca 5 x 3 mm, enlarged in fruit to 6 mm, subentire at margin, brown-striate, yellow or pinkish-yellow in colour; disc-saucer shaped, margin flattened, wavy, obscurely 6 lobed, tuberculate. Ovary globose, ca 1.3 mm across; styles 3, above half bifid, erect, reflexed. Capsules ca 6 mm across, brown-striate, smooth; seeds triangular, ca 3.3 x 1 mm, appressed hairy.

Common, in evergreen forests between 600-1200 m.

Fl.: & Fr.: August - October.

Specimens examined: XCH 10099, 10686, 12437, 14079, 14864, 17195, 17357, 17369, 19524, 19592.

Distribution: Peninsular India.

12. Phyllanthus maderaspatensis L., Sp. Pl. 982. 1753; Hook.f., Fl. Brit. India 5: 292. 1887; Gamble, Fl. Pres. Madras 2: 1289. 1925; Rani in Matthew, Fl. Tamil. Carnatic 3: 1466. 1983; Chandrab. in A.N. Henry *et al*., Fl. Tamil Nadu 1: 2: 239. 1987; Philcox in Dassan. & Clayton, Rev. Handb. Fl. Ceylon 11: 209. 1997; Matthew, Fl. Palni Hills 2: 1138. 1999; N.P. Balakr. & T.Chakrab., Fam. Euphorb. India 378. 2007; T.Chakrab., M.Gangop. & N.P. Balakr. in N.P. Balakr. *et al.* Fl. India 23: 469. 2012. **(Fig. 200).**

Tamil: Melanelli

Glabrous, erect or decumbent herbs to 60 cm high; stem often much branched, spreading, wiry, slender. Leaves subsessile, oblanceolate, 1-1.3 x 4-0.5 cm, cuneate at base, obtuse at apex, apiculate, sometimes retuse, entire at margin; lateral nerves ca 4 pairs, obscure; stipules lanceolate, to 1 mm. Male flowers: In distal axils, fascicled with one female flower. Pedicels minute. Sepals 6, obovate to spathulate, ca 0.75 x 0.5 mm, 1 veined, equal; disc 6 glandular; glands globose, minute. Filaments to 0.2 mm; anthers oblong, longitudinally dehiscent. Female flowers: Pedicels to 0.3 mm. Sepals 6, obovate to spathulate, 1.25 x 0.8 mm, biseriate, equal, persistent, 1 veined; disc 6 glandular; glands ca 0.2 x 0.1 mm. Ovary ca 1 mm across. Capsules ca 3.3 mm across; seeds trigonous, ca 2 x 0.6 mm, striated, muriculate, brown.

Very common in plains upto 100 m.

Fl.: & Fr.: Throughout the year.

Specimens examined: XCH 11528, 17831 (Papanasam). MH 13656, 3336.

Distribution: Tropical Africa and Asia east to Australia

13. Phyllanthus narayanswamii Gamble in Kew Bull. 1925. 329. 1925 *et* Fl. Pres. Madras 2: 1289. 1925; Nayar & Sastry, Red Data Book of Indian plants 1: 145. 1987; N.P. Balakr., & T.Chakrab., Fam. Euphorb. India 379. 2007; T.Chakrab., M.Gangop. & N.P. Balakr. in N.P. Balakr. *et al.* Fl. India 23: 470. 2012. **(Fig. 201).**

Glabrous undershrubs to 50 cm; branches many, arising from stout root stock, wiry, reddish brown, terete, scabrous with stipular scars. Leaves elliptic-oblong to obovate, 1-1.6 x 0.4-0.7 cm, rounded at base, obtuse at apex, often apiculate, entire at margin, often unequal; lateral nerves ca 5 pairs, arcuate, raised beneath; petioles to 3 mm; stipules peltate, subsagittate, ca 2 x 0.7 mm. Bracts subsagittate, ca 1 mm, lacerate. Male flowers: Pedicels to 3.6 mm. Sepals 6, oblong-ovate or ovate, subequal, 1.8-2.2 x 1-1.2 mm, sinuate at margin; disc 6 glandular; glands globose, ca 0.45 x 0.5 mm. Stamens 3, free, erect, exserted; filaments to 0.9 mm; anthers subglobose, longitudinally dehiscent, transverse like. Female flowers: Solitary. Pedicels to 5.8 mm, pulvinate at apex; disc saucer-shaped. Ovary globose, ca 2.2 mm across, verrucose, glandular; styles 3, bifid in the upper half; arms slightly recurved, spreading. Capsules not seen.

In moist rocky places in evergreen forests.

Fl.: & Fr.: April - May.

Specimen examined: XCH 19276 (Kothayar).

Distribution: Peninsular India.

Note: This species is included in the "Red Data Book of Indian Plants" and recorded from Eastern Ghats. New record for Agasthiyamalai hills.

14. Phyllanthus pinnatus (Wight) G.L.Webster in J. Arnold Arbor. 38: 52. 1957. in *ad not*.; Chandrab. in A.N. Henry *et al*., Fl. Tamil Nadu 2: 238. 1987; N.P. Balakr. & T.Chakrab., Fam. Euphorb. India 373. 2007. *Chorisandra pinnata* Wight, Ic. t. 1994. 1853. *Reidia floribunda* Wight, Ic. t. 1903. 1852, non *Phyllanthus floribundus* (Baill.) Muell.Arg. in Linnaea 32. 14. 1863; Gamble, Fl. Pres. Madras 2: 1293. 1925; Rani in Matthew, Fl. Tamil. Carnatic 3: 1470. t. 1000. 102K. 1983; T.Chakrab., M.Gangop. & N.P. Balakr. in N.P. Balakr. *et al.* Fl. India 23: 478. 2012. *Flueggea microcarpa* sensu Hook.f., Fl. Brit. India 5: 328. 1887, p.p. non Blume 1826. **(Fig. 202).**

Subshrubs to 90 cm, tomentose; branches reddish, rough with persistent stipules, scabrous, reddish-brown on drying, rufous-tomentose when young, later glabrous to pubescent; branchlets 15-20 cm, spreading towards the apex, crowded at summit of branches, hirsute. Leaves falcately elliptic-oblong, 1-2 x 0.5-1 cm; glaucous beneath, glabrous, obliquely cuneate at base, acute at apex, apiculate, sometimes hairy at apex, entire at margin; lateral nerves ca 6 pairs, forked near margin, obscure; petioles minute, hirsute; stipules subsagittate, to 1.5 mm, ciliate. Flowers axillary, solitary. Male flowers: In proximal axils. Pedicels to 9 mm, filiform, tomentose. Sepals 4, ovate, ca 3.8 x 2 mm, sparsely tomentose without, glabrous except central line within, reddish in colour; disc 4 glandular; glands truncate, ca 2 x 2.5 mm, tuberculate. Stamens 2, exserted; filaments united in a column to 2 mm; anthers hemispheric, ca 1.7 mm across; pistillode to 0.45 mm. Female flowers: In distal axils. Pedicels to 1.2 cm, pulvinate at apex, tomentose. Sepals 6, oblong-ovate, ca 3.5 x 1.8 mm, sparsely tomentose without, glabrous except central line within, reddish; disc 6 glandular; glands truncate, tuberculate. Ovary globose, ca 0.5 mm across, papillate; styles 3, bifurcate, ca 0.75 mm. Capsules globose, ca 5 mm across, subsquamosely hirsute with rusty, short, unequal bristles; seeds 6, trigonous, ca 3 x 1.5 mm, appressed hairy.

Rare, in semi evergreen forests, in rocky places near streams.

Fl.: & Fr.: June - August.

Specimens examined: XCH 19838. 19901 (Sivagiri), MH 98302.

Distribution: Peninsular India and Sri Lanka.

Note: The specimens are very close to Sri Lankan species *Phyllanthus cinereus* Muell.Arg. Both the species may be same.

15. Phyllanthus polyphyllus Willd., Sp. Pl. 4: 586. 1805; Hook.f., Fl. Brit. India 5: 290. 1887; Gamble, Fl. Pres. Madras 2: 1288. 1925; Rani in Matthew, Fl. Tamil. Carnatic 3: 1468. t. 106 d. 1983; Chandrab. in A.N. Henry *et al.*, Fl. Tamil Nadu 2: 238. 1987; Philcox in Dassan. & Clayton, Rev. Handb. Fl. Ceylon 11: 220. 1997; Matthew, Fl. Palni Hills 2: 1140. 1999; N.P. Balakr. & T.Chakrab., Fam. Euphorb. India 379. 2007; T.Chakrab., M.Gangop. & N.P. Balakr. in N.P. Balakr. *et al.* Fl. India 23: 472. 2012. **(Fig. 203).**

Malayalam: Kattunelli, Kilarannelli

Glabrous shrubs to 3 m; branches terete, woody, close-grained; branchlets at base sometimes puberulous. Leaves subsessile, linear-oblong, 1.4 x 0.35 cm, obtuse at base, obtuse at apex, apiculate, entire at margin; lateral nerves ca 8 pairs, faint; stipules lanceolate, minute. Male flowers: In proximal axils, glomerulate. Pedicels ca 2 mm, filiform. Sepals 6, oblong-obtuse, ca 1.8 x 0.7 mm, free, greenish-white; disc 6 glandular; glands globose, margin crenulate. Stamens united in a column; filaments connate to 1.5 mm; anthers oblong, longitudinally dehiscent. Female flowers: In distal axils, solitary. Pedicels to 2.5 mm. Sepals 6, oblong-obtuse, ca 3.5 x 1.5 mm; disc annular, crenulate. Ovary ca 2 mm across; styles 3, basally connate to 1 mm, above free to 1.5(-2) mm long, each bifid at apex and reflexed. Capsules ca 5 mm across; seeds trigonous, ca 3 x 1.5 mm, minutely foveolate.

Common in deciduous forests, mostly in rocky places.

Fl.: & Fr.: April - July.

Specimens examined: XCH 10660, 11115, 11393, 13056, 13239, 13364, 13772, 13861, 15645, 15681, 15729, 15807, 15975, 15246, 16372, 16549, 16737, 17143, 17809. MH 8397, 33898, 27916.

Distribution: Peninsular India and Sri Lanka.

16. Phyllanthus reticulatus Poir., Encycl. 5: 298. 1804; Hook.f., Fl. Brit. India 5: 288. 1887; Airy shaw, Euphorbiaceae Borneo 185. 1975; Rani in Matthew, F1. Tamil. Carnatic 3: 1458. tt. 100 m, 122 i, 109 n. 1983; Chandrab., in A.N. Henry *et al.*, Fl. Tamil Nadu 2: 238. 1987; Philcox in Dassan. & Clayton, Rev. Handb. Fl. Ceylon, 217. 1997; Matthew, Fl. Palni Hills 2: 1140. 1999; N.P. Balakr. & T.Chakrab., Fam. Euphorb. India 382. 2007; T.Chakrab., M.Gangop. & N.P. Balakr. in N.P. Balakr *et al.* Fl. India 23: 479. 2012. *P. multiflorus* Willd., Sp. Pl. 4: 581. 1805. *Anisonema multiflora* (Willd.) Wight,

Ic. Pl. Ind. Or. 5: 26. t. 1899. 1852. *Kirganelia reticulata* (Poir.) Baillon, Etud. Gen. Euphorb. 614.1858; Gamble, Fl. Pres. Madras 2: 1294. 1925 (repr. ed.). *Cicca microcarpa* Benth., Fl. Hongk. 312. 1861. *Kirganelia multiflora* Baillon ex Thwaites, Enum. Pl. Zeyl. 282. 1861. *Kirganelia lineata* sensu Alston in Trimen, Handb. Fl. Ceylon. 6: 259. 1931, non (Willd.) Alston which is *Berchemia lineata* (L.) DC. (Rhamnaceae).

(Fig. 204).

Straggling shrubs to 2 m; branchlets slender, puberulous to glabrous, terete, reddish brown on drying. Leaves elliptic-oblong or elliptic-obovate, 1-2.5 x 0.5-0.8 cm, puberulous beneath,acute at both ends, entire at margin; lateral nerves ca 8 pairs, arcuate, faint; petioles to 3 mm, puberulous; stipules linear-lanceolate, to 3 mm. Male flowers: Pedicels to 5.5 mm, capillary, puberulous. Sepals 5, elliptic-oblong or obovate, 2-2.4 x 1-1.4 mm, puberulous without at base, imbricate, dull pinkish in colour; disc 5 glandular; glands ca 0.3 mm, minutely pitted. Stamens 5, in 2 series; outer 3 with free filaments; inner two united into a column, ca 0.75 mm; anthers oblong, longitudinally dehiscent. Female flowers: Pedicels to 5.5 mm, filiform, puberulous. Sepals 5, unequal, elliptic or spathulate or retusely obovate, 0.75-1.2 x 0.3-0.75 mm, veined; disc 5 glandular; glands smaller; ovary elliptic-globose, ca 2 x 1.5 mm, 10-12 locular; ovules superposed; styles 0; stigma sessile, minute, persistent. Berries subglobose, ca 5 x 6 mm, dark purplish or bluish black.

Common, in foot hills to dry deciduous forests, mostly near stream banks.

Fl.: & Fr.: Throughout the year.

Specimens examined: XCH 16290, 16291, 17690, 19936. MH 438, 3362, 8373, 32213.

Distribution: India, Sri Lanka, Africa to Southeast Asia and Northern Australia.

17. Phyllanthus rheedei Wight, Ic. Pl. Ind. Or. 5, 2: 25. 1852; Hook.f., Fl. Brit. India 5: 293. 1887; Gamble, Fl. Pres. Madras 2: 1289. 1925; Fyson, Fl. S. Ind. Hill Stat. 528. t. 459. 1932; Chandrab. in A.N. Henry *et al*., Fl. Tamil Nadu 2: 238. 1987; Philcox in Dassan. & Clayton, Rev. Handb. Fl. Ceylon 11: 223. 1997; Matthew. Fl. Palni Hills 2: 1140. 1999; N.P. Balakr. & T.Chakrab., Fam. Euphorb. India 379. 2007; T.Chakrab., M.Gangop. & N.P. Balakr. in N.P. Balakr. *et al.* Fl. India 23: 474. 2012. **(Fig. 205).**

Glabrous herbs to 30 cm; branches terete, striate, flexible; branchlets compressed, basal half leafless, smooth. Leaves elliptic or elliptic-ovate, 1.4-2.2 x 0.8-1.2 cm, subglaucous beneath, greenish or blackish brown above when dry, acute at both ends, entire at margin; lateral nerves 6-8 pairs, arcuate, prominent, raised beneath, obscure above; petioles filiform, to 2 mm; stipules linear-lanceolate, to 1.8 mm, margin dentate. Bracts linear to oblong-lanceolate, ca 1.3 mm. Male flowers: Pedicels to 1.2 mm, filiform. Sepals 6, oblong, ca 2 x 1 mm, cucullate at apex; disc 6 glandular; glands globose, margin lobed. Stamens connate in to a column; filaments connate to 0.25 mm, above free to 0.12 mm; anthers oblong, longitudinally dehiscent. Female flowers: Pedicels to 2 mm, winged. Sepals 6, biseriate, unequal, oblong-ovate, 2.2-2.7 x 0.9-1.2 mm, enlarged in fruit; disc 6 glandular; glands linear, to 0.75 mm. Ovary ovate, ca 1.5 x 1.3 mm; styles 3, erect to 0.5 mm, above forked, slightly curved at margin. Capsules ca 3.75 x 4 mm; seeds trigonous, ca 3 x 0.6 mm, with deep vertical striae and faint transverse ribs, often laxly appressed hairy.

Occasional, in semi evergreen forests and montane forests between 700-1200 m.

Fl.: & Fr.: June – August.

Specimens examined: XCH 14913, 16770, 17251, 18883, 19995, 20065, 20175. MH 17417.

Distribution: India and Sri Lanka.

18. Phyllanthus rotundifolius Klein ex Willd., Sp. Pl. 4: 584. 1805; Hook.f., Fl. Brit. India 5: 299. 1887; Gamble, Fl. Pres. Madras 2: 1290. 1925; Chandrab. in A.N. Henry *et al.*, Fl. Tamil Nadu 2: 238. 1987; Philcox in Dassan. & Clayton, Rev. Handb. Fl. Ceylon 11: 225. 1997; N.P. Balakr. & T.Chakrab., Fam. Euphorb. India 389. 2007; T.Chakrab., M.Gangop. & N.P. Balakr. in N.P. Balakr. *et al.* Fl. India 23: 489. 2012. **(Fig. 206).**

Glabrous, erect herbs to 60 cm, terete to angular. Leaves obovate or elliptic-oblong, 0.7-1.3 x 0.3-0.7 cm, pale yellow to green when dry, cuneate at base, obtuse at apex, apiculate, entire at margin; lateral nerves ca 5 pairs, faint beneath, obscure above; stipules lanceolate, ca 1.2 x 0.5 mm. Bracts linear-lanceolate, ca 0.6 mm. Male flowers: Solitary. Pedicels to 0.75 mm. Sepals 6, biseriate, obovate, ca 0.25 x 0.5-0.6 mm; disc 6 glandular; glands triangular, ca 0.15 x 0.15 mm. Stamens connate in a column; filaments to 0.24 mm; anthers transversely dehiscent. Female flowers: Pedicels to 0.65 mm. Sepals biseriate, subequal, oblanceolate, ca 0.7-1 x 0.5-0.75 mm, broadly white margined, enlarged in fruit; disc saucer-shaped, margin shallowly 3 lobed, crenulate. Ovary ca 0.6 mm across, smooth, sometimes white punctate; styles 3, apex forked; arms recurved. Capsules ca 1.7 mm across; seeds trigonous, ca 0.9 x 0.35 mm, with equal slender ribs and transverse striae, sometimes puberulous.

Common near Dam sites, occasionally in cultivated lands.

Fl.: & Fr.: Throughout the year.

Specimens examined: XCH 16353, 16362, 16572, 16983, 17389, 17666, 18123, 18528, 19761, 20020. MH 13688, 88661.

Distribution: India, Sri Lanka, Tropical Africa and Arabia.

19. Phyllanthus simplex Retz., Obs. Bot. 5 : 29. 1788; Hook.f., Fl. Brit. India 5: 295. 1887; Gamble, Fl. Pres. Madras 2: 1289. 1925; Philcox in Dassan. &

Clayton, Rev. Handb. Fl. Ceylon 11: 213. 1997; N.P. Balakr. & T.Chakrab., Fam. Euphorb. India 380. 2007; T.Chakrab., M.Gangop. & N.P. Balakr. in N.P. Balakr. *et al.* Fl. India 23: 476. 2012. *P. virgatus* auct. non G. Forst. 1786; T.Chakrab. & N.P. Balakr. in J. Econ. Taxon. Bot., Add. ser. 9: 98. 1992; N.P. Balakr. & T.Chakrab., Fam. Euphorb. India 380. 2007. *Macraea oblongifolia* Wight, Ic. Pl. Ind. Or. 5: 27. T. 1902-1. 1852. *Phyllanthus simplex* var. *oblongifolius* (Wight) Muell.Arg. in Linnaea 32: 32. 1863 *et* in DC., Prodr. 15(2): 391. 1866; Hook.f., Fl. Brit. India 5: 295. 1887.

Key to the varieties

1a. Branchlets flattened, winged; leaves elliptic-oblong; small at end of branchlets; male flowers glomerulate in proximal axils, female flowers solitary in distal axils. .... 1. var. **gardnerianus**

1b. Branchlets terete; leaves linear-oblong, evenly arranged; male and female flowers glomerulate in proximal axils. .... 2. var. **virgatus**

1. var. **gardnerianus** (Wight) Muell.Arg. in Linnaea 32: 33. 1863; Hook.f., Fl. Brit. India 5: 295. 1887. *Macraea gardneriana* Wight, Ic. Pl. Ind. Or. 5: 27. t. 1902 – 3.1852. *M. ovalifolia* Wight, Ic. Pl. Ind. Or. 5: 27. t. 1892 - 4.1852 (non *Phyllanthus ovalifolius* Forssk., 1776). *P. gardnerianus* (Wight.) Baill., Etude Gen. Euphorb. 628. 1858; Gamble, Fl. Pres. Madras 2: 1290. 1925; Chandrab. in A.N. Henry *et al.*, Fl. Tamil Nadu 2: 237. 1987; Philcox in Dassan. & Clayton, Rev. Handb. Fl. Ceylon 11: 212. 1997; Matthew, Fl. Palni Hills 2: 1139. 1999; N.P. Balakr. & T.Chakrab., Fam. Euphorb. India 380. 2007. non. *P. gardnerianus* (Wight) Baillon; T.Chakrab., M.Gangop. & N.P. Balakr. in N.P. Balakr. *et al.* Fl. India 23: 477. 2012. *Phyllanthus virgatus* var. *gardnerianus* (Wight) Govaerts & Radcl.-sm. in Kew Bull. 51: 177. 1996. **(Fig. 207).**

Glabrous herbs-subshrubs to 60 cm; branchlets spreading, flattened, winged, rough with stipules. Leaves on main stem elliptic-oblong, ca 2.5 x 0.8 cm, on branchlets, ca 1.7 x 0.5 cm, at ends of lateral branchlets 1 x 0.3 cm, obtuse to acute at base, obtuse or sometimes subacute at apex, entire at margin, chartaceous, pale brownish on drying; lateral nerves ca 5 pairs; petioles to 1 mm; stipules subsagitatte to lanceolate, ca 2.4 mm, persistent. Bracts broadly oblong to 1 x 7 mm, deeply lacinate at apex. Male flowers: axillary glomerulate at proximal axils; pedicels filiform to 4 mm, sepals 6, biseriate, linear-obovate, subequal, ca 1.5-2 x 1 mm, margin has no shape, cucullate at apex, hyaline; disc 6 glandular, glands subspherical, ca 0.5 x 0.35 mm; stamens free; filaments to 1.2 mm, anthers globose. Female flowers: Solitary at distal axils; pedicels to 0.5 cm, 1 cm in fruit, pulvinate at apex, filiform; sepals 6, oblong or oblong-ovate, ca 2 x 0.7-1.2 mm, acute at apex, sometimes cuculate, slightly connate at base; disc cushion-like, wavy, ovary ca 1 mm, densely tuberculate, styles 3, connate at base to 0.5 mm; deeply bifid above, arms recurved. Capsules ca 4 mm, prominently tuberculate; seeds trigonous, ca 3 x 1 mm, verruculose.

Common in montane forests.

Fl. & Fr.: January – March.

Specimens examined: XCH 17926, 17960, 17987, 19082, 20154, 17213. MH 88709, 90696, 93261.

Distribution: Peninsular India and Sri Lanka

2. var. **simplex** **(Fig. 208).**

Glabrous woody herbs to 30 cm high; branchlets prostrate to erect, spreading. Leaves linear-oblong, 0.7-1.9 x 0.2-0.5 cm, obtuse at base, acute at apex, sometimes apiculate, entire, ciliate at margin; lateral nerves ca 7 pairs, parallel, arcuate, faint; petioles minute, to 0.6 mm. Bracts deltoid. Male flowers: few, glomerulate with female flowers, sessile. Sepals 6, obovate, ca 0.5 x 0.4 mm, cucullate, imbricate; disc 6 glandular; glands tuberculate. Stamens free; filaments to 0.2 mm; anthers globose, transversely dehiscent. Female flowers: Pedicels to 4 mm, up to 6 mm in fruit, filiform. Sepals 6, oblong, ca 0.8 x 0.2-0.3 mm, apex acute, often apiculate; disc cupular, thin, crenate at margin. Ovary ca 0.5 x 0.5 mm, papillate; styles 3, free, deeply bifid to ca 0.2 mm; stigma capitate. Capsules ca 1.5 x 1.5 mm, papillate-verrucose; seeds trigonous, ca 9 x 6.5 mm, covered loosely with minute tubercles in regular lines.

Occasional, in dried rocky places in low altitude grasslands.

Fl.: & Fr.: Throughout the year.

Specimens examined: XCH 10827, 11132, 12055, 13842, 16412, 17162, 17171, 17788, 19251, 19759, 20053. MH 10925, 3597.

Distribution: India to China and Indonesia and Sri Lanka.

20. Phyllanthus singampattianus (Sebastine & A.N. Henry) Kumari & Chandrab. in A.N. Henry *et al.*, Fl. Tamil Nadu 2: 238. 1987; N.P. Balakr. & T.Chakrab., Fam. Euphorb. India 374. 2007; T.Chakrab., M.Gangop. & N.P. Balakr. in N.P. Balakr. *et al.* Fl. India 23: 458. 2012. *Reidia singampattiana* Sebastine & A.N. Henry in Bull. Bot Surv. India 2: 43. 1960. **(Fig. 209).**

Glabrous, woody undershrubs to 1 m; stems woody, rough, terete, slender; branchlets spreading towards top, crowded at summit of stem, upto 13 cm, reddish green. Leaves elliptic-lanceolate, ca 7 x 4 mm, obtusely acute at base, oblique, acute at apex, entire at margin, glaucous beneath; lateral nerves ca 5 pairs, inconspicuous; subsessile; stipules linear-subulate, to 1 mm, dark brown. Male flowers: In proximal axils, arising from minute clustered bracts. Pedicels to 3 mm, filiform, pulvinate at apex. Sepals 4, broadly ovate, deeply lacerate at margin, ca 2.3 x 1.3 mm, equal, brick red; disc 4 glandular; glands globose, minutely tuberculate. Stamens 4; filaments united in a column to 1 mm; anthers oblong, transverse in position. Female flowers: Pedicels to 4.5 mm, filiform, pulvinate at apex. Sepals 6, broadly ovate, ca 2.3 x 1.3 mm, brick-red; disc annular. Ovary globose, ca 1.5 mm across; styles 3, to 5 mm, each deeply bifid, spreading. Capsules depressed globose, ca 3 mm across, brick red; seeds 6, trigonous, ca 0.4 x 0.2 mm, smooth, dark brown.

Rare, in moist deciduous to semi evergreen forests in rocky places and on river banks.

Fl.: & Fr.: July – September (Peak).

Specimens examined: XCH 16711, 18890, 18975 (Papanasam).

Distribution: Peninsular India.

Note: **Endemic** to Tirunelveli Hills.

21. Phyllanthus tenellus Roxb., Fl. Ind. 3: 668. 1832; Muell.Arg. in DC. Prodr. 15, 2: 337. 1866; Hook.f. in Hook., Ic. Pl. 1569. 1866; Radcl.-sm. in Kew Bull. 50: 814. 1995; Philcox in Dassan. & Clayton, Rev. Handb. Fl. Ceylon 11: 218. 1997; N.P. Balakr. & T.Chakrab., Fam. Euphorb. India 382. 2007; T.Chakrab., M.Gangop. & N.P. Balakr. in N.P. Balakr. *et al.* Fl. India 23: 480. 2012. **(Fig. 210).**

Glabrous herbs to 30 cm; stems terete, granular, woody; branchlets to 3 cm with ca 10 pairs of leaves. Leaves elliptic-ovate, ca 8 x 5 mm, acute at both ends, entire at margin; lateral nerves ca 5 pairs; petioles to 5 mm; stipules lanceolate, acuminate. Flowers in axillary glomerules, bracteate. Bracts linear-lanceolate, to 0.6 mm. Male flowers: Pedicels to 1.6 mm. Sepals 5, obovate, ca 0.7 x 1 mm, cucullate at apex; disc 5 glandular, ca 0.12 x 0.15 mm. Stamens 5, free; filaments to 0.25 mm; anthers globose, didymous, transversely dehiscent. Female flowers: Pedicels to 1 mm. Sepals 5, oblong-ovate, ca 0.7 x 0.3 mm, sometimes slightly cucullate at apex; disc saucer shaped, thin, wavy. Ovary obovate, ca 0.4 mm across, granular; styles 3, ca 0.15 mm, deeply bifid, spreading. Capsules ca 1 cm across; seeds trigonous, ca 0.5 x 0.3 mm, regularly vertically muriculate.

Rare, in dry to moist deciduous forests.

Fl.: & Fr.: January - March.

Specimen examined: XCH 15410 (Kalakad).

Distribution: Native to Mascarene Islands, widely introduced into tropical and subtropical areas. India and Sri Lanka.

Note: Apparently introduced in India after 1980. Our finding is a new record for Peninsular India. It might have been introduced to these hills from Sri Lanka.

22. Phyllanthus urinaria L., Sp. Pl. 982. 1753; Hook.f., Fl. Brit. India 5: 293. 1887; Gamble, Fl. Pres. Madras 2: 1289. 1925; Matthew, Fl. Tamil. Carnatic 3: 1469. t. 106 e. 1983; Chandrab. in A.N. Henry *et al.*, Fl. Tamil Nadu 2: 239. 1987; Philcox in Dassan. & Clayton, Rev. Handb. Fl. Ceylon 11: 227. 1997; Matthew, Fl. Palni Hills 2: 1140. 1999; N.P. Balakr. & T.Chakrab.,

Fam. Euphorb. India 389. 2007; T.Chakrab., M.Gangop. & N.P. Balakr. in N.P. Balakr. *et al.* Fl. India 23: 490. 2012. *P. leprocarpus* Wight, Ic. t. 1895. f. 4. 1852. **(Fig. 211).**

Glabrous herbs to 60 cm; branchlets flattened, winged, rough with stipules. Leaves oblong, hirsutulous to glabrous, glaucous beneath, obtuse to subcordate at base, acute at apex, apiculate, entire at margin, ca 1.8 x 0.6 cm; lateral nerves ca 6 pairs, arcuate; subsessile to minutely petiolate; stipules subulate, ca 2.2 mm. Bracts linear-lanceolate, ca 0.3 mm. Male flowers: Glomerulate in proximal axils, sessile. Sepals 5-(6), oblong-ovate, ca 0.5 x 0.3 mm, cucullate at apex; disc 5-(6) glandular; glands minute; filaments minute; anthers oblong, longitudinally dehiscent. Female flowers: In distal axils, solitary, sessile. Sepals 6, slightly connate at base, biseriate, ovate or subspathulate, ca 0.5-0.6 x 0.25 mm; disc saucer shaped, sinuate; ovary oblate, ca 0.5 mm across; styles 3, forked at apex; arms recurved. Capsules ca 3 mm across, scurfy verrucouse; seeds trigonous, ca 1.5 x 0.5 mm, with transverse ridges.

Occassional in low altitude grasslands, rare in forest undergrowth.

Fl.: & Fr.: October - December.

Specimens examined: XCH 13173, 13946, 16680, 17111, 20051. MH 3326, 12036, 3697, 19241.

Distribution: Southern India and Sri Lanka

34. PUTRANJIVA Wallich

Around 3 species distributed in India, Sri Lanka, Bangladesh, Myanmar, Thailand, Indochina, Taiwan, Ryukyu islands, Java, Molluccas and New Guinea; 1 in India.

Putranjiva roxburghii Wallich, Tent. Fl. Nep. 61.1826; Wight, Ic. Pl. Ind. Or. 5: t. 1876. 1852; Hook.f., Fl. Brit. India 5: 336. 1887; Gamble, Fl. Pres. Madras 2: 1311. 1925; N.P. Balakr. & T.Chakrab., Fam. Euphorb. India 335. 2007; T.Chakrab. & N.P. Balakr. in N.P. Balakr. *et al.* Fl. India 23: 405. 2012. *Nagea putranjiva* Roxb., Fl. Ind. 3: 766. 1832. *Drypetes roxburghii* (Wallich) Hurus in J. Fac. Sci. Univ. Tokyo, Ser. II Sect. 3 Bot. 6: 335. 1954. **(Fig. 212).**

Trees to 4 m; branches tawny-puberulous. Leaves oblong to oblong-elliptic, 4-8 x 1.5-2.8 cm, pubescent beneath, thinly pubescent to glabrous above, blackish on drying, obtuse to acute at base, acute at apex, closely serrulate at margin; lateral nerves ca 12 pairs, arcuate; petioles to 1 cm, thin. Male flowers: Calyx 5 lobed, lobes oblong, ca 0.7 x 0.4 mm, basally connate, puberulous without. Stamens 2 or 4; filaments to 1.5 mm; anthers ca 0.6 mm, hairy. Female flowers: Pedicels ca 1 cm, pubescent. Calyx 5 lobed; lobes ovate or elliptic, unequal, 1.3-2 x 0.5-1 mm, pubescent. Ovary ellipsoid, ca 4 x 3.6 mm; stigmas flabellate, each to 3.6 mm across. Drupes ellipsoid, ca 1 x 0.5 cm, white tomentose.

Occasional, in moist-decidusous forests between 300-600 m.

Fl.: February - April. *Fr.:* June onwards.

Distribution: India, Sri Lanka, Myanmar, Thailand, Indo-china, Java, Moluccas, Lesser Sunda Islands and New Guinea.

Specimens examined: XCH 11363 (Papanasam), 15483, 15925 (Courtallam). MH. 38830, 43966.

35. SAUROPUS Blume

The genus *Sauropus,* with about 40 species, is Indo-Malaysian and South-East Asian in distribution, extending to Australia. Of the 19 species occurring in the Indian subcontinent, *S. androgynus* and *S. bacciformis* are widespread from Sri Lanka and India to S. China, SE. Asia, through Malaysia to Celebes and Moluccas (Chakrabarty & Gangopadhyay, 1996). 12 species in India.

Key to the species

1a. Leaves 2 cm long or less 2
1b. Leaves more than 2 cm long 3

2a. Calyx glandular in male flowers .... 2. **S. bacciformis**
2b. Calyx eglandular in male flowers .... 4. **S. quadrangularis**

3a. Calyx of male flowers cupular .... 3. **S. gour-maitii**
3b. Calyx of male flowers discoid .... 1. **S. androgynus**

1. **Sauropus androgynus** (L.) Merr. in Bull. Bur. Forest Philipp. Islands 1: 30. 1903; Gamble, Fl. Pres. Madras 2: 1303. 1925; Chandrab. in A.N. Henry *et al.,* Fl. Tamil Nadu 2: 239. 1987; T.Chakrab. & M.Gangop. in J. Econ. Taxon. Bot. 20: 519. 1996; N.P. Balakr. & T.Chakrab. Fam. Euphorb. India 393. 2007; T.Chakrab. & M.Gangop. in N.P. Balakr. *et al.* Fl. India 23: 493. 2012. *Clutia androgyna* L., Mant. 1: 128. 1767. *Sauropus albicans* Blume, Bijdr. 596. 1826; Hook.f., Fl. Brit. India 5: 332. 1887. *S. gardnerianus* Wight, Ic. Pl. Ind. Or. 6: t. 1951. 1853. **(Fig. 213).**

Shrubs to 3 m. Leaves elliptic or elliptic-lanceolate to lanceolate or triangular - lanceolate, 3.3-6.5 x 1.5-2.3 cm, rounded to acute at base, acute or sometimes acuminate at apex to 5 mm long, often mucronate, entire at margin; lateral nerves 5-6 pairs; petioles to 2 mm; stipules triangular-lanceolate, to 3 mm long. Male flowers: Pedicels to 3 mm. Calyx discoid, flat, ca 7 mm across, shallowly 6 lobed, thick, reddish. Stamens 3; filaments combined in 3-gonous column. Female flowers: Pedicels to 5 mm. Sepals 6, 3 + 3, spathulate-obovate, ca 4 x 3.2 mm, reddish. Ovary turbinate, ca 1.5 mm across. Capsules globose, slightly depressed, ca 1 cm across, 6 valved, reddish white.

Rare, in moist deciduous forests at ca 500 m, often cultivated.

Fl.: & Fr.: August - October.

Specimens examined: XCH 17578, 18532 (Mahendragiri).

Distribution: Peninsular India, Sri Lanka, Bangladesh, Myanmar, Thailand, S. China, Malay Peninsula, W. Malaysia to Celebes, Moluccas and New Guinea.

2. **Sauropus bacciformis** (L.) Airy Shaw in Kew Bull. 35: 685. 1980 & 36: 343. 1981; Rani in Matthew, Fl. Tamil. Carnatic 3: 1472. 1983; Chandrab. in A.N. Henry *et al.*, Fl. Tamil Nadu 2: 239. 1987; T.Chakrab. & M.Gangop. in J. Econ. Taxon. Bot. 20: 523. 1996; Matthew, Fl. Palni Hills 2: 1143. 1999; N.P. Balakr. & T.Chakrab., Fam. Euphorb. India 393. 2007; T.Chakrab. & M.Gangop. in N.P. Balakr. *et al.* Fl. India 23: 494. 2012. *Phyllanthus bacciformis* L., Mant. 298. 1767. *Agyneia bacciformis* (L.) Juss., Euphorb. Gen. Tent. 2: 24. t. 6. 1824; Wight, Ic. t. 1893. 1852; Hook.f., Fl. Brit. India 5: 285. 1887; Gamble, Fl. Pres. Madras 2: 1284. 1925. *Synostemon bacciformis* (L.) Webster in Taxon 9: 26. 1960. **(Fig. 214).**

Herbs to 30 cm, sometimes woody; Leaves elliptic or elliptic-oblong or elliptic-orbicular, 0.8-2 x 0.5-1 cm, glaucous beneath, obtuse to subacute at base, obtuse to acute at apex; lateral nerves 3-5 pairs, obscure to faint; petioles to 2 mm, filiform. Male flowers: Pedicels to 1.5 mm, filiform. Calyx 6, 3 + 3, subflabellate, ca 1 x 0.4-1 mm, emarginate to bilobulate at apex, white-margined, glandular. Stamens 3; filaments combined in a gonophore, extrorse; anthers parallel, horizontal. Female flowers: Pedicels to 2.5 mm, filiform. Calyx 6, 3 + 3, obovate, ca 1.4 x 1 mm, acute at apex, minutely apiculate, greenish-white. Ovary ovoid, ca 0.75 mm across; styles 3, sunk on ovary, bifid; arms recurved. Capsules ovoid-ellipsoid, ca 7.5 x 6 mm, beaked; seeds ca 4.5 x 1.4 mm.

Occasional, in grasslands between 300-500 m.

Fl.: & Fr.: September - November.

Specimens examined: XCH 16256, 16348, 17647, 17648, 17686 (Thirukurungudi.). MH 13668.

Distribution: India to SE Asia, S. China, Malaya, Java, Celebes, Ceylon and Mauritius.

3. **Sauropus gour-maitii** T.Chakrab. & M.Gangop. in J. Econ. Taxon. Bot. 20: 529. 1996; N.P. Balakr. & T.Chakrab., Fam. Euphorb. India 394. 2007; T.Chakrab. & M.Gangop. in N.P. Balakr. *et al.* Fl. India 23: 496. 2012. **(Fig. 215).**

Shrubs 2-3 m; upper leafy branchlets 2 winged, yellowish green or dark greenish brown when dry. Leaves distichous, alternate, ovate-lanceolate or elliptic-lanceolate or triangular-lanceolate, 4-7 x 1.7-2.8 cm, glabrous, truncate or acute at base, acute or acuminate at apex, entire at margin; lateral nerves 5-6 pairs; petioles to 2 mm; stipules lanceolate, ca 2 mm. Male flowers: Pedicels linear-filiform, ca 7.4 mm, slender. Calyx deeply 6 lobed, concave, cupular; lobes oblong-oblanceolate, ca 2 x 1.5 mm, white, incurved and shortly bifid at apex. Stamens 3; filaments combined in a trigonous column ca 1.5 x 1 mm; anthers seated at angles, linear-oblong, ca 0.5 mm, longitudinally dehiscent. Female flowers: Pedicels capillary, ca 5 mm, slender. Calyx 5 lobed; lobes broadly obovate or spathulate or sub-orbicular, 2.8-3 x 2-2.3 mm, subcoriaceous, glabrous, rounded at apex, persistent. Ovary elliptically globose, ca 2.5 x 1 mm, exserted, 3 lobed, 3 locular; ovules 6, 2 in each locule, pendulous; styles erect, ca 0.7 mm, with deeply incurved arms at apex. Fruits capsular, depressed globose, ca 1 x 0.8 cm, 6 valved, white in colour, apically dehiscent; seeds 6, trigonous, ca 0.75 x 0.3 mm, brown, smooth; pedicel ca 1 cm.

Rare, in evergreen forests between 900-1200 m.

Fl.: & Fr.: March - July.

Specimens examined: XCH 18858 (Papanasam), 19003, 19695 (Kalakad).

Distribution: Peninsular India.

4. **Sauropus quadrangularis** (Willd.) Muell.Arg. in Linnaea 32: 73. 1863; Hook.f., Fl. Brit. India 5: 335. 1887; Gamble, Fl. Pres. Madras 2: 1303. 1925; Chandrab. in A.N. Henry *et al.*, Fl. Tamil Nadu 2: 239. 1986; T.Chakrab. & M.Gangop. in J. Econ. Taxon. Bot. 20: 534. 1996; N.P. Balakr. & T.Chakrab., Fam. Euphorb. India 395. 2007; T.Chakrab. & M.Gangop. in N.P. Balakr. *et al.* Fl. India 23: 494. 2012. *Phyllanthus quadrangularis* Willd., Sp. Pl. 4: 585. 1805. **(Fig. 216).**

Woody herbs or shrubs to 50 cm; branchlets compressed. Leaves elliptic or oblong, sometimes ovate, 0.7-2 x 0.5-0.65 cm; coppery-brown on drying, acute or subacute at both ends, sometimes mucronulate at apex; lateral nerves ca 6 pairs, faint; petioles ca 0.1 mm, filiform; stipules lanceolate-acuminate, to 1.6 mm. Male flowers: Pedicels to 1.5 mm, filiform. Calyx 6, 3 + 3; lobes subflabellate, 0.5 x 0.5-0.7 mm. Stamens 3; column erect; anthers parallel, vertical in position. Female flowers: Pedicels to 3 mm. Calyx 6, 3 + 3; lobes obovate, 2-3 x 1.8-2.4 mm, narrowed at base, acute to apiculate at apex, greenish white. Ovary turbinate, ca 0.7 x 1 mm; styles 3, bifid, above half exserted. Capsules ovoid-globose, ca 5 mm across, beaked, greenish white.

Occasional in deciduous forests in rocky crevices between 150-400 m.

Fl.: & Fr.: September - March.

Specimens examined: XCH 13863 (Courtallam), 18385 (Sivagiri). MH 3173.

Distribution: Peninsular India, extending to Bihar.

36. SUREGADA Roxb. ex Rottler

A paleotropical genus of ca 40 species, the majority in tropical Asia; 2 in India.

1. **Suregada angustifolia** (Muell.Arg.) Airy Shaw in Kew Bull. 23: 128. 1969; Matthew, Fl. Tamil. Carnatic 3: 1478. 1983; Chandrab. in A.N. Henry *et al.*, Fl. Tamil Nadu 2: 240. 1987; N.P. Balakr. & T.Chakrab., Fam. Euphorb. India 219. 2007; T.Chakrab. & N.P. Balakr. in N.P. Balakr. *et al.*, Fl. India 23: 247. 2012. *Gelonium angustifolium* Muell.Arg. in DC., Prodr. 15(2): 1128. 1866. *G. lanceolatum* auct. non Willd. 1806; Hook.f., Fl. Brit. India 5: 459. 1887. **(Fig. 217).**

Dioecious, glabrous or minutely puberulous shrubs to 2 m, much branched; branches terete, with stipular lines, grayish white. Leaves alternate, obovate or oblanceolate, 5-8.5 x 2-3.5 cm, acute to cuneate at base, obtuse or subacute at apex, usally entire at margin, penninerved; lateral nerves 3-8 pairs, arcuate, obscure above, faint beneath; petioles to 5 mm, minutely puberulous. Flowers axillary or extra-axillary, clustered, leaf-opposed. Male flowers: fascicled. Pedicels to 1.4 mm. Calyx 5 lobed; lobes broadly ovate-orbicular, ca 2.5 x 2 mm, membranous at margin, obscurely laciniate, ciliate, imbricate. Stamens many, exserted; filaments unequal, 3-5 mm, free; anthers oblong, 1.8 mm, apiculate. Female flowers: mostly solitary. Pedicels to 1 mm, upto 3.5 mm in fruit. Calyx 5 lobed; lobes rotund, ca 1.5 mm across, puberulous to glabrous without, ciliate. Ovary rotund, ca 8 mm across, 3 locular; ovules 3, pendulous. Fruit capsular, depressed-globose, ca 1 x 1 cm, 3 lobed; seeds subglobose, ca 7.5 x 4 mm, smooth.

Common along the borders of deciduous forests.

Fl. & Fr.: March – May.

Specimens examined: XCH 16429, 16943, 17603, 17811, 18845, 18846.

Distribution: Peninsular India and Sri Lanka.

37. TRAGIA Plum. ex L.

A large and diverse genus of ca 125 species, mainly in Africa and America (Webster, 1994); 7 in India.

Tragia involucrata L., Sp. Pl. 980. 1753; Muell.Arg. in DC., Prodr. 15(2): 943. 1866; Hook.f., Fl. Brit. India 5: 465. 1888; Gamble, Fl. Pres. Madras 2: 1332. 1925; Rani in Matthew, Fl. Tamil. Carnatic 3: 1479. t. 100r. 1983; Chandrab. in A.N. Henry *et al.*, Fl. Tamil Nadu 2: 240. 1987; Philcox in Dassan. & Clayton, Rev. Handb. Fl. Ceylon 11: 140. 1997; Matthew, Fl. Palni Hills 2: 1145. 1999; N.P. Balakr. & T.Chakrab., Fam. Euphorb. India 181. 2007; T.Chakrab. & N.P. Balakr. in N.P. Balakr. *et al.*, Fl. India 23: 196. 2012. **(Fig. 218).**

Monoecious, hispid-pubescent vines. Leaves alternate, broadly ovate-cordate or elliptic-ovate, 4-9 x 1.6-3.7 cm; densely hirsute-pubescent beneath, less so above, with scattered, stinging bristles; pale to dark green on drying, acute-cordate at base, acuminate at apex, dentate-serrate at margin, 3-5 ribbed at base, above penninerved; lateral nerves 4-5 pairs; cross nervules prominent beneath; petioles 0.3-1 cm, densely hispid; stipules linear-lanceolate, to 4 mm. Inflorescence axillary or terminal racemes, 1-2 cm, androgynous. Male flowers: in distal axils. Pedicels very short, hispid. Calyx 3 lobed; lobes ovate, ca 1.5 x 1 mm, acute at apex, glabrous, valvate, greenish-yellow. Stamens 3, exserted; filaments connate, very short, to 0.3 mm; anthers oblong, to 0.7 mm, parallel, longitudinally dehiscent. Female flowers: subsessile. Calyx 6 lobed, to 4 x 2.5 mm, deeply pinnatifid, spreading, hardened, densely strigosely-hispid without, less so within, hardened, imbricate, yellowish, persistent. Ovary globose-ovoid, ca 1.5 x 2.5 mm, 3 lobed, hispid; ovules 3; styles 3, united in a column to 1 mm, spreading at apex, persistent. Capsules subsessile, depressed-globose, ca 6.3 x 6.6 mm, hispid.

Occasional, in shady places in moist deciduous to evergreen forests between 300-700 m.

Fl.: & Fr.: Throughout the year.

Specimens examined: XCH 13367, 14044, 17878, 19787. MH. 3890.

Distribution: India, Sri Lanka, Myanmar and Southern China.

20. URTICACEAE Juss.

Herbs, shrubs or small trees some with stinging hairs. Leaves opposite or alternate, 3 nerved or penninerved, entire or lobed. Flowers small, greenish, monoecious or dioecious, in open or capitate cymes or spikes or aggregated on fleshy receptacle, sometimes involucrate. Perianth of 4-5 (rarely 2-3) free or united sepals, usually united in female and sometimes accrescent and fleshy in fruit. Stamens as many as the tepals, usually inflexed in bud with reversed anthers, later elastically reflexed. Pistillode usually present, various. Ovary superior, 1 locular; ovule single, basal, erect; style simple or 0; stigma papillose, plumose or penicillate. Fruit a drupe or an achene, usually enclosed in and adnate to perianth; seeds with a membranous testa; embryo straight.

53 genera and 2625 species have been reported mainly in the tropical and subtropical regions; 29 genera and 155 species are in India.

Key to the genera

1a. Plants with stinging hairs 2
1b. Plants without stinging hairs 4

2a. Stipules foliaceous; stinging hairs stout .... 4. **Girardinia**
2b. Stipules not foliaceous; stinging hairs not stout 3

3a. Leaves membranous, margin coarsely serrate; flowers in pseudospikes, shorter than leaves .... 5. **Laportea**
3b. Leaves chartaceous, margin entire or sinuate or crenulate; flowers in cymes, cymes often longer than leaves .... 2. **Dendrocnide**

4a. Herbs or undershrubs 5
4b. Small trees or large shrubs 9

5a. Flowers, at least female flowers aggregated on a fleshy receptacle 6
5b. Flowers not aggregated on a fleshy receptacle 7

6a. Stem succulent; leaf margin serrate in the upper half; stamens 4 ... ... 3. **Elatostema**
6b. Stem not succulent; leaf margin serrate throughout; stamens 5 .... 10. **Procris**

7a. Flowers cymose; cystoliths of leaves linear 8
7b. Flowers in sessile clusters; cystoliths of leaves punctiform .... 9. **Pouzolzia**

8a. Leaves opposite, margin coarsely serrate .... 8. **Pilea**
8b. Leaves alternate or subopposite, margin entire or slightly crenulate ...... 7. **Pellionia**

9a. Leaves white tomentose beneath; stigma sessile, penicillate .... 1. **Debregeasia**
9b. Leaves not white tomentose beneath; stigma discoid, ciliate-fimbriate ..... 6. **Oreocnide**

1. DEBREGEASIA Gaud.

Ca 9 species have been reported from N.E. Tropical Africa, Tropical and Subtropical Asia to S. W. Pacific; 6 in India.

Debregeasia longifolia (Burm.f.) Wedd. in DC., Prodr. 16(1): 235. 1869; Rani & Matthew in Matthew, Fl. Tamil. Carnatic. 3: 1489. 1983; Chithra in A.N. Henry *et al.*, Fl. Tamil Nadu 2: 243. 1987. *Urtica longifolia* Burm.f., Fl. Ind. 197. 1768. *Debregeasia velutina* Gaud., Voy. Bonite. Bot. Atlas t. 90 1844; Hook.f., Fl. Brit. India 5: 590. 1888; Trimen, Handb. Fl. Ceylon 4: 118. 1898; Fischer in Gamble, Fl. Pres. Madras 3: 1389. 1928; Wilmot Dear in Kew Bull. 43(4): 675. 1988; Mohanan & Sivadasan, Fl. Agasthyamala 639. 2002. **(Fig. 219).**

Tamil: Kattu nochi.

Shrubs to 3 m; branchlets terete, tender parts prominently hispid-pubescent. Leaves alternate, lanceolate or oblong-lanceolate, 5.5-16 x 1-4 cm, chartaceous, white-pilose or ashy-tomentose beneath, pubescent or scabrid above, black when dry; cystoliths punctiform; base rounded to subacute, margin closely, regularly serrulate, 3 ribbed at base; lateral nerves above ca 5 pairs, distally arranged, reticulations prominent beneath; petioles 1-5 cm, scabrid with white-hispid hairs; stipules connate, intrapetiolar. Flowers monoecious or dioecious, in dense, stalked, globose heads in short, dichotomous, paniculate cymes from the axils along branchlets, subsessile. Male flowers: Tepals 4, ovate, concave, ca 1.5 x 0.5 mm, pubescent without, basally connate. Stamens 4; filaments ca 2 mm. Pistillode ellipsoid, ca 1 mm. Female flowers: Tepals 3 or 4, free, unequal, ovate, ca 1 x 0.5 mm; outer one larger, sparsely pubescent without. Ovary obovoid, ca 0.7 x 0.3 mm, 1 locular, included; ovule 1, basal, erect; stigma sessile, penicillate. Achenes minute, orange in colour.

Occasional, in semi-evergreen to evergreen forests between 800-1500 m.

Fl.: October – December; *Fr.:* February – April.

Specimens examined: XCH 10548, 12136 (Mancholai); 12504, 13548, 20132 (Kothayar); 18200 (Courtallam).

Distribution: Sri Lanka, India, Himalaya, Myanmar, Indo-China, W. China, Malaysia.

2. DENDROCNIDE L.

Ca 42 species distributed in Tropical and Subtropical Asia to Pacific; 1 in India.

Dendrocnide sinuata (Blume) Chew in Gard. Bull. Straits Settlem 21: 206. 1965; Chithra in A.N. Henry *et al.*, Fl. Tamil Nadu 2: 244. 1987. *Urtica sinuata* Blume, Bijdr. 505. 1826. *Laportea crenulata* (Roxb.) Gaud. in Freyc. Voy. Bont. Bot. 498. 1826; Hook.f., Fl. Brit. India 5: 550. 1888; Trimen, Handb. Fl. Ceylon 4: 105. 1898; Fischer in Gamble, Fl. Pres. Madras 3: 1373. 1928; Mohanan & Sivadasan, Fl. Agasthyamala 642. 2002. **(Fig. 220).**

Tamil: Anai-choriyan.

Large shrubs or small trees to 4 m, with stinging hairs; branchlets subsucculent, smooth, soft. Leaves alternate, broadly elliptic or elliptic-lanceolate or obovate, 8.5-21 x 2.3-7.3 cm, chartaceous, with minute stinging hairs on nerves, base rounded, rarely narrowly acute, margin entire or sinuate or crenulate, apex acute to shortly acuminate; lateral nerves ca 10 pairs, distally arranged; petioles 2-5.5 cm, with stinging hairs, flexuous. Flowers monoecious or dioecious, in paniculate cymes or racemes; cymes often equal to or longer than leaves, flexuous, dichotomously branched; flower clusters remote, often unilateral, flowers subsessile. Male: Tepals 4, oblong, ca 3 x 1.5 mm. Stamens 4; filaments to 2 mm. Pistillode clavate. Female: Tepals 4, subequal, subcampanulate; lobes oblong, ca 2 x 1.5 mm, acute. Ovary oblique; ovule erect; style linear, papillose on one side. Achenes oblique, ventricose, seated on minute perianth, crowned with style.

Rare, in evergreen forests between 900-1200 m.

Fl.& Fr.: September – January.

Specimens examined: XCH 11022, 11631, 14357 (Kalakad).

Distribution: Indomalaysia.

3. ELATOSTEMA J.R.Forst. & G.Forst.

Ca 630 species distributed in the tropical and subtropical Africa, Asia, and Oceania; ca 27 in India.

Key to the species

1a. Male receptacles with peduncles >3 cm 2
1b. Male receptacles sessile or with peduncles <1 cm 3

2a. Small herb; leaves upto 10 cm; apex acuminate .... 5. **E. wightii**
2b. Gregarious; leaves upto 6 cm; apex obtuse - acute .... 4. **E. monandrum**

3a. Leaves membranous, without cystoliths, coarsely crenate dentate in the upper two-third .... 1. **E. acuminatum**
3b. Leaves chartaceous or subcoriaceous, with large conspicuous cystoliths, entire or undulately toothed in the upper half 4

4a. Leaves acute at apex .... 2. **E. cuneatum**
4b. Leaves acuminate or caudate at apex .... 3. **E. lineolatum**

1. **Elatostema acuminatum** (Poir.) Brongn. in Duperr. Voy. Bot. 211. 1834; Hook.f., Fl. Brit. India 5: 566. 1888; Fischer in Gamble, Fl. Pres. Madras 3: 1377. 1928; Chithra in A.N. Henry *et al.*, Fl. Tamil Nadu 2: 244. 1987; Mohanan & Sivadasan, Fl. Agasthyamala 640. 2002. *Procris acuminata* Poir. in Lam., Encycl. 5: 629. 1804. **(Fig. 221).**

Much branched undershrubs to 1 m; branchlets 4 angular, glabrous. Leaves alternate, obliquely oblanceolate or oblong, 7-13 x 2-3.5 cm, membranous, glabrous, when dry greenish-brown above, paler beneath, base acute, margin coarsely crenate-dentate, apex caudate to 2.5 cm; lateral nerves ca 8 pairs; subsessile. Receptacles axillary, solitary, subsessile. Male: Tepals 5, ovate, ca 2.5 x 1.5 mm; stamens 5; filaments ca 0.75 mm. Pistillode ellipsoid, to 0.5 mm. Female: Tepals 5, ovate, ca 1.5 x 1 mm, persistent; ovary larger than tepals; stigma sessile, penicillate. Achenes ellipsoid, minute.

Occasional, in semi-evergreen forests between 800-1000 m.

Fl. & Fr.: May – September.

Specimens examined: XCH 10374, 10473, 13588 (Kakachi), 12334 (Papanasam); 19146 (Courtallam).

Distribution: India, Sri Lanka, Myanmar, Java.

2. **Elatostema cuneatum** Wight, Ic. t. 2091. f. 3. 1853, "*cuniata*"; Hook.f., Fl. Brit. India 5: 568. 1888; Fischer in Gamble, Fl. Pres. Madras 3: 1377. 1928; Chithra in A.N. Henry *et al.*, Fl. Tamil Nadu 2: 244. 1987. **(Fig. 222).**

Dwarf, procumbent herbs to 30 cm, often rooting at lower nodes; stem more or less tomentose. Leaves alternate, falcately lanceolate, 1.5-3 x 0.5-1 cm, chartaceous, sparsely tomentose when young, when dry brownish, base cuneate, inequilateral, margin crenate, chiefly in the upper half, apex acute; lateral nerves ca 4 pairs; subsessile. Receptacles minute, in axillary fascicles. Male: Bracteoles ovate-lanceolate, to 2 mm, ciliate. Tepals 5, unequal, ovate or rotund or lanceolate, ca 3 x 1-2.5 mm, ciliate. Stamens 5; filaments to 1.75 mm. Pistillode globose, stalked, ca 1.5 mm. Female: Bracteoles ovate-lanceolate, ca 1 mm, long ciliate. Tepals 5, unequal, minute, ca 0.75 mm, ciliate; ovary subglobose, ca 0.75 mm across. Achenes minute, ellipsoid, ribbed.

In semi-evergreen forests between 900-1000 m, rivarian.

Fl.: August.

Specimen examined: XCH 16775 (Papanasam).

Distribution: Java, India. **Vulnerable.**

3. Elatostema lineolatum Wight, Ic. t. 1984. 1853; Wedd., Monogr. Urtic. 312. 1856; Hook.f., Fl. Brit. India 5: 565. 1888; Fischer in Gamble, Fl. Pres. Madras 3: 1376. 1928; Matthew & Rani in Matthew, Fl. Tamil. Carnatic 3: 1490. 1983; Chithra in A.N. Henry *et al.*, Fl. Tamil Nadu 2: 244. 1987; Matthew, Fl. Palni Hills 2: 1154. 1999; Mohanan & Sivadasan, Fl. Agasthyamala 641. 2002.

Monoecious herbs. Leaves subsessile, alternate, distichous variable in shape, thick, chartaceous, 3 nerved at base, base obliquely acute or cuneate, apex acuminate or caudate. Flowers in short, unisexual clusters, often mixed with spathulate bracteoles. Bracts ovate, rarely spurred, basally connate into a lobed disc. Male: Tepals 4, spurred at back. Stamens 4. Pistillode linear, clavate. Female: Tepals 3-5, persistent, often spurred on back. Ovary obovoid; stigma sessile, penicillate. Achenes ellipsoid, ribbed; seeds oblong; albumen 0.

Key to the varieties

1a. Leaves linear, ribs and nerves obscure ....2. var. **linearis**

1b. Leaves other than linear, ribs and nerves prominent **2**

2a. Leaves 4-15 cm long, broader, margin coarsely crenate toothed or serrate above middle, cystoliths very crowded and conspicuous ....3. var. **lineolatum**

2b. Leaves 1.5-4 cm long, narrower, margin entire or with one or two large teeth on each side, cystoliths not crowded and slightly or not conspicuous .... 1. var. **falcigera**

1. var. **falcigera** Thwaites, Enum. Pl. Zeyl. 260. 1861; Hook.f., Fl. Brit. India 5: 565. 1888; Fischer in Gamble, Fl. Pres. Madras 3: 1377. 1928; Mohanan & Sivadasan, Fl. Agasthyamala 642. 2002. **(Fig. 223).**

Much branched herbs; stem grey-tomentose. Leaves small, falcately linear-lanceolate, 1.5-4 x 0.3-0.6 cm, when dry brownish, pubescent, base cuneate to attenuate, margin entire or with one or two large teeth, apex caudate-acuminate.

Very rare, in evergreen forests, above 1000 m.

Fl. & Fr.: August – November.

Specimens examined: XCH 10431, 10475 (Kakachi).

Distribution: Peninsular India, Sri Lanka.

2. var. **linearis** Wedd. in Arch. Mus. Hist. Nat. 9: 313. 1856, "lineare"; Hook.f., Fl. Brit. India 5: 565. 1888; Fischer in Gamble, Fl. Pres. Madras 3: 1377. 1928; Chithra in A.N. Henry *et al.*, Fl. Tamil Nadu 2: 244. 1987. **(Fig. 224).**

Herbs to 50 cm; branchlets woody, pubescent. Leaves linear, 2.5-21 x 0.3-1 cm, chartaceous, pubescent on midrib, when dry greenish above, brownish beneath, base obtuse, margin entire, or with one or two large teeth at apex, apex acuminate; nerves obscure. Receptacles small, pisiform, solitary or crowded. Involucratal bracts confluent. Male flowers: Tepals 4. Stamens 4. Female flowers: Tepals 5, oblong, ca 1.5 mm, long ciliate. Ovary ovoid, ca 0.5 mm, stalked.

Very rare in evergreen forests between 900-1200 m.

Fl. & Fr.: July – October.

Specimens examined: XCH 16777 (Papanasam); 17201, 17224 (Kothayar).

Distribution: Sri Lanka, India, Myanmar.

3. var. **lineolatum** **(Fig. 225).**

Herbs or subshrubs, 50 cm to 1 m; branchlets woody, slightly zig-zag; bark creamy. Leaves very variable in shape and size, oblong or oblong-lanceolate or narrowly linear-oblong, or obovate, 4-15 x 1-3.5 cm, glabrous above, glabrous or hairy beneath; cystoliths crowded and conspicuous; base cuneate, coarsely crenate toothed or serrate above middle, apex caudate or acuminate; lateral nerves distant, distinct beneath, lower pair nearly straight; stipules lanceolate, usually hairy. Receptacles sessile, small, solitary or crowded in the axils.

Frequent, in semi-evergreen to evergreen forests between 700-1500 m.

Fl. & Fr.: Throughout the year, peak in March.

Specimens examined: XCH 19610 (Kalakad); 12951, 15569, 19668 (Papanasam); 19857 (Sivagiri); 11214, 15741, 15757, 16106 (Courtallam); 16023 (Naraikadu); 16668 (Sivasilam); 10005, 10085, 10429, 10585, 13560 (Kakachi); 12609 (Kothayar); 12778, 18630, 18634 (Mahendragiri).

Distribution: Sri Lanka, India, Myanmar.

4. Elatostema monandrum (Buch.-Ham. ex D. Don) H.Hara in H.Ohashi, Fl. E. Himalaya, 3rd Report 21. 1975. *E. surculosum* Wight, Ic. t. 2091. f. 4. 1853, "*surculosa*"; Hook.f., Fl. Brit. India 5: 572. 1888; Fischer in Gamble, Fl. Pres. Madras 3: 1377. 1928; Chithra in A.N. Henry *et al.*, Fl. Tamil Nadu 2: 245. 1987; Matthew, Fl. Palni Hills 2: 1155. 1999. **(Fig. 226).**

Herbs to 30 cm, often succulent. Leaves subsessile, alternate, subfalcately ovate or lanceolate, chartaceous, glabrous, when dry dark-brownish above, paler beneath, base inequilateral, acute or subauricled, margin deeply crenate-serrate in the upper half or sometimes with two teeth on either side of apical one, apex obtuse – acute; cystoliths conspicuous above; lateral nerves 4 pairs, lower pair long; narrowed towards apex; stipules minute. Receptacles axillary, subtended by bracts and bracteoles. Male receptacles long peduncled. Peduncles ca 3 cm. Involucral bracts free, broad, membranous, all round or one or more with broad spreading or recurved, acute, laterally compressed spur or horn. Flowers few. Male: Tepals 4, ovate, basally connate. Stamens 4. Female: very much smaller, sessile. Bracteoles ciliate. Tepals 5. Ovary 1 locular; ovule 1. Achenes ovoid.

Very rare, in shady, marshy rocky areas in grasslands, above 1000 m.

Fl.: September.

Specimen examined: XCH 17319 (Kothayar).

Distribution: India, Sri Lanka.

5. **Elatostema wightii** Hook.f., Fl. Brit. India 5: 570. 1888; Fischer in Gamble, Fl. Pres. Madras 3: 1377. 1928; Chithra in A.N. Henry *et al.*, Fl. Tamil Nadu 2: 245. 1987. **(Fig. 227).**

Succulent herbs to 60 cm, spreading; branchlets often zig-zag, ribbed. Leaves alternate, subfalcately oblong-lanceolate, 4-8 x 1-2 cm, subchartaceous, when dry dark brownish above, greenish beneath, base inequilateral, acute, sometimes auricled, margin coarsely toothed from middle, apex caudate-acuminate; lateral nerves obscure, lower long pair prominent; cystoliths mostly only on midrib and lower pairs subsessile. Male: Receptacles long peduncled, subtended by bracteoles. Peduncles ca 3 cm, filiform. Bracteoles linear, ca 3.5 mm, ciliate. Male: Tepals 4, oblong, ca 3 x 1 mm, apiculate. Stamens 4. Female: smaller, sessile, silvery. Tepals 5. Ovary 1 locular, ovule 1.

Vey rare, in marshy rocky places in grasslands, above 1000 m.

Fl.: September.

Specimen examined: XCH 17329 (Kothayar).

Distribution: Peninsular India.

4. GIRARDINIA Gaud.

Two species distributed in the tropical and subtropical Old World to Russian far east; 1 in India.

Girardinia diversifolia (Link) Friis in Kew Bull. 36: 145. 1981; Matthew & Rani in Matthew, Fl. Tamil. Carnatic 3: 1490. 1983; Chithra in A.N. Henry *et al.*, Fl. Tamil Nadu 2: 245. 1987; Matthew, Fl. Palni Hills 2: 1155. 1999. *Urtica diversifolia* Link, Enum. Pl. Hort. Berol. 2: 385. 1822. *Girardinia heterophylla* Decne. in Jacq., Voy. 4: 151. t. 153. 1844; Hook.f., Fl. Brit. India 5: 551. 1888 incl. var. *palmata. G. leschenaultiana* Decne. in Jacq., Voy. 4: 152. 1844; Wight, Ic. t. 1976. 1853; Fischer in Gamble, Fl. Pres. Madras 3: 1372. 1928. *G. heterophylla* Decne. var. *zeylanica* (Decne.) Hook.f., Fl. Brit. India 5: 551. 1888. **(Fig. 228).**

Strong herbs to 60 cm; branchlets with ca 5 mm long stinging hairs, furrowed. Leaves alternate, ca 12 x 15 cm, pedately 3-7 lobed, chartaceous, scabrid above, mingled with sparsely spreading stinging hairs, densely scabrid along nerves beneath, base truncate to subcordate, margin inciso-crenate, apex acuminate, 3 nerved from base; petioles to 9 cm, with stinging hairs; stipules connate, foliaceous. Flowers monoecious or dioecious, clustered on simple or panicled spikes or heads, armed with numerous stinging hairs. Male: Flowers in dense clusters. Peduncles 1.5-5 cm. Tepals 4, ovate, ca 1 x 0.5 mm, cucullate, hairy without. Stamens 4. Pistillode globose. Female: inflorescence much branched. Tepals 4, connate into a ventricose tube, split on one end, strigose without. Ovary ovate, ca 0.5 x 0.3 mm, 1 locular, flattened; ovule 1; style filiform, ca 1.5 mm. Achenes compressed, ca 3 mm across.

Very rare, in shady places in semi-evergreen forests, ca 1000 m.

Fl. & Fr.: March.

Distribution: Sri Lanka, India, China, Malaysia.

Specimen examined: XCH 18285 (Courtallam).

5. LAPORTEA Gaud.

Ca 36 species distributed in the tropical and subtropical regions; ca 5 in India.

Laportea interrupta (L.) Chew in Gard. Bull. Straits Settlem. 21: 200. 1965; Matthew & Rani in Matthew, Fl. Tamil. Carnatic 3: 1491. 1983; Chithra in A.N. Henry *et al.*, Fl. Tamil Nadu 2: 245. 1987. *Urtica interrupta* L., Sp. Pl. 985. 1753. *Fleurya interrupta* (L.) Gaud. Voy. Uranie 12: 497. t. 8. 1830; Wight, Ic. t. 1975. 1853; Hook.f., Fl. Brit. India 5: 548. 1888; Fischer in Gamble, Fl. Pres. Madras. 3: 1372. 1928. **(Fig. 229).**

Herbs to 50 cm; branchlets with scattered stinging hairs, ribbed. Leaves alternate, ovate or elliptic-ovate, 4.5-9 x 2-4 cm, membranous, with sparse stinging hairs, more so on nerves beneath, greenish when dry, base acute, margin coarsely serrate, apex shortly acuminate, 3 ribbed at base; lateral nerves ca 3 pairs; petioles 1-4 cm. Flowers monoecious or dioecious, clustered in axillary, unisexual or androgynous spikes or cymes. Male: Tepals 5, ovate, ca 2 x 1.6 mm. Stamens 5, unequal; filaments ca 2 mm. Pistillode globose, stalked. Female: Tepals 5, connate into a cup at base; lobes ovate or lanceolate, 0.6-1.1 mm, margin irregular or minutely glandular. Ovary obliquely oblong, ca 0.75 mm, 1 locular; ovule 1; style obliquely linear, to 1.5 mm. Achenes oblique, compressed, membranous.

Very rare, in moist-deciduous forests, ca 700 m.

Fl. & Fr.: June.

Specimen examined: XCH 19787 (Sivagiri).

Distribution: Africa, Abyssinia, Sri Lanka, India, Japan, China through S. E. Asia to Queensland and Pacific islands.

6. OREOCNIDE L.

Ca 16 species distributed in the tropical and subtropical Asia; 3 in India.

Oreocnide integrifolia (Gaud.) Miq., Ann. Mus. Lugd.-bat. 4: 306. 1869; Matthew & Rani in Matthew, Fl. Tamil. Carnatic 3: 1491. 1983; Chithra in A.N. Henry *et al.*, Fl. Tamil Nadu 2: 246. 1987; Matthew, Fl. Palni Hills 2: 1157. 1999. *Villebrunea integrifolia* Gaud. Voy. Bonite Bot. t. 91. 1844; Hook.f., Fl. Brit. India 5: 589. 1888; Fischer in Gamble, Fl. Pres. Madras 3: 1388. 1928. **(Fig. 230).**

Trees or large shrubs to 3 m; tender parts grey-pubescent. Leaves alternate, lanceolate or oblanceolate, 7.5-20 x 2.8-5.8 cm, chartaceous; cystoliths punctiform; when dry greenish brown above, paler beneath, base acute to cuneate, margin entire or the upper half crenulate, apex acuminate or subcaudate; lateral nerves 6-8 pairs, looping, raised beneath; transverse nervules prominent beneath, reticulate; petioles 0.5-5 cm; stipules bifid. Flowers monoecious or dioecious, clustered in axils, sometimes in short cymes. Bracts minute. Male: Globose. Tepals 4, ovate, ca 1 mm long, ciliate. Stamens 4, ca 1 mm. Pistillode clavate, ca 0.75 mm, pilose. Female: Tepals connate into a tube, ca 2 mm, narrowed into a toothed mouth. Ovary adnate to tepals, obovoid, ca 2 mm across; ovule erect; stigma discoid, ciliate-fimbriate. Achenes free or adnate to tepals, embraced below by marcescent, fleshy bracteoles.

Occasional, in semievergreen to evergreen forests between 600-1200 m.

Fl.: Throughout the year, peak in February – April.

Fr.: November – January.

Specimens examined: XCH 11624, 15363, 19517 (Kalakad); 16202, 18152 (Courtallam); 12875, 16786 (Papanasam); 17740 (Thirukurungudi); 19836 (Sivagiri); 10528, 12124, 12133 (Mancholai); 11914 (Kakachi); 12561 (Kothayar).

Distribution: India, tropical Himalaya, Myanmar, Indo-China, S. China.

7. PELLIONIA Gaud.

39 species in the tropical and subtropical Asia to S. W. Pacific; ca 4 in India.

Pellionia heyneana Wedd., Monog. Urtic. 287. t. 5. 1856; Hook.f., Fl. Brit. India 5: 561. 1888; Fischer in Gamble, Fl. Pres. Madras 3: 1380. 1928; Chithra in A.N. Henry *et al.*, Fl. Tamil Nadu 2: 246. 1987; Mohanan & Sivadasan, Fl. Agasthyamala 643. 2002. **(Fig. 231).**

Herbs to 60 cm; stem woody, creeping below, ribbed; tender parts pubescent. Leaves alternate or subopposite, inequilateral, linear to broadly lanceolate, falcate, 8.5-19 x 3.5-9 cm, chartaceous, glabrous or the ribs beneath pilose, unequally cordate at base, margin entire or slightly crenulate, apex acuminate or caudate, 3 ribbed at base; lateral nerves distally 3 or 4, on upper side only, all ribs raised beneath; transverse nervules prominently reticulate; cystoliths numerous, prominent; petioles 0-1 cm, hirsute-pubescent; stipules large, narrow-ensiform. Flowers monoecious or dioecious, pinkish or violet in colour, in open or contracted or capituliform, pedunculate axillary cymes; male in long peduncled lax cymes; female in axillary heads. Male: Peduncles to 12 cm. Tepals 5, oblong, ca 2 x 0.5 mm, pubescent without, ciliate. Stamens 5; filaments to 1.5 mm; anthers ovate, ca. 0.5 mm. Pistillode conical. Female: Calyx deeply 5; lobes equal or unequal, mucronate, oblong, ca 1.5 x 0.3 mm. Ovary elliptic, 1 locular. Achenes elliptic-ovate, compressed.

Occasional, in evergreen forests between 500-700 m.

Fl.: Throughout the year, peak in April.

Specimens examined: XCH 11217, 15493, 15750, 16105, 16201, 19157 (Courtallam); 12822, 12939, 16665 (Papanasam); 17749 (Thirukurungudi); 16665 (Sivasilam); 19883 (Sivagiri).

Distribution: Peninsular India, Sri Lanka.

8. PILEA Lindl.

604 species mainly distributed in the Asia, Africa, Pacific and the America; ca 15 in India.

Pilea melastomoides (Poir.) Blume. Mus. Bot. Lugduno-Batavum 2: 54. 1856; Matthew & Rani in Matthew, Fl. Tamil. Carnatic 3: 1493. 1983; Chithra in A.N. Henry *et al.*, Fl. Tamil Nadu 2: 246. 1987; Matthew, Fl. Palni Hills 2: 1159. 1999; Mohanan & Sivadasan, Fl. Agasthyamala 644. 2002. *Urtica melastomoides* Poir. in Lam., Encycl. Suppl. 4: 223. 1816 *Urtica trinervia* Roxb., Fl. Ind. 3: 582. 1832. *Pilea trinervia* (Roxb.) Wight, Ic. t. 1973. 1853; Hook.f., Fl. Brit. India 5: 557. 1888; Fischer in Gamble, Fl. Pres. Madras 3: 1379. 1928. **(Fig. 232).**

Robust succulent herbs to 60 cm. Leaves opposite, broadly elliptic or oblong-lanceolate, 9-18 x 3-7.5 cm, membranous, glabrous, when dry greenish-black or brown above, paler beneath, obtuse at base, margin coarsely sharply-serrulate, apex acuminate or caudate, 3 ribbed at base; primary nerves numerous, regular, nearly straight, parallel between ribs; petioles 1.5-8.5 cm; stipules connate, intrapetiolar. Flowers minute, monoecious or dioecious in axillary, peduncled, dichotomous cymes. Bracts small or 0. Peduncles ca 9 cm. Male: Pedicels ca 1 mm. Tepals 4, lanceolate, ca 1 x 0.5 mm, unequal; outer larger. Stamens 4, ca 2.5 mm. Female: Tepals 4, unequal. Ovary ca 1 cm; ovules erect; stigma sessile; penicillate. Achenes ovoid, compressed.

Very rare, in open places of evergreen forests above 700 m.

Fl. & Fr.: September - December.

Specimens examined: XCH 14312 (Kalakad); 10432, 10480, 11917 (Kakachi); 12163 (Mancholai); 17252 (Kothayar).

Distribution: Peninsular India, Sri Lanka.

9. POUZOLZIA Gaud.

Ca 57 species in the tropics and subtropics; ca 14 in India.

Key to the species

1a. Lateral ribs not forked .... 4. **P. zeylanica**
1b. Lateral ribs forked 2

2a. Flowers in terminal leafy spikes; upper leaves reduced to floral bracts ... 3. **P. wightii**
2b. Flowers in axillary clusters or panicles; upper leaves not reduced 3

3a. Leaves alternate; tepals in male flowers 5 .... 1. **P. auriculata**
3b. Leaves opposite or ternate; tepals in male flowers 4 .... 2. **P. bennettiana**

1. Pouzolzia auriculata Wight, Ic. t. 1980. f. 2. & t. 2099. f. 37. 1853; Hook.f., Fl. Brit. India 5: 582. 1888; Fischer in Gamble, Fl. Pres. Madras 3: 1383. 1928; Matthew & Rani in Matthew, Fl. Tamil. Carnatic 3: 1495. 1983; Chithra in A.N. Henry *et al*., Fl. Tamil Nadu 2: 247. 1987; Matthew, Fl. Palni Hills 2: 1160. 1999; Mohanan & Sivadasan, Fl. Agasthyamala 645. 2002. **(Fig. 233).**

Subshrubs to 2 m; tender parts white-hairy. Leaves alternate, elliptic-lanceolate or linear-lanceolate, 2-7 x 1-2 cm, chartaceous, appressed-hairy, more so beneath, obtuse at base, margin entire, apex acuminate, ribbed at base; lateral nerves forked; petioles 0.3-2 cm, pilose. Male flowers: buds rounded. Tepals 5, oblong-obovate, ca 1.5 x 0.5 mm, cucullate, pubescent without. Stamens 5, ca 1.5 mm. Pistillode ellipsoid. Female flowers: Tepals lanceolate, ca 1.5 x 0.3 mm, connate, ciliate, pubescent without. Ovary ellipsoid, ca 1.5 mm; stigma hairy.

Frequent near stream banks in moist deciduous forests between 300-600 m.

Fl. & Fr.: Throughout the year, peak September – November.

Specimens examined: XCH 12415, 14213, 14442, 17064 (Papanasam); 11167, 13784, 13858 (Courtallam); 18355, 18456 (Sivagiri); 17692 (Thirukurungudi); 12079 (Manimuthar).

Distribution: India, Sri Lanka.

2. Pouzolzia bennettiana Wight var. **macrophylla** Hook.f., Fl. Brit. India 5: 586. 1888; Fischer in Gamble, Fl. Pres. Madras 3: 1384. 1928; Chithra in A.N. Henry *et al*., Fl. Tamil Nadu 2: 247. 1987. **(Fig. 234).**

Shrubs to 2 m; tender parts grey-pubescent. Leaves sessile, lanceolate or narrowly elliptic-lanceolate, 5.5-11 x 1-2.5 cm, chartaceous, scaberulous, pubescent, more so beneath on nerves, obtuse or acute or cordate at base, apex acuminate, mucronate, 3-5 ribbed; ribs raised beneath; transverse nervules parallel, reticulate. Bracts and bracteoles ovate, ciliate. Male flowers: Pedicels to 1 mm, hispid. Tepals 4, oblanceolate, ca 2.5 x 2 mm, hispidulous without, cucullate. Stamens 4, ca 3 mm. Female flowers: Tepals connate, pubescent, acute. Ovary ellipsoid, to 3 mm; style hairy. Achenes ellipsoid, winged.

Very rare near stream sides in moist-deciduous forests between 300-600 m.

Fl. & Fr.: Throughout the year.

Specimens examined: XCH 14378 (Papanasam); 14281 (Kalakad); 11148, 13977 (Courtallam); 12194 (Mancholai).

Distribution: Peninsular India, Sri Lanka.

3. **Pouzolzia wightii** Benn. & Br., Pl. Jav. Rar. 1: 66. 1838; Wight, Ic. t. 2093. f. 8. 1853; Hook.f., Fl. Brit. India 5: 584. 1888; Fischer in Gamble, Fl. Pres. Madras 3: 1383. 1928; Matthew & Rani in Matthew, Fl. Tamil. Carnatic 3: 1496. 1983; Chithra in A.N. Henry *et al.*, Fl. Tamil Nadu 2: 248. 1987; Matthew, Fl. Palni Hills 2: 1161. 1999; Mohanan & Sivadasan, Fl. Agasthyamala 646. 2002. **(Fig. 235).**

Shrubs to 2 m; tender parts pubescent; stem tapering into leafy spikes. Leaves opposite to ternate or alternate, very variable in shape, size and indumentum, elliptic-lanceolate, or oblong-ovate, obovate, 1.5-8 x 0.5-2.5 cm, chartaceous, scabrid, pubescent, base rounded, truncate or acute, apex acute, upper leaves gradually reduced to floral bracts, ca 6 mm; petioles 0-8 mm. Male flowers: Buds truncate. Tepals 4, ovate, ca 1 x 0.5 mm, ciliate, apiculate, cucullate. Stamens 4, ca 1 mm. Female flowers: Tepals connate, acute. Ovary ellipsoid, ca 1 mm; style hairy. Achenes ca 1.2 mm, prominently winged; wings ca 1 mm.

Occasional along stream banks of moist-deciduous to semievergreen forests between 500-1000 m.

Fl. & Fr.: Throughout the year, peak in September to November.

Specimens examined: XCH 18442, 19968 (Sivagiri); 16120, 13199 (Courtallam); 12148 (Mancholai); 18660 (Mahendragiri).

Distribution: **Endemic** to Peninsular India.

4. **Pouzolzia zeylanica** (L.) Benn., Pl. Jav. Rar. 1: 67. 1838 Matthew & Rani in Matthew, Fl. Tamil. Carnatic 3: 1497. 1983; Chithra in A.N. Henry *et al.*, Fl. Tamil Nadu 2: 249. 1987; Mohanan & Sivadasan, Fl. Agasthyamala 646. 2002. *Parietaria zeylanica* L., Sp. Pl. 1052. 1753. *Parietaria indica* L., Mant. Pl. 128. 1767. *Pouzolzia indica* (L.) Gaud., Voy. Uranie 12: 563. 1830; Wight, Ic. t. 1980. f. 1. & 2100. f. 40. 1853; Hook.f., Fl. Brit. India 5: 581. 1888; Fischer in Gamble, Fl. Pres. Madras 3: 1382. 1928. **(Fig. 236).**

Shrubs to 50 cm. Leaves opposite or alternate, very variable in shape, linear, lanceolate or elliptic or ovate-lanceolate or broadly ovate, 2-7 x 0.5-1 cm, chartaceous, pubescent, obtuse or truncate or acute at base, apex acute or shortly acuminate, mucronate, 3 ribbed at base; petioles 0-1 cm. Male flowers: Pedicels ca 1 mm, pubescent. Tepals 4, oblong-ovate, ca 2.5 x 1 mm, cucullate, pubescent without. Stamens 4, ca 3 mm. Female flowers: Tepals connate, pubescent without. Ovary ellipsoid; ovule erect; stigma linear. Achenes 2 winged or not.

Frequent, from deciduous forests to semievergreen forests between 200-800 m.

Fl. & Fr.: Throughout the year, peak in January – March.

Specimens examined: XCH 11942, 12035, 15102, 15121 (Mancholai); 11592, 11748 (Kalakad); 1130, 13995 (Courtallam); 14387, 16568, 15556 (Papanasam); 17283, 17321 (Kothayar); 17443 (Mahendragiri).

Distribution: Sri Lanka, India, Himalaya, Myanmar, Indo Malaysia, China to Japan.

10. PROCRIS Juss.

24 species distributed in the Asia, Africa and Pacific; 3 in India.

Procris crenata Robinson in Philipp. J. Sci. 5: 507. 1911; Matthew & Rani in Matthew, Fl. Tamil. Carnatic 3: 1498. 1983; Chithra in A.N. Henry *et al.*, Fl. Tamil Nadu 2: 249. 1987; Matthew, Fl. Palni Hills 2: 1162. 1999. *P. wightiana* Wallich ex Wedd., Monog. Urtic. 336. 1856, nom. illeg.; Fischer in Gamble, Fl. Pres. Madras 3: 1378. 1928. *P. laevigata* sensu Hook.f., Fl. Brit. India 5: 575. 1890, non Blume 1825-1826. **(Fig. 237).**

Succulent herbs to 50 cm. Leaves alternate, linearly oblanceolate or lanceolate, 9-12 x 1.5-2.5 cm, glabrous, when dry greenish-black, base attenuate tapering into petiole, oblique, margin crenate-serrate throughout, apex acuminate or caudate; lateral nerves ca 7 pairs; petioles ca 1 cm. Flowers monoecious, axillary, mostly in the axils of fallen leaves. Male: In cymes. Tepals 5, ovate, ca 1 x 0.5 mm, fleshy. Stamens 5. Female: Tepals 5, ovate, ca 0.5 x 0.2 mm, cucullate. Ovary ellipsoid, 1 locular; ovule 1; stigma sessile.

Very rare, in evergreen forests above 1000 m.

Fl.: March.

Specimens examined: XCH 17965 (Papanasam); 11908 (Kakachi); 14993 (Kothayar).

Distribution: Africa, Sri Lanka, India, Indo-China, Philippines.

21. ULMACEAE Mirb.

Trees or shrubs. Leaves simple, alternate, bifarious, gland-dotted; stipules lateral or intrapetiolar. Flowers usually bisexual or polygamous, small, in axillary clusters or cymes, female sometimes solitary. Perianth calycine, usually 4-5-merous, polyphyllous or gamophyllous. Stamens as many or twice as many as the perianth-lobes; anthers erect in bud, dehiscing laterally. Ovary of 2 carpels, rarely 2 locular; ovule 1 in each cell, pendulous; stigmas 2, bifid, branches stigmatic on the inner side; pistillode in male usually present. Fruit indehiscent, a nut, drupe or samara; seeds with thin testa; albumen 0 or scanty; embryo straight or curved; cotyledons flat or folded.

Ca 15 genera and 200 species in the tropical and subtropical regions; ca 7 genera and 21 species in India.

Key to the genera

1a. Flowers polygamous; perianth lobes imbricate .... 1. **Celtis**

1b. Flowers dioecious; perianth lobes induplicate-valvate .... 2. **Trema**

1. CELTIS L.

65 species cosmopolitan in distribution; ca 7 in India.

Key to the species

1a. Twigs reddish-brown, lenticellate; reticulation on leaves indistinct ... 2. **C. timorensis**

1b. Twigs grey-brown, not lenticellate; reticulation on leaves distinct ..... 1. **C. philippensis**

1. Celtis philippensis Blanco in Steenis, Fl. Males. Ser. 1. 8: 62. 1977; Matthew & Rani in Matthew, Fl. Tamil. Carnatic 3: 1501. 1983; Kumari in A.N. Henry *et al.*, Fl. Tamil Nadu 2: 249. 1987; Matthew, Fl. Palni Hills 2: 1164. 1999. *C. wightii* Planch. in Ann. Sci. Nat. Bot. Ser. 3. 10: 307. 1848; Wight, Ic. t. 1969. 1853; Hook.f., Fl. Brit. India 5: 483. 1888; Fischer in Gamble, Fl. Pres. Madras 3: 1349. 1928. **(Fig. 238).**

Trees to 7 m. Leaves oblong or broadly ovate or oblong-lanceolate, 3-9.5 x 1.2-3.7 cm, coriaceeous, glabrous, shining, base rounded to truncate, margin entire, apex obtuse or acute or shortly acuminate, strongly 3 nerved at base; nerves running almost to the apex, flattened above, prominently raised beneath; petioles thick, hard, to 2 cm. Panicles short, ca 2 cm. Male flowers in lower parts of new shoots, bisexual ones above. Pedicels ca 5 mm, puberulous. Perianth 5 lobed; lobes suborbicular, unequal, ca 2 x 1 mm, concave, membranous, ciliate, acute, obtuse or emarginate at apex, puberulous without. Stamens 5, subequal, ca 1.5 mm; anthers oblong. Pistillode small. Ovary oblong-ovoid, ca 2.5 mm; style-arms broadly bifurcate, papillose. Drupes ellipsoid, ca 1 x 0.6 cm; epicarp leathery-coriaceous, beaked when young.

Frequent from foot hills to semi-evergreen forests between 50-700 m.

Fl.: Mostly throughout the year, flowers peak during November – March.

Specimens examined: XCH 11420, 13864 (Courtallam); 12417, 14184, 15703, 16598 (Papanasam); 19713 (Kalakad); 18005 (Thirukurungudi); 18357, 18444 (Sivagiri); 18569 (Mahendragiri).

Distribution: Tropical Africa to Madagascar, India, Myanmar, S.E. China, Hong Kong, Taiwan, Indo-China, Thailand, Malaysia to N.E. Australia, Solomon islands.

2. Celtis timorensis Spanoghe in Linnaea 15: 343. 1841; Soep. in Steenis, Fl. Males. Ser. 1. 8: 65. 1977; Matthew & Rani in Matthew, Fl. Tamil. Carnatic 3: 1503. 1983; Kumari in A.N. Henry *et al*., Fl. Tamil Nadu 2: 250. 1987; Matthew, Fl. Palni Hills 2: 1165. 1999. *C. cinnamomea* Lindl. ex Planch. in Ann. Sci. Nat. Bot. Ser. 3. 10: 303. 1848; Hook.f., Fl. Brit. India 5: 482. 1888; Fischer in Gamble, Fl. Pres. Madras 3: 1349. 1928. **(Fig. 239).**

Trees to 15 m; tender parts pubescent. Leaves broadly elliptic, subcoriaceous, glabrous, base rounded, margin entire or serrate in the upper half, apex acute or acuminate, 4-9 x 1-4 cm, 3 nerved at base; nerves running almost to the apex, raised and prominent beneath; petioles filiform, to 1 cm. Drupes ovoid-globose, ca 1 x 0.5 cm, reddish when ripe.

Rare, in semi-evergreen forests between 500-800 m.

Fr.: October.

Distribution: Sri Lanka, India, Myanmar, Indo-China, Malaysia.

Specimens examined: XCH 11165, 16185 (Courtallam); 10695 (Kakachi).

2. TREMA Lour.

20 species distributed mainly in tropical and subtropical regions; 6 in India.

Trema orientalis (L.) Blume, Mus. Bot. Lugduno-Batavum 2: 62. 1856; Hook.f., Fl. Brit. India 5: 484. 1888; Fischer in Gamble, Fl. Pres. Madras 3: 1350. 1928; Matthew & Rani in Matthew, Fl. Tamil. Carnatic 3: 1504. 1983; Kumari in A.N. Henry *et al*., Fl. Tamil Nadu 2: 250. 1987; Matthew, Fl. Palni Hills 2: 1166. 1999; Mohanan & Sivadasan, Fl. Agasthyamala 623. 2002. *Celtis orientialis* L., Sp. Pl. 1044. 1753; Wight, Ic. tt. 602 & 603. 1842. *Sponia wightii* Planch. in Ann. Sci. Nat. Bot. Ser. 3, 10: 322. 1848; Wight, Ic. t. 1971. 1853; Bedd., Fl. Sylv. t. 311. 1873. **(Fig. 240).**

Trees to 5 m, dioecious; tender parts appressed-pubescent. Leaves broadly lanceolate, 5-9 x 1-3 cm, subcoriaceous, strigose above, appressed-hirsute beneath, base obliquely subcordate, margin coarsely serrulate, apex acuminate, 3-nerved at base, lateral nerves 4-5 pairs, impressed above, raised beneath; petioles filiform, to 1 cm, pubescent. Cymes axillary, in clusters. Flowers unisexual, subsessile. Male: Perianth 5 lobed; lobes oblong, ca 2 x 1 mm, concave, appressed pubescent without. Stamens 5, ca 2 mm, introrse. Pistillode oblong. Female: Perianth 4 lobed; lobes ovate, ca 1.5 x 1 mm, ciliate. Ovary obovoid, ca 2.5 x 2 mm; style arms 2, erect, to 1 mm, hairy. Drupes globose, ca 4 x 3 mm, tipped with style.

Common in ravines and foot hills to evergreen forests.

Fl & Fr.: Throughout the year, flowers peak during March – June.

Specimens examined: XCH 12288, 13063, 13079, 15554 (Papanasam); 10789, 11051, 11673, 14363 (Kalakad); 11151, 15732 (Courtallam); 13242 (Thirukurungudi); 19937 (Sivagiri); 15114, 10558, 10603, 10607 (Mancholai); 12782 (Mahendragiri); 12660 (Balamore); 14659 (Kallar).

Distribution: Tropical Africa, Sri Lanka, India, Himalaya, Indo-China, W & S. China, Australia, Polynesia.

22. MORACEAE Link

Trees, shrubs or herbs, often with milky juice. Leaves usually alternate, rarely opposite, simple, frequently dotted (cystoliths); stipules deciduous, often sheathing and leaving an annular scar. Flowers small, monoecious or dioecious, usually cymose but sometimes becoming capitate or spiciform, the female sometimes solitary. Perianth single, calycine, usually 4-merous, free or connate, often accrescent in fruit. Stamens isomerous and opposite the sepals or reduced in number. Ovary superior, 1 locular; styles 1 or 2, simple or 2-partite; ovule solitary, pendulous. Fruit an achene or drupe or sub-2 valvate, often aggregate into fleshy pseudocarps or anthocarps. Embryo usually curved; cotyledons often unequal, the larger enfolding the smaller.

A family of 53 genera and ca 1400 species; 15 genera and 115 species in India.

Key to the genera

1a. Thorny trees or shrubs .... 6. **Plecospermum**
1b. Thornless trees, shrubs or herbs (except *Streblus taxoides*) 2

2a. Plants with watery juice 3
2b. Plants with milky juice 4

3a. Trees; flowers spicate or subcapitate .... 5. **Morus**
3b. Herbs; flowers crowded on the upper surface of a flat receptacle .... 3. **Dorstenia**

4a. Male flowers in pedunculate heads .... 7. **Streblus**
4b. Male flowers on the interior or exterior surface of a receptacle 5

5a. Male and female flowers crowded on the inner surface of the fleshy receptacle ..... 4. **Ficus**
5b. Male flowers crowded on the exterior surface of the fleshy receptacle 6

6a. Male flowers crowded on the upper surface of the disciform receptacle, female solitary .... 1. **Antiaris**
6b. Male and female flowers crowded around an oblong or globose receptacle .... 2. **Artocarpus**

1. ANTIARIS Lesch.

Monotypic genus naturally distributed in Tropical Africa to SW. Pacific.

Antiaris toxicaria (Pers.) Lesch. in Ann. Mus. Natl. Hist. Nat. 16: 478. t. 22. 1810; Hook.f., Fl. Brit. India 5: 537. 1888; Fischer in Gamble, Fl. Pres. Madras 3: 1367. 1928; Chithra in A.N. Henry *et al.*, Fl. Tamil Nadu 2: 251. 1987. *Ipo toxicaria* Pers., Syn. Pl. 2: 566. 1807. *Antiaris saccidora* Dalzell in Hooker's J. Bot. Kew Gard. Misc. 3: 232. 1851. **(Fig. 241).**

Monoecious trees, 40-50 m high; latex milky; branchlets lenticellate, spreading, rusty hairy when young. Leaves oblong-oblanceolate, 6-13 x 4-6 cm, scabrid, apex acuminate, margin entire-hardly denticulate, puberulous on either surfaces, base rounded, penninerved; lateral nerves ca 8 pairs; petioles, ca 3 mm; stipules caducous, connate, hairy. Flowers unisexual. Male crowded on the upper surface of receptacle, axillary, ca 1.5 cm across; peduncles to 5 mm, surrounded by confluent, imbricate bracts. Tepals 4, spathulate. Stamens 5-7. Female: Minute, solitary. Bracts many. Tepals 0. Ovary adnate to involucre, 1-locular; ovule 1, pendulous; style arms 2, ca 1 cm, recurved. Drupes fleshy, deep purple.

Rare, in moist deciduous forests at low elevations.

Fl & Fr.: February – April.

Specimens examined: XCH 15475 (Courtallam).

Distribution: Africa, Asia, Australia.

2. ARTOCARPUS Forst.

Ca 57 speices distributed in Tropical & Subtropical Asia to W. Pacific; ca 10 in India.

Artocarpus heterophyllus Lam., Encycl. 3: 210. 1789; Chithra in A.N. Henry *et al.*, Fl. Tamil Nadu 2: 251. 1987; Matthew & Rani in Matthew Fl. Tamil Carnatic 3: 1170. 1999. *A. integrifolia* auct. non L.f. 1781; King in Hook.f., Fl. Brit. India 5: 541. 1888; Fischer in Gamble, Fl Pres. Madras 3: 1369. 1928. **(Fig. 242).**

Trees, 15-20 m high; branchlets glabrescent. Leaves alternate, broadly ovate-elliptic, 5-15 x 3-10 cm, coriaceous, slightly puberulous, glossy above, apex shortly acute-obtuse, margin entire, base acute-subobtuse, rarely cuneate; lateral nerves 4-8 pairs; petioles stout, to 2.5 cm; stipules amplexicaul. Inflorescence cauliflorous, also from main branches, puberulous when young; peduncles stout, cylindric, fleshy, ca 4 cm. Flowers crowded on globose-cylindrical receptacles. Male: Heads 3-6.5 x 1-2 cm. Perianth 2-lobed, ca 0.5 mm, puberulous. Stamen 1, ca 0.8 mm; filament stout, to 0.5 mm; anthers oblong, to 0.3 mm. Female heads 2.5-7 x 1-3 cm. Ovary obovoid; ovule 1; stigma spathulate. Anthocarps fleshy, tubercled, oblong-ellipsoid; seeds elliptic-oblong, 2-3.5 cm.

Occasional, in evergreen forests between 800-1100 m.

Fl. & Fr.: January – June.

Specimen examined: XCH 12384.

Distribution: South West India.

3. DORSTENIA L.

Ca 120 species distributed in the tropics; 2 in India.

Dorstenia indica Wight, Ic. Pl. Ind. Orient. t. 1964. 1853; Hook.f., Fl. Brit. India 5: 494. 1888; Fischer in Gamble, Fl. Pres. Madras 3: 1370.1928; Matthew & Rani in Matthew, Fl. Tamil. Carnatic 2: t. 675. 1982 & 3: 1510. 1983; Chithra in A.N. Henry *et al.*, Fl. Tamil Nadu 2: 251. 1987. **(Fig. 243).**

Unarmed, monoecious, succulent herbs to 30 cm high. Leaves simple, alternate, oblong-oblanceolate, 4-9.5 x 1-3 cm, apex acuminate, margin toothed, base acute, minutely hairy beneath, less so above, chartaceous, penninerved; lateral nerves 7-10 pairs; petioles to 3 cm. Inflorescence greenish, discoid, ca 2 cm across, margin lobed; lobes 12-20, larger ones alternate to shorter, ca 0.4-1.2 cm. Flowers minute, embedded in flattened, fleshy receptacles. Male: Tepals 2, to 0.5 mm, base connate, hairy without, apex acute-slightly obtuse, boat shaped. Stamens 2, ca 0.5 mm; filaments ca 0.5 mm; anthers ca 0.2 mm, Female: Tepals obscurely 2 lobed, toothed, connate and adnate to receptacle. Ovary included, stalked, globose, ca 1 mm across, 1 locular; ovule pendulous; style ca 1 mm, laterally 2-fid, unequal. Achenes ovoid, verrucose.

Frequent, as shady undergrowth of evergreen forests between 600-900 m.

Fl. & Fr.: January – May.

Specimens examined: XCH 19829 (Sivagiri); 10364, 17245 (Kothayar); 15309, 19365 (Kalakad); 17711 (Thirukarankudi); 12929, 18871 (Papanasam).

Distribution: Sri Lanka and Peninsular India.

4. FICUS L.

Ca 875 speices in the tropical and subtropical regions of the world; ca 90 in India.

Key to the species

1a. Leaves opposite, very rarely alternate .... 9. **F. hispida**
1b. Leaves alternate 2

2a. Leaves neither scabrid nor hispid above 3
2b. Leaves more or less scabrid above 16

3a. Erect trees, often epiphytic first 4
3b. Stranglers 15

4a. Receptacles sessile or subsessile; peduncle if present <0.5 cm long 5
4b. Receptacles distinctly pedunculate; peduncle >0.5 cm long 12

5a. Leaves caudate .... 2. **F. arnottiana**
5b. Leaves not caudate 6

6a. Leaves tomentose or puberulous beneath, at least when young 7
6b. Leaves glabrous 8

7a. Leaves grey tomentose beneath, with a gland at the base of the midrib 12. **F. mollis**
7b. Leaves puberulous; without gland .... 6. **F. drupacea**

8a. Petioles d"1 cm 9
8b. Petioles e"2 cm 10

9a. Branches drooping; leaves glossy, drying pale green/brown .... 3. **F. benjamina**
9b. Branches not drooping; leaves not glossy, drying dark brown .... 11. **F. microcarpa**

10a. Figs subsessile; ripening greenish-white .... 17. **F. virens**
10b. Figs sessile; ripening pale yellow, red or purple 11

11a. Receptacles 2-6 together on tubercles from leafless axils, ripening pale yellow .... 16. **F. tsjahela**
11b. Receptacles in pairs on leafy axils, ripening red or purple .... 1. **F. amplissima**

12a. Receptacles on short leafless branchlets or from larger branches or trunk .... 14. **F. racemosa**
12b. Receptacles axillary on leafy twigs 13

13a. Leaves up to 8 cm, upto 2.5 cm across; figs with small bracts at the top of the stalk .... 13. **F. nervosa**
13b. Leaves up to 18 cm, upto 9.5 cm across; figs with small bracts at the base of the stalk 14

14a. Leaves ovate-elliptic, apex rounded or acute .... 4. **F. callosa**
14b. Leaves oblong, apex shortly acuminate .... 5. **F. costata**

15a. Leaves more or less gibbose one side; primary nerves ca 5 pairs, distantly arranged, prominently looping .... 15. **F. tinctoria** ssp. **gibbosa**
15b. Leaves not gibbose; primary nerves ca 10 pairs, distantly arranged, not looping .... 10. **F. lawesii**

16a. Weak shrubs or trees with solid twigs .... 8. **F. heterophylla**
16b. Erect, strong shrubs with hollow twigs .... 7. **F. exasperata**

1. Ficus amplissima J.E.Smith in Rees, Cyclop. 14: n.1. 1810; Corner in Dassan. & Fosberg, Rev. Handb. Fl. Ceylon 3: 342. 1981; Rani & Matthew in Matthew, Fl. Tamil. Carnatic 3: 1515. 1983; Chithra in A.N. Henry *et al.*, Fl. Tamil Nadu 2: 251. 1987; Matthew, Fl. Palni Hills 2: 1178. 1990. *F. tsiela* Roxb. ex Buch.-Ham. in Trans. Linn. Soc. London 15: 149. 1826; Wight, Ic. t. 668. 1840; King in Hook.f., Fl. Brit. India 5: 515. 1888; Fischer in Gamble, Fl. Pres. Madras 3: 1362. 1928. **(Fig. 244).**

Trees to 6 m high; branchlets yellowish when dry. Leaves alternate, broadly ovate-elliptic, 8-12 x 4.5-5.2 cm, coriaceous, glabrous, pale green or yellowish-green when dry; cystoliths present on upper surface; base rounded to subtruncate, margin entire, often undulate, apex acuminate; lateral nerves 6-9 pairs, faint; petioles up to 2.5 cm, blackish when dry. Receptacles in axillary pairs or solitary, sessile, globose, ca 5 mm across. Basal bracts 4 lobed, broader than long, ovate. Male flower: Few, sessile. Tepals 3, ovate. Stamen 1; filament thicker; anther oblong. Female flower: Subsessile. Tepals 3, subequal, oblanceolate, ca 1.25 mm. Style long; stigma simple. Gall flowers similar.

Occasional, in moist deciduous forests at ca 300 m.

Fl.: March.

Specimen examined: XCH 15106 (Mancholai).

Distribution: Peninsular India, Sri Lanka, Maldives.

2. **Ficus arnottiana** (Miq.) Miq. in Ann. Mus. Bot. Lugduno-Batavum 3: 287. 1867; King in Hook.f., Fl. Brit. India 5: 513. 1888; Fischer in Gamble, Fl. Pres. Madras 3: 1363. 1928; Corner in Dassan. & Fosberg, Rev. Handb. Fl. Ceylon 3: 244. 1981; Rani & Matthew in Matthew, Fl. Tamil. Carnatic 3: 1515. 1983; Chithra in A.N. Henry *et al.*, Fl. Tamil Nadu 2: 252. 1987. *Urostigma arnottiana* Miq. in London J. Bot. 6: 564. 1847; Mohanan & Sivadasan, Fl. Agasthyamala 629. 2002. *Urostigma courtallense* Miq., ibid. 564. *U. caulobotryum* Miq., ibid. 568. *Pogonotrophe emodi* Miq., in London J. Bot. 7: 73 1848. *Ficus emodi* (Miq.) Miq. in Ann. Mus. Bot. Lugduno Batavum 3: 278, 293. 1867. *F. arnottiana* (Miq.) var. *courtallensis* (Miq.) Miq. in Ann. Mus. Bot. Lugduno Batavum 3: 287. 1867. **(Fig. 245).**

Trees to 8 m high. Leaves spirally arranged, ovate-cordate, 4-9 x 3-6 cm, subcoriaceous, glabrous, base deeply cordate, margin entire, slightly sinuate, 2-3 nerved at base; lateral nerves ca 5-6 pairs, prominent on both surfaces; nerves and petioles pinkish; petioles 2-6 cm. Receptacles in axillary pairs and also on twigs beneath leaves, glabrous, subobovoid, ca 4 mm across, ripening yellowish-brown, red or purple with spots. Peduncles ca 3 mm. External bracts 3, triangular, hairy. Tepals 3, obtuse, ca 1.5 mm. Male: Few, sessile. Stamen 1; filaments thick, adnate; anthers 2 celled. Female: Sessile or shortly stalked. Ovary subglobose, ca 0.5 mm; style lateral, slender; stigmas cylindrical.

Occasional, in deciduous and evergreen forests above 700 m.

Fl. & Fr.: March – June.

Specimens examined: XCH 19754 (Kalakad); 13726 (Courtallam).

Distribution: Peninsular India

3. **Ficus benjamina** L., Mant. Pl. 129. 1767; Matthew & Rani in Matthew, Fl. Tamil Carnatic 3: 344. 1981. **(Fig. 246).**

Trees to 8 m high. Leaves alternate, ovate, 4-6.5 x 2.5-4.5 cm, subcoriaceous, base rounded to broadly acute, apex acute to shortly acuminate, not prominently nerved from base; lateral nerves 6-15 pairs, prominent; petioles 8 mm, glabrous. Receptacles axillary, paired, subsessile, globose or obovoid. Tepals 3-4, free or shortly united, ovate-lanceolate, 1-1.25 x 0.5-0.75 mm, brownish, glabrous. Male: Abundant, dispersed, shortly pedicellate. Stamen 1; filament short; anther oblong; Female: Sessile. Ovary ovate, ca 1 mm across, brownish; style ca 1 mm, tip curved, dilated, stigmatiferous. Achenes smooth.

Rare, in deciduous forests at ca 200 m.

Fl. & Fr.: June – August.

Specimens examined: XCH 11312 (Courtallam) 11176.

Distribution: India and South China.

4. Ficus callosa Willd. in Mem. Acad. Roy. Sci. Hist. (Berlin) 102. 1798; King in Hook.f., Fl. Brit. India 5: 516. 1888; Fischer in Gamble, Fl. Pres. Madras 3: 1364.1928; Corner in Dassan. & Fosberg, Rev. Handb. Ceylon 3: 263. 1981; Chithra in A.N. Henry *et al.*, Fl. Tamil Nadu 2: 253. 1987. **(Fig. 247).**

Trees to 30 m high. Leaves alternate, broadly ovate-elliptic, 8-17 x 3.5-9.5 cm, apex rounded-acute, base rounded, margin entire, 3 nerved from base; lateral nerves ca 7-11 pairs, prominent on both surfaces; petioles to 5 cm, glabrous; stipules triangular, caducous. Receptacles axillary, solitary, globose, pubescent, scabrid, to 2.5 cm across. Peduncles to 1.5 cm. External bracts 3. Tepals 5, oblong, glabrous, to 1.5 mm, apex acute. Stamen 1, 1-3 mm; filament short; anther 2 celled. Ovary globose; ovule 1; stigmas bifid. Achenes smooth, ca 2 mm, obovoid.

Occasional, in deciduous and evergreen forests between 500-700 m.

Fl. & Fr.: March – June.

Distribution: Sri Lanka, India, Myanmar.

Specimen examined: XCH 19919 (Sivagiri).

5. Ficus costata Ait., Hort. Kew 3: 452. 1789; Corner in Dassan. & Fosberg, Rev. Handb. Fl. Ceylon 3: 246. 1981. *Ficus venusta* Kunth et Bouche, Ind. Sem. Hort. Berol. 16. 1846. *Urostigma wightianum* Miq. var. *majus* Thwaites, Enum. Pl. Zeyl. 265. 1864. *F. caudiculata* Trimen in J. Bot. 23: 242. 1885; King in Hook.f., Fl. Brit. India 5: 497. 1888. *F. mooniana* King in Ann. Roy. Bot. Gard. Calcutta 1: 57. pl. 69. 1887. **(Fig. 248).**

A large tree to 30 m; branchlets thick, when dry blackish-brown. Leaves spirally arranged, oblong, 15-18 x 5-6 cm, glabrous, when dry brownish, coriaceous, stiff, slightly shining, base rounded to cuneate, margin entire, slightly subrevolute when dry, shortly acuminate at apex; lateral nerves 11-15 pairs; transverse nervules prominent, raised beneath; lateral veins and midrib reddish-brown beneath when dry, minutely brown-areolate beneath; petioles ca 2 cm, thick, brown when dry. Receptacles (immature) axillary, solitary, globose, ca 1.5 cm across, shortly pedunculate, reddish, minutely dark brownish-scurfy. Peduncle to 8 mm, thick. Tepals 3, lanceolate, red.

Very rare, in evergreen forests at ca 700 m.

Fr.: April.

Specimen examined: XCH 15610 (Papanasam)

Distribution: Sri Lanka, India, Myanmar.

Note: New record for India.

6. **Ficus drupacea** Thunb., Diss. Ficus 6, 11. 1786. *F. mysorensis* B. Heyne ex Roth in Roem. & Schult., Syst. Veg. 1 : 508. 1817; King in Ann. Roy. Bot. Gard. Calcutta 1: 19. tt. 14 & 81 D. 1887 & in Hook.f., Fl. Brit. India 5: 500. 1888; Fischer in Gamble, Fl. Pres. Madras 3: 1361. 1928. **(Fig. 249).**

Trees to 7 m; branchlets blackish when dry. Leaves broadly ovate-elliptic, rarely obovate, 9-11 x 4-5 cm, coriaceous, glabrous or minutely puberulous beneath, base rounded, subtruncate, margin entire, apex obtuse or shortly acuminate; lateral nerves 10-13 pairs, strongly and very prominently raised beneath, parallel, prominently looped just below margin; transverse nervules prominent beneath; petioles to 2 cm, thick. Receptacle in axillary pairs, sometimes solitary, sessile, subglobose, pale yellowish, ca 1 cm across.

Rare, in moist deciduous forest at ca 600 m.

Fl.: October.

Specimen examined: XCH 17760 (Mahendragiri)

Distribution: Sri Lanka, India, Bangladesh, Laos.

7. **Ficus exasperata** Vahl, Enum. Pl. 2: 197 & 402. 1806; Rani & Matthew in Matthew, Fl. Tamil. Carnatic 3: 1522. 1983; Matthew, Fl. Palni Hills 2: 1181. 1999; Chithra in A.N. Henry *et al.*, Fl. Tamil Nadu 2: 253 1987. *F. serrata* Forssk., Fl. Aeg.-Arab. 179. 1775. *F. asperrima* Roxb., Fl. Ind. 3: 554. 1832; Fischer in Gamble, Fl. Pres. Madras 3: 1366.1928. **(Fig. 250).**

Trees to 4 m; branchlets hairy. Leaves alternate, opposite above, oblong-elliptic, subcoriaceous, apex acute-acuminate, base rounded-subacute, margin sparingly toothed; petioles to 1 cm; stipules paired. Receptacles axillary, solitary, globose, ca 0.5 cm across, scabrid, pubescent without, Peduncles to 1 cm. External bracts 3, lateral. Tepals 3-5, linear-spathulate, white hairy. Male: Stamen 1; filament ca 0.3 mm; anther oblong, to 0.6 mm. Female: Sessile. Ovary obovoid, ca 0.3 mm; style filiform; stigma obscurely 2-fid. Achenes oblong, reticulate.

Common, in deciduous and evergreen forests between 500-1000 m.

Fl. & Fr.: December – May.

Specimens examined: XCH 12688, 14722 (Balamore); 11266, 14822 (Mahendragiri); 16950 (Papanasam).

Distribution: Arabia, Central and South India, Sri Lanka, and East Africa.

8. Ficus heterophylla L.f., Suppl. Pl. 442. 1781; Roxb., Fl. Ind. 3: 532. 1832; Hook.f., Fl. Brit. India 5: 518. 1888; Fischer in Gamble, Fl. Pres. Madras 3: 1366. 1928; Matthew & Rani in Matthew, Mat. Fl. Tamil Carnatic., 2: 344. 1981 & Fl. Tamil. Carnatic., 3: 1521. 1983; Chithra in A.N. Henry *et al.*, Fl. Tamil Nadu 2: 254. 1987. *F. scabrella* Roxb., Fl. Ind. 3: 532. 1832. **(Fig. 251).**

Shrubs to 3 m; branchlets rusty hairy. Leaves alternate, elliptic-lanceolate or oblong, 5-11 x 2-5.2 cm, scabrid on both surfaces, subcoriaceous, base acute, margin dentate, apex acute-shortly acuminate, 3 nerved at base; lateral nerves ca 6-8 pairs, prominent beneath; petioles to 8 mm. Receptacles axillary solitary, ellipsoid-globose, ca 1.5 cm across. Basal bracts ovate, subacute. Tepals 3, linear, lanceolate, to 1 mm. Male: Stamen 1; anther oblong, parallel. Female: Ovary ca 0.8 mm; ovule 1; style to 1 mm, dilated above. Achenes compressed, slightly keeled.

Occasional, in evergreen forests above 700 m.

Fl. & Fr.: October – December.

Specimens examined: XCH 19617 (Kalakad); 15842, 14021, 15844 (Courtallam).

Distribution: Borneo, Myanmar, Sri Lanka, India and South China.

9. Ficus hispida L.f., Suppl. Pl. 442. 1781; Hook.f., Fl. Brit. India 5: 522. 1888; Fischer in Gamble, Fl. Pres. Madras 3: 1367.1928; Matthew, Fl. Tamil Carnatic. 3: 1522. 1983 & Fl. Palni Hills 2: 1181. 1999; Chithra in A.N. Henry *et al.*, Fl. Tamil Nadu 2: 254. 1987. *Ficus oppositifolia* Roxb., Pl. Cor. t. 124. 1799 & Fl. Ind. 3. 561. 1832. *Ficus daemona* J. Koenig ex Vahl, Enum. Pl. 2: 198. 1806; Roxb., Fl. Ind. 3: 562. 1832. **(Fig. 252).**

Trees to 7 m; branchlets hispid. Leaves opposite, elliptic oblong-oblanceolate, 8-14 x 3-6 cm, hispid beneath, slightly hairy above, apex abruptly acuminate, base rounded, subtruncate, margin minutely toothed; 3-nerved from the base; lateral nerves ca 5 pairs, prominent beneath; petioles ca 1 cm, densely hairy; stipules lanceolate, to 7 mm. Receptacles on stem and special shoots, globose, ca 1.5 cm across, pubescent without. Peduncle stout, to 5 mm. External bracts ovate-triangular, ca 1.5 mm; pubescent without. Tepals 0, glabrous. Male: Stamen 1, subsessile; anther oblong, parallel, unequal, to 1 mm. Female: Sessile or stalked. Ovary globose, ca 0.9 mm, 1-ovuled; style to 1.3 mm; slightly hairy, apex clavate. Achenes lenticular, ca 1.3 mm, keeled, with prominent hilum.

Common, in moist deciduous forests upto 1200 m especially near rivers and streams.

Fl. & Fr.: December – May.

Specimens examined: XCH 10531, 10673, 10957, 12157, 12191, 11962 (Kalakad); 14059, 15731, 15893, 18197 (Courtallam); 12674, 14628, 14756 (Balamore).

Distribution: Sri Lanka, India, Queensland and South China to New Guinea.

10. Ficus lawesii King in J. Asiat. Soc. Bengal, Pt. 2, Nat. Hist. 55(2): 403 1887. *F. glaberrima* Blume var. *bracteata* Corner in Gard. Bull. Straits settlem 17: 388. 1959 (1960); Chithra in A.N. Henry *et al.*, Fl. Tamil Nadu 2: 254. 1987. *F. travancorica* King in Ann. Roy. Bot. Gard. Calcutta 1: 28. tt. 26 & 820. 1887 & in Hook.f., Fl. Brit. India 5: 503. 1888; Fischer in Gamble, Fl. Pres. Madras 3: 1365. 1928. **(Fig. 253).**

Straggling to 15 m, glabrous. Leaves oblong-lanceolate, 5-18 x 1-4 cm, chartaceous, chocolate-brown when dry, paler beneath, base acute, margin entire, undulate, apex acuminate; acumen to 1 cm; 3 ribbed at base; lateral nerves ca 14 pairs, arched just below margin, prominent beneath, obscure above; petioles 2-2.5 cm, blackish when dry. Receptacles in axillary pairs, globose, ca 1 cm across, coppery-brown when dry. Male flowers: Scattered, sessile. Tepals 4 or 5, unequal. Stamen 1, ca 1 mm. Gall and female flowers subsessile. Tepals 4 or 5-lobed. Ovary of gall flowers elongate, ovate; style short; in female flower ovary ovate; style long.

Occasional, in deciduous forests, at ca 150 m.

Fr.: February

Specimens examined: XCH 14639, 14648 (Petchiparai)

Distribution: Southern Western Ghats

11. Ficus microcarpa L.f., Suppl. Pl. 442. 1781; Corner in Dassan. & Fosberg, Rev. Handb. Fl. Ceylon 3: 258. 1981; Rani & Matthew in Matthew, Fl. Tamil. Carnatic 3: 1522. 1983; Chithra in A.N. Henry *et al.*, Fl. Tamil Nadu 2: 254. 1987; Matthew, Fl. Palni Hills 2: 1182. 1999; Mohanan & Sivadasan, Fl. Agasthyamala 634. 2002. *F. retusa* auct. non L. 1767; King in Hook.f., Fl. Brit. India 5: 511. 1888; Fischer in Gamble, Fl. Pres. Madas 3: 1362. 1928. *F. retusa* var. *nitida* auct. non Thunb. 1781. 1801; Hook.f., Fl. Brit. India 5: 511. 1888; Fischer in Gamble, Fl. Pres. Madras 3: 1362. 1928.

Trees to 10 m; branchlets grayish brown when dry, hard. Leaves obovate, 4.5-7 x 2-3.5 cm, coriaceous, glabrous, drak brown when dry, base acute to cuneate, margin entire, apex obtuse, obtusely shortly acuminate, 3 ribbed at base; lateral nerves ca 5 pairs, close, parallel, looping; petioles to 1 cm, blackish when dry. Receptacles in axillary pairs, depressed-globose, ca 8 mm across, yellowish or orange when mature, sessile. Basal bracts 4, ovate, ca 2 mm, minutely puberulous or glabrous, obtuse, persistent. Male flowers: Sessile or subsessile. Tepals 3, subspathulate. Stamen 1; filaments thick; anthers cordate-apiculate. Female flowers: Sessile, pedicellate. Tepals 4, spathulate, unequal, ca 0.5 mm. Ovary obovoid; style short; stigma cylindric. Gall flowers: similar. Perianth lobes much smaller.

Frequent in deciduous forests between 150-300 m, mostly near stream banks.

Fl. & Fr.: Throughout the year.

Specimens examined: XCH 12277, 13057, 15636, 15676, 17021 (Papanasam); 11322, 11364, 11390, 11474, 13716 (Courtallam); 13262 (Thirukurungudi); 10666 (Mancholai); 19662 (Kalakad).

Distribution: Sri Lanka, India, South China, Ryukuy to New Britain islands.

12. Ficus mollis Vahl, Symb. Bot. 1: 82. 1790; Corner in Dassan. & Fosberg, Rev. Handb. Fl. Ceylon 3: 249. 1981; Rani & Matthew in Matthew, Fl. Tamil Carnatic 3: 1524. 1983; Chithra in A.N. Henry *et al.* Fl. Tamil Nadu 2: 254. 1987. *Ficus tomentosa* Roxb. ex Willd., Sp. Pl. 4: 1136. 1806; King in Hook.f., Fl. Brit. India 5: 501. 1888; Fischer in Gamble, Fl. Pres. Madras 3: 1361. 1928. **(Fig. 254).**

Trees to 5 m high; branchlets fulvous-tomentose. Leaves alternate-subopposite, oblong-oblanceolate or elliptic, 4-8.5 x 2-4 cm, apex acute, sometimes bluntly apiculate-emarginate, base rounded, subcordate, margin entire, subcoriaceous, 3-5 nerved from base; lateral nerves ca 3 pairs, raised beneath; petioles to 3.5 cm, tomentose; stipules lanceolate. Receptacles axillary, paired, or clustered, globose, ca 6 mm across, densely tomentose. Outer bracts 3, ovate-orbicular, tomentose, ca 0.3 mm across, concave. Tepals 3-5, lanceolate, ca 0.8 mm, glabrous. Male flowers: Stamen 1, to 1.2 mm; filament ca 0.4 mm; anther oblong, to 0.8 mm. Female: Ovary globose, to 0.7 mm; ovule 1; style ca 1.3 mm, simple. Achenes smooth.

Rare, in deciduous forests upto 800 m.

Fl. & Fr.: June – August.

Specimens examined: XCH 12361 (Papanasam); 13291 (Thirukurunkudi); 18053 (Courtallam); 12391, 13928, 14171 (Kalakad).

Distribution: Sri Lanka and India.

13. Ficus nervosa B.Heyne ex Roth in Roem. et Schult., Syst. Veg. 1: 513. 1817; King in Hook.f., Fl. Brit. India 5: 512. 1888; Fischer in Gamble, Fl. Pres. Madras 3: 1364. 1928; Corner in Dassan. & Fosberg, Rev. Handb. Fl. Ceylon 3: 264. 1981; Rani & Matthew in Matthew, Fl. Tamil. Carnatic 3: 1525. 1983; Chithra in A.N. Henry *et al.*, Fl. Tamil Nadu 2: 254. 1987; Matthew, Fl. Palni Hills 2: 1183. 1999; Mohanan & Sivadasan, Fl. Agasthyamala 634. 2002. *F. angustifolia* Roxb., Fl. Ind. 3: 554. 1832; Wight, Ic. t. 660. 1840. **(Fig. 255).**

Tamil: Neer-aal

Trees to 3 m high; branchlets thick, ribbed, black when dry. Leaves spirally arranged, 4-8 x 1.5-2.5 cm, chartaceous, glabrous, shining, chocolate-brown beneath, greenish-brown above when dry; cystoliths present beneath; base rounded or truncate, margin undulate, apex acuminate; lateral nerves ca 7 pairs; lateral nerves, midrib and transverse nervules prominently raised beneath; petioles slender, to 2 cm, blackish when dry. Receptacles axillary, ca 5 mm across, solitary or paired, pedunculate. Peduncle to 15 mm.

Rare, in semi-evergreen forests, ca 1000 m.

Fl.: September.

Specimen examined: XCH 10500 (Kakachi).

Distribution: Sri Lanka, India, Myanmar, South China, Taiwan, Tonkin, Annam.

14. Ficus racemosa L., Sp. Pl. 1060. 1753; Rani & Matthew in Matthew, Fl. Tamil. Carnatic 3: 1526. 1983; Matthew, Fl. Palni Hills 2: 1184. 1999. *Ficus glomerata* Roxb., Pl. Cor. t. 123. 1799; King in Hook.f., Fl. Brit. India 5: 535. 1888; Fischer in Gamble, Fl. Pres. Madras 3: 1364.1928. **(Fig. 256).**

Trees to 4 m high. Leaves alternate, oblong-oblanceolate, 8-10 x 3-5 cm, faint, apex acute, base subacute-rounded, margin entire, subcoriaceous, glabrous, 3 nerved from base; lateral nerves ca 5 pairs; stipules lanceolate, hairy, to 1 cm. Receptalecs on leafless branchlets, globose, white-pilose, axillary, ca 1.5 cm across. Peduncles 5-10 mm, hairy. Basal bracts ovate, ca 1.5 mm. Tepals 4, free, ovate-oblanceolate, scarious, ca 1 mm. Male flowers: Stamens 2, exserted, ca 1.6 mm; filaments 1. 2 mm; anthers oblong, ca 0.4 mm. Female: Ovary ovoid-oblong, ca 1.2 mm; style ca 1.5 mm, simple. Achenes smooth.

Common, in deciduous forest at low attitudes.

Fl. & Fr.: Throughout the year.

Specimens examined: XCH 13210 (Kalakad); 13288, 11402 (Nanguneri); 13803, 15547, 15740 (Courtallam).

Distribution: Sri Lanka, India, New Guinea, Pakistan, Queensland and South China.

15. Ficus tinctoria Forst.f. subsp. **gibbosa** (Blume) Corner, Gard. Bull. Singapore 17: 476 1960. *F. parasitica* Koen. ex Willd. in Mem. Acad. Roy. Sci. Hist. (Berlin) 2: 102. t. 3. 1798; King in Hook.f., Fl. Brit. India 5: 497. 1888. *F. gibbosa* Blume var. *parasitica* (Koen. ex Willd.) King in Ann. Roy. Bot. Gard. Calcutta 1: 6. t. 2 bN. 1887; Fischer in Gamble, Fl. Pres. Madras 3: 1365. 1928. *F. tinctoria* subsp. *parasitica* (Willd.) Corner in Gard. Bull. Straits Settlem. 17: 476. 1960 et in Dassan. & Fosberg, Rev. Handb. Fl. Ceylon 3: 276. 1981; Rani & Matthew in Matthew, Fl. Tamil. Carnatic 3: 1528. 1983. *F. tinctoria* subsp. *cuspidifera* (Miq.) Chithra in A. N. Henry *et al.*, Fl. Tamil Nadu 2: 255. 1987; Mohanan & Sivadasan, Fl. Agasthyamala 636. 2002. **(Fig. 257).**

Straggling to 5 m; branchlets pale yellowish when dry. Leaves alternate, broadly subrhomboid or ovate-elliptic, 6.5-20.1 x 3-4 cm, chartaceous, glabrous, when dry grayish-green above, paler beneath, base acute to cuneate, usually prominently inequilateral with one side more or less gibbous, margin entire, sinuate, apex acute or shortly obtusely acuminate; lateral nerves ca 5 pairs, distantly arranged, looping just below margin, faint to prominent beneath, obscure above; petioles up to 1 cm; stipules lanceolate, long acuminate, ca 5 mm. Receptacle in axillary pairs or solitary, globose, pedunculate, yellowish, ca 8 mm across. Peduncle slender, to 6 mm. Basal bracts 0. Male flowers: Tepals 4, linear-lanceolate, ca 1.5 mm, subequal, pubescent without. Stamen 1, attached by its base to pistil (abortive); filament to 1 mm; anther oblong, to 1 mm. Female: Tepals 4, linear-oblong or lanceolate, hyaline, ca 1.5 mm, subequal, pubescent without. Ovary obliquely ovoid; style lateral, elongate. Gall flowers like female flowers.

Rare, in moist deciduous to semi-evergreen forests between 300-600 m.

Fl.: Throughout the year.

Specimens examined: XCH 13996, 14845 (Courtallam); 12199 (Mancholai).

Distribution: Sri Lanka, India, Myanmar, South China, Indo-China, Andaman and Nicobar islands.

16. Ficus tsjahela Rheede ex Burm.f., Fl. Ind. 227. 1768; Hook.f., Fl. Brit. India 5: 514. 1888; Fischer in Gamble, Fl. Pres. Madras 3: 1362.1928; Rani & Matthew in Matthew, Fl. Tamil. Carnatic 3: 1530. 1983 & 4: t. 609. 1988; Matthew, Fl. Palni Hills 2: 1185 1999; Chithra in A.N. Henry *et al.*, Fl. Tamil Nadu 2: 256. 1987. **(Fig. 258).**

Trees to 10 m. Leaves alternate, broadly elliptic-oblong, 4-7 x 2.5-4.5 cm, subcoriaceous, glabrous, apex abruptly acuminate, base rounded-slightly cordate, margin entire, cusp ca 1 cm long, 3-5 nerved from base; lateral nerves ca 6 pairs, prominent beneath; petioles articulate, ca 4 cm; stipules to 2 cm, leafy. Receptacles in axils of present or fallen leaves, paired or clustered, depressed-globose, ca 4 mm across. Tepals 3 in male, free or basally jointed, 3-4 in female, oblanceolate, ca 1.2 mm, glabrous. Male flowers: Stamen 1, to 1.2 mm; filament to 0.4 mm; anther oblong, to 0.8 mm. Female: Sessile. Ovary obovoid, ca 0.9 mm; narrow at base, ovule 1; style to 1.2 mm, simple. Achenes smooth.

Common, in moist deciduous forests upto 800 m.

Fl. & Fr.: September – March.

Specimens examined: XCH 11018, 13041, 14486, 15576 (Kalakad); 11324, 11415, 13726, 13770 (Courtallam); 13240, 16417, 16467 (Thirukurungudi): 18537 (Mahendragiri).

Distribution: Sri Lanka and Peninsular India.

17. Ficus virens Ait., Hort. Kew. ed. 1. 3: 451. 1789. Corner in Dassan. & Fosberg, Rev. Handb. Fl. Ceylon 3: 238. 1981; Rani & Matthew in Matthew, Fl. Tamil. Carnatic 3: 1531. 1983; Chithra in A.N. Henry *et al.*, Fl. Tamil Nadu 2: 256. 1987; Matthew, Fl. Palni Hills 2: 1185. 1999. *F. infectoria* sensu Roxb., Fl. Ind. 3: 551. 1832, non Willd. 1806; Wight, Ic. t. 665. 1840; King in Hook.f., Fl. Brit. India 5: 515. 1888; Fischer in Gamble, Fl. Pres. Madras 3: 1362. 1928. **(Fig. 259).**

Trees to 10 m; branchlets pale yellowish when dry, terete. Leaves spirally arranged, elliptic-ovate or oblong-ovate, 2-7.5 x 3-4 cm, chartaceous, glabrous, when dry grayish-brown, base rounded, subcordate, subtruncate to cuneate, margin entire, apex obtusely shortly acuminate, 3-ribbed at base; lateral nerves ca 8 pairs, raised beneath with midrib, arched; transverse nervules prominent beneath; petioles long, slender, 5.5-6 cm, glandular at apex. Receptacles axillary or on twigs just at base of leaves, paired, globose, pale white, ca 5 mm across, 12-15 mm when mature, sessile or shortly pedunculate. Basal bracts 3, ovate, ca 5 x 3 mm, sparsely puberulous, persistent. Peduncle ca 3 mm. Male flowers: Ostiolar, sessile. Tepals 3, ovate, unequal, free or slightly gamophyllous. Stamen 1; filament ca 0.5 mm; anther oblong, ca 0.6 mm. Female flower: Sessile. Tepals 3, lanceolate, unequal, ca 2.5 mm, free. Ovary oblong, ca 0.4 mm, stalked; style to 0.75 mm, filiform. Gall flowers similar, pedicellate.

Occasional, in moist-deciduous forests between 300-600 m.

Fl. & Fr.: Throughout the year.

Specimens examined: XCH 19541 (Kalakad); 11324, 11415, 13770 (Courtallam); 12081, 13240 (Thirukurungudi)

Distribution: Sri Lanka, India, Solomon islands.

5. MORUS L.

Ca 19 speices in the tropical and temperate Asia, Africa, Southern and Northen America; ca 6 in India.

Morus alba L., Sp. Pl. 986. 1753; Hook.f., Fl. Brit. India 5: 492. 1888; Fischer in Gamble, Fl. Pres. Madras 3: 1370. 1928. **(Fig. 260).**

English: Mulberry.

Dioecious, shrubs or small trees. Leaves alternate, ovate, 3-10.5 x 1.8-6.3, acuminate, subcoriaceous, 2-4 nerved from base; base subcordate, margin serrate, apex acute, veins prominent; lateral nerves ca 7 pairs, raised beneath, softly pubescent; petioles 2-3 cm; stipules minute, caducous. Male flowers not seen. Female flowers: In spike to 2 cm. Perianth lobes 4, obovate, decussately imbricate, 2 x 1.7 mm. Ovary 1 locular; ovule 1, pendulous; styles branched towards apex; stigma divergent, feathery. Fruit not seen.

Cultivated.

Fl. & Fr.: February – March.

Specimen examined: 14898 (Kothayar).

Distribution: East Asia.

6. PLECOSPERMUM Trec.

Ca 10 species distributed in the tropical and temperate Asia, Africa, New Caledonia, Southern and Northern America; 2 in India.

Plecospermum spinosum Trec. in Ann. Sci. Nat. Bot. Ser 3. 8: 124. 1842; Hook.f., Fl. Brit. India 5: 491. 1888; Fischer in Gamble, Fl. Pres. Madras 3: 1353.1928; Rani & Matthew in Matthew, Fl. Tamil. Carnatic 3: 346. 1981; Chithra in A.N. Henry *et al.*, Fl. Tamil Nadu 2: 256. 1987. *Batis spinosa* Roxb., Fl. Ind. 3: 762. 1832. **(Fig. 261).**

Thorny, straggling shrubs 1-3 m high; branchlets glabrous. Leaves alternate, ovate-elliptic or oblanceolate, 4-8 x 2-3 cm, subcoriaceous, apex acute-subobtuse, base narrowly acute, penninerved; lateral nerves ca 7 pairs, faint; petioles to 1 cm; stipules minute. Flowers dioecious, in axillary 1-3-nate, pedunculate, capitate clusters. Peduncles to 1 cm. Male flowers: Bract basal, ovate, ca 5 mm; bracteoles dispersed with flowers. Perianth 4 lobed, concave, lanceolate, ca 2 mm, slightly hairy. Stamens 4, to 6 mm; filaments ca 5 mm; anthers to 1 mm. Pistillode minute, hairy. Female: tepals fleshy, 4 toothed. Ovary obovoid, sunken, 1 locular; ovule 1; style to 2.8 mm, slightly puberulous. Fruits irregularly shaped, anthocarps to 8 mm across.

Occasional, in deciduous forests between 400-700 m.

Fl. & Fr.: March – May.

Specimens examined: XCH 14165, 14822, 16589, 16950 (Kalakad); 11421 (Courtallam); 17611, 16323 (Thirukurunkudi).

Distribution: Sri Lanka, Peninsular India.

7. STREBLUS Lour.

Ca 27 species distributed in tropical and temperate Asia, Africa, Pacific, and Australasia; 5 in India.

Key to the species

1a. Armed .... 2. **S. taxoides**
1b. Unarmed .... 1. **S. asper**

1. **Streblus asper** Lour., Fl. Cochinch. 615. 1790; Hook.f., Fl. Brit. India 5: 489. 1888; Fischer in Gamble, Fl. Pres. Madras 3: 1353.1928; Rani & Matthew in Matthew, Fl. Tamil. Carnatic 2: t. 679. 1982 & 3: 1534. 1983; Chithra in A.N. Henry *et al.*, Fl. Tamil Nadu 2: 256. 1987. *Trophis aspera* Retz., Obs. 5: 30.

1789. *Epicarpurus orientalis* Blume, Bijdr. 488. 1826. *Streblus lactescens* Blume, Ann. Mus. Bot. Lugduno-Batavum. 2: 80. 1856. **(Fig. 262).**

Small trees to 3 m; branchlets glabrous. Leaves elliptic-obovate or rhomboid, minutely hairy beneath, less so above, apex acute, margin toothed, base cuneate; petioles to 2 mm; stipules caducous. Flowers axillary. Male: Clustered, stalked. Peduncles to 1.2 cm. Tepals 4, free, lanceolate, concave, ca 2.2 mm, glabrous within, puberulous without, apex subobtuse. Stamens 4, ca 3 mm; filaments to 2 mm; anthers to 1 mm. Female: Stalk to 1.2 cm. Bracts ovate; bracteoles ovate-elliptic, to 2.2 mm, hairy. Tepals 4, ovate-elliptic, to 3.5 cm, concave, hairy. Ovary ovoid, to 2.8 mm; ovule 1; style to 6.5 mm. Drupes included in perianth.

Occasional, in deciduous forests upto 500 m.

Fl. & Fr.: February – April.

Specimen examined: XCH 14514, 15678, 18378 (Kalakad); 14824 (Mahendragiri); 18398 (Sivagiri).

Distribution: India, Myanmar, Celebes, Sri Lanka, Haenan, Indo-China, Java, Malaya, Philippines, South China, Sumatra, Sumbowa, and Thailand.

2. Streblus taxoides (Heyne ex Roth) Kurz, For. Fl. Burma 2: 465. 1877; Rani & Matthew in Matthew, III Fl. Tamil. Carnatic t. 680. 1982. *Trophis taxoides* Heyne ex Roth, Nov. Pl. Sp. 368. 1821 *T. spinosa* Roxb., Fl. Ind. 3: 762. 1832. *Phyllochlamys spinosa* (Roxb.) Bureau in DC., Prodr. 17: 218. 1873; Hook.f., Fl. Brit. India 5: 488. 1888; Rani & Matthew in Matthew, Fl. Tamil. Carnatic 3: 346. 1981. *Epicarpurus spinosus* Wight, Icon. Pl. Ind. Orient. t. 1962. 1853. **(Fig. 263).**

Armed shrubs to 3 m high; branchlets glabrous. Leaves alternate, elliptic-oblanceolate, 2-8 x 1-4 cm, coriaceous, glabrous, apex acuminate, base cuneate; lateral nerves ca 5 pairs; petioles to 2 mm. Flowers axillary. Male: Minute, clusters subsessile, ca 1 cm across. Bracts ovate, concave, ca 2.5 mm, apex acute. Tepals 4, elliptic-obovate, concave, to 2 mm, slightly puberulous without, herbaceous. Stamens 4, ca 5.5 mm; filaments to 4 mm; anthers to 1.5 mm. Female: Axillary, solitary, stalked. Pedicel to 5 mm. Bracts lanceolate, ca 2 mm; bracteoles lanceolate, ca 3 mm. Tepals 4, oblanceolate, ca 3.5 mm, free, apex acute. Ovary compressed, to 1.5 m, 1 locular; ovule 1; styles lateral, ca 1 mm, slightly hairy, 2-partite. Drupes ellipsoid, included in the foliaceous, persistent perianth.

Occasional, in moist deciduous and semi evergreen forests between 300-700 m.

Fl. & Fr.: February – June.

Specimens examined: XCH 13044, 15521, 16878 (Papanasam).

Distribution: Myanmar, Celebes, Sri Lanka, Haenan, India, Indo-China, Java, Malaya, Meadoro, Palawas, Thailand and Timor.

REFERENCES

Balakrishnan, N.P., and Chakrabarty, T. 2007. *The Family Euphorbiaceae in India: A Synopsis of its Profile, Taxonomy and Bibliography*. Bishen Singh Mahendra Pal Singh, Dehra Dun, India.

Balakrishnan, N.P., Chakrabarty, T., Sanjappa, M., Lakshiminarsimhan, P., Singh, P. 2012. *Flora of India* Vol. 23 *(Loranthaceae – Daphniphyllaceae)*, Botanical Survey of India, Kolkata.

Beddome, R.H. 1868-1874. *Icones Plantarum Indiae Orientalis*. Gantz Brothers, Madras.

Beddome, R.H. 1869-1874. *The Flora Sylvatica for Southern India*. Gantz Brothers, Madras.

Blasco, F. 1970. Aspects of the flora and ecology of savannas of South Indian hills. *J. Bombay Nat. Hist. Soc.* **67**: 305-311.

Bourdillon, T.F. 1908. *The Forest Trees of Travancore*. Govt. Press, Trivandrum.

Champion, H.G. and Seth, S.K. 1968. *A revised survey of the forest types of India*. C.S.I.R., New Delhi.

Dassanayake, M.D. and Fosberg, F.R. (eds.) 1980-1998. A Revised Handbook to the Flora of Ceylon, Vols. 1-XII, Amerind Publ. Co., New Delhi.

Davis, S.D. Heywood, V.H. Hamilton A.C. (Eds.). 1994. *Centres of Plant Diversity: A Guide and Strategy for Their Conservation.* Vol. Europe, Africa, South West Asia and the Middle East, WWF/IUCN. IUCN Publications Unit, Cambridge.

Fyson, P.F. 1932. *The Flora of South Indian Hill Stations*. Govt. Press, Madras.

Gamble, J.S. 1915-1936. *Flora of the Presidency of Madras*, Adlard & Son. Ltd., London.

Gopalan, R. and Henry, A.N. 2000. *Endemic plants of India, camp for the strict endemics of Agasthiyamalai Hills, Southern Western Ghats*. Bishen Singh Mahendra Pal Singh, Dehra Dun, India.

Henry, A. N. and Gopalan, R. 1995. Agastyamalai hills, India. In: Davis, S. D., Heywood, V. H. and Hamilton, A. C (Eds.), *Centres of Plants Diversity - A guide and strategy for their Conservation.* Vol. **2** (Asia, Australia and The Pacific): CPD site IS 7. Pp.111-113. The World Wide Fund for Nature (WWF) and IUCN (The World Conservation Union), Switzerland.

Henry, A.N., Chandrabose, M., Swaminathan, M.S., and Nair, N.C. 1984. Agastyamalai and its environs: A potential area for a biosphere reserve. *J. Bombay Nat. Hist. Soc.*, **81**: 282-290.

Hooker, J.D. 1872-1897. *The Flora of British India.* Vols. I-11. Reeve & Co. Ltd. London.

Jothi, G.J., and Manickam, V.S. 2005. Intraspecific variation in some species of Euphorbiaceae from Tirunelveli hills of southern Western Ghats, Tamil Nadu. *Trop. Ecol.* **46**(2):145-150.

Jothi, G.J., Sundaresan, V., and Manickam, V.S. 2002. Taxonomic and Nomenclatural notes on *Claoxylon wightii* Hook.f. and its allies (Euphorbiaceae) from Southern Western Ghats, Tamil Nadu, India. *Rheedea*, **12**(2):150.

Jothi, G.J. and Manickam, V.S. 2005. Intraspecific variation in some species of Euphorbiaceae from Tirunelveli hill of Southern Western Ghats, Tamil Nadu. *Trop. Ecology* **46**(2): 145-150.

Jothi, G.J., Manickam, V.S., Sundaresan, V. and Josephine M.M. 2002. New species of *Glochidion* Forst. (Euphorbiaceae) from southern India. *J. Econ. Taxon. Bot.* **26**(1): 114 -116.

Jothi, G.J., Manickam, V.S., Sundaresan, V., and Murugan, C. 2001. Addition to description of *Dimorphocalyx beddomei* (Benth.) Airy Shaw (Euphorbiaceae) from India. *J. Econ. Taxon. Bot.* **25**(3): 721-726.

Keshava Murthy, K.R., and Yoganarasimhan, S.N. 1990. *Flora of Coorg (Kodagu), Karnataka, India*. Vimsat publishers, Bangalore – 560 010.

Manickam, V.S. and Irudayaraj, V. 1992. *Pteridophyte flora of the Western Ghats, South India,* B.I. Publications, New Delhi – 110 002.

Manickam, V.S., and Irudayaraj, V. 2003. *Pteridophyte flora of Nilgiris, South India,* Bishen Singh Mahendra Pal Singh, Dehra Dun.

Manickam, V.S. Murugan, C., Jothi, G.J., and Sundaresan, V. 2008. *Flora of Tirunelveli Hills* (Southern Western Ghats) Vol. I. Bishen Singh Mahendra Pal Singh, Dehra Dun.

Manickam, V.S., Murugan, C., Jothi C.J., and Sundaresan, V. 2005. A new species of *Beilschmiedia* (Lauraceae). *Nord. J. Botany* **24**(4): 407.

Manickam, V.S., Sundaresan, V., Murugan, C., and Jothi, G.J. 2001. *Thottea dinghoui* Swarup, family Aristolochiaceae, A new record for Tamil Nadu. *J. Bombay Nat. Hist. Soc.* **98**(2): 318-320.

Matthew, K.M. 1981-1984. *The Flora of the Tamil Nadu Carnatic.* Vols. I-III. The Rapinat Herbarium, St. Joseph's College, Tiruchirapalli.

Matthew, K.M. 1999. *The flora of the Palni Hills, South India.* Vol.I-III. The Rapinat Herbarium, St. Joseph's College, Tiruchirapalli.

Mohanan, M. and Henry, A.N. 1994. *Flora of Thiruvananthapuram*. Botanical Survey of India. Calcutta.

Mohanan, N. and Sivadasan, M. 2002. *Flora of Agasthyamala*. Bishen Singh Mahendra Pal Singh, Dehra Dun.

Mudaliar, C.R. and Sundararaj, D.D. 1954. A short account of the flora of Tirunelveli district. Madras State. Herb. Cent. Souv. Pp. 57-63.

Nair, N.C. and Daniel, P. 1986. The floristic diversity of the Western Ghats and its conservation: A review. *Proc. Indian Acad. Sci.* (Animal Sci/Plant Sci) Suppl, 127-163.

Nair, N.C. and Henry, A.N. 1983-1989. *Flora of Tamil Nadu, India. Analysis,* Vols. 1-3. Botanical Survey of India, Coimbatore.

Nayar, M.P. 1996. *Hot spots of Endemic Plants of India, Nepal and Bhutan*. Tropical Botanic Garden and Research Institute, Palode.

Raghavan, R.S. and Singh, N.P. 1983. Endemic and Threatened Plants of Western India, pp. 8–11. In: Jain, S.K. & Sastry, A.R.K. (eds.). *Plant Conservation Bulletin, Project on the Study, Survey and Conservation of Endangered Flora (POSSCEF)*. Botanical Survey of India, Calcutta.

Ramachandraan, V.S. and Nair, V.J. 1988. *Flora of Cannanore*. Botanical Survey of India, Coimbatore.

Ramaswamy, M.S. 1914. A botanical tour in the Tirunelveli Hills. *Rec. Bot. Surv. India* **6**(5): 105-171.

Rangacharya, K. 1919. A note on the flora of Tirunelveli District. *Madras Agri. Dept. year book* 95-109.

Saldanha, C.J. 1984. *Flora of Karnataka*, Vol. 1. Oxford & IBI-1, New Delhi.

Saldanha, C.J. and Nicolson, D.H. 1976. *Flora of Hassan District, Karnataka, India*. Amerind Publishing Co. Pvt. Ltd., New Delhi.

Santapau, H. 1953. *The Flora of Khandala on the Western Ghats of India*. Botanical Survey of India, Calcutta.

Sasidharan, N. and Sivarajan, V.V. 1996. *Flowering Plants of Thrissur Forest (Western Ghats, Kerala, India)*. Scientific Publishers, Jodhpur.

Sebastian, K.M. and Henry, A.N. 1960. A new species of *Reidia* from South India. *Bull. Bot. Surv. India* **2**: 437-439.

Shankarnarayan, K.A. 1960. The vegetation of Tirunelveli district. *J. Indian Bot. Soc.* **39**. 474-479.

Sharma, B.D., Shetty, B.V., Karthikeyan, S. and Chandrabose, M., 1973. Studies on the vascular flora of Mahendragiri hill and the surrounding regions, Kanyakumari and Tirunelveli districts, Tamil Nadu. *Bull. Bot. Surv. India* **15**: 45-70.

Subramanyam, K., and Nayar, M.P. 1974. Vegetation and phytogeography of the Western Ghats. *Ecol. Biogeogr. India* **23**:178-196.

Vajravelu, E. 1990. *Flora of Palghat district, including Silent Valley National Park*. Botanical Survey of India, Coimbatore.

INDEX OF BOTANICAL NAMES